VISTAS IN PHYSICAL REALITY

VISTAS IN PHYSICAL REALITY

Festschrift for Henry Margenau

Edited by
Ervin Laszlo
State University of New York
College of Arts and Science at Geneseo

and
Emily B. Sellon
Editor, *Main Currents in Modern Thought*

Plenum Press · New York and London

Library of Congress Cataloging in Publication Data

Main entry under title:

Vistas in physical reality.

"Publications by Henry Margenau": p.
Includes bibliographical references and index.
CONTENTS: Vistas in physics: Wigner, E. P. On some of physics' problems. Grünbaum, A. Is the coarse-grained entropy of classical statistical mechanics an anthropomorphism? Törnebohm H. Aspects of the special theory of relativity. Park, J. L. The measurement act in quantum physics. Lindsay, R. B. Acoustics, old and new [etc.]
1. Science–Philosophy–Addresses, essays, lectures. 2. Physics–Philosophy–Addresses, essays, lectures. 3. Education–Philosophy–Addresses, essays, lectures. 4. Margenau, Henry, 1901- I. Margenau, Henry, 1901- II. Laszlo, Ervin, 1932- III. Sellon, Emily B., 1908-
Q175.3.V57 530'.1 75-34356
ISBN 0-306-30884-3

A Division of Plenum Publishing Corporation
227 West 17th Street, New York, N.Y. 10011

United Kingdom edition published by Plenum Press, London
A Division of Plenum Publishing Company, Ltd.
Davis House (4th Floor), 8 Scrubs Lane, Harlesden, London, NW10 6SE, England

Printed in the United States of America

CONTRIBUTORS

Eugene P. Wigner, Department of Physics, Princeton University

Adolf Grünbaum, Department of Philosophy, University of Pittsburgh

Håkan Törnebohm, Insititute for Theory of Science, University of Gothenburg

James L. Park, Department of Physics, Washington State University

R. Bruce Lindsay, History of Science and Theoretical Physics, Brown University

Wolfgang Yourgrau, Department of Physics, University of Denver

Ervin Laszlo, Department of Philosophy, State University of New York at Geneseo

Peter Caws, Department of Philosophy, Hunter College of the City University of New York

Siegfried Müller-Markus, Television Commentator, formerly University of Fribourg

John Bacon, Department of History and Philosophy, York College (CUNY)

Lee Thayer, Department of Communication Studies, Simon Fraser University

Harold G. Cassidy, Professor-at-Large, Hanover College

Emily B. Sellon, Editor, *Main Currents in Modern Thought*

PREFACE

Festschriften, when they are haphazard collections of pieces written by colleagues and well-wishers on the occasion of a major anniversary in the life of a distinguished man, tend to be tedious. One can more profitably go directly to the writings of the celebrant, as well as other, more voluntary publications of his well-wishers. However, the editors wish to claim that this Festschrift is different. This is so first of all because of the almost unique combination of interests and competence of Henry Margenau. He is at once a distinguished physicist, an equally distinguished educator, and a prominent philosopher. These broad areas of his extraordinarily active and fruitful career are each represented in this volume in his honor, and this constitutes the particular interest of the collection. Without limiting themselves to paraphrases or empty compliments, the contributors to this book range over the scope of interest of Margenau's work, and, acknowledging its influence and significance, present their own viewpoints and conclusions. Since they include some of the most distinguished men in science and philosophy today, the privilege of having them speak to some broadly defined common concerns in a single volume is a rare one, for which our thanks must go to Henry Margenau, who inspired the papers.

Although the editors have divided the contents into three sections which correspond to the three principal areas of interest represented, the contributions in all sections manifest a comparable breadth of philosophical vision and scientific sophistication. In Section I, some of the most intriguing problems of contemporary physics are dealt with. Wigner considers both external and internal problems: problems of communication and purpose in a science which has continued to grow and diversify rapidly, and questions of knowledge peculiar to the present state of physics. Cosmology and epistemology are among the most fascinating problems with which physicists find themsevles confronted, and Wigner reflects the best in the scientific tradition when he concludes that they are essential to science: what is to be feared is not that we will run out of them but that we may one day get tired of them.

Both Grünbaum and Törnebohm present original contributions to the cause of defending contemporary physics from the charge of instrumentalism, formalism, and anthropomorphic theory construction. Grünbaum demonstrates that human choice and decision cannot invalidate the physical credentials of the statistical entropy principle of classical mechanics. Törnebohm defends the special theory of relativity from a somewhat analogous charge: that it is a theory about clocks and measuring sticks. He distinguishes between theories that map a part or aspect of the real world (*X*-ological maps) and theories that map the behavior of instruments gathering information about the world (*X*-ometric maps). Törnebohm then shows that special relativity theory can be interpreted as a primarily *X*-ological map.

The controversy between Einstein and Bohr in the interpretation of quantum mechanics is the context of Park's contribution. He shows that the epistemology advanced by Margenau already in the 1930s, and restated in the latency theory of observables in 1954, does not require of quantum measurements to come up with possessed observables in the form of measured observables, yet it also does not require the problematic conceptual burden of duality, complementarity, and subjectivism. Finally, Lindsay raises the intriguing question of whether acoustics is, or ever has been, a true physical science. He argues impressively that acoustics is a valid science, and that it is a technology and an art besides.

Section II explores still wider philosophical vistas. It opens somewhat drastically with Yourgrau's radical denial that there is, or can be, *a* philosophy of science. Instead, Yourgrau wishes to treat philosophy of science as a pluralistic venture, consisting of a philosophy of physics, of mathematics, chemistry, and the other natural and social sciences. Laszlo's paper takes a contrary standpoint. Not a monolithic unity of all sciences is presupposed, but the unity of science with respect to epistemology, ideals, and method. Elaborating concepts offered by Margenau, Laszlo attacks the problem of discerning the nature of scientific progress: one which contemporary philosophy of science is on the whole at a loss to provide with cogent meaning. The exploration of philosophical vistas in science is set forth by Caws. The world of experience is closed in on itself, and all our knowledge of the physically real is found within the circle. Can we break out of it? Caws suggests that the isomorphy of well-defined conceptual systems in the sciences present in the minds of all qualified individuals allows us to to so by arguing for the existence of another, underlying system, which is isomorphic with all of them.

Next, Müller-Markus raises an essentially psychological problem and sheds new light on it by drawing on comparisons and analogies with the structure of physical theory. The problem is that of explaining how physicists come by the insights which present major and minor innovations in physics. The science explaining such creativity is not physics but "protophysics," and Müller-Markus lays down its foundations through detailed analyses of discoveries, such as Planck's of the quantum constant and Leibniz's anticipation of it two centuries earlier.

Bacon shifts the discussion to the level of ethics. He subscribes to Margenau's suggestion that a method like that of science can be fruitfully applied to ethics, appraises Margenau's ethics, and then proceeds to outline his own variant of the hypotheticodeductive approach to the problem of constructing and validating a system of normative ethics. Thayer's contribution furnishes an immediate contrast to rational constructivism in matters of general human concern: he argues that in complex societies such as ours there comes a point at which the functions of incompetence exceed those of competence. Competence (specialization to the extent where one man can perform only one kind of task) is a danger not only in complex societies but also in complex social and cognitive institutions, such as science.

This takes us to the third and last section of this volume. Margenau has been an articulate and long-standing advocate of integration and reform in education, especially in the sciences and the philosophy of science. Section III has these issues as its focus. Here Cassidy reports on his sustained experience with using Margenau's epistemology of science as a "philosophy by example" that offers a frame of reference within which cognitive knowledge, subjective experience, and action in applying these to practice can unify the task of the university and give meaning to education and life. The concluding paper by Sellon brings together some of Margenau's own ideas on integrative education as he has presented them over many decades of writing and teaching, and indicates some of the possibilities and directions toward which it may be tending.

E. L.
E. B. S.

CONTENTS

I

VISTAS IN PHYSICS

1

Eugene P. Wigner

ON SOME OF PHYSICS' PROBLEMS

The problems of physics that I would like to discuss have not created a crisis situation. It may be well to have a look at them, nevertheless, and to think about solutions for them. Problems are in the foreground of interest these days, and it may be refreshing to contemplate some which can be considered with a degree of detachment.

Similar to most other institutions, physics has both external and internal problems. The external are the more serious in this case and they are less pleasant to contemplate. Some of them are, indeed, on the verge of being threatening. Let me therefore begin with them. My discussion will not be scientifically ordered: I will talk about those questions which bother me particularly, which are in my mind most of the time. This applies to the discussion of both external and internal problems.

The First External Problem: The Problem of Communication

The problem of communication is not new. I myself first spoke about it almost 30 years ago, and it was not new then. Dr. Alvin Weinberg, Director of Oak Ridge National Laboratory, dealt with it most incisively in his *Reflections on Big Science*, and much of what I will say will be a personalized version of what I have learned from him.

Eugene P. Wigner • Department of Physics, Princeton University.

In short, science begins to form such a large body that it threatens to fall to pieces. It has been, for some time, difficult to explain the intricacies of quantum theory to a chemist. The language we have developed and the many concepts we have created are difficult to assimilate for someone whose principal interest is in another field, with questions and with a special language of its own. Nevertheless, chemists, and in particular theoretical chemists, are trying to think in terms of quantum mechanical theory. Sometimes, it seems to us, they rediscover pictures which have long been familiar to us, and the novelty of their discovery is the projection of this picture in a language which we find cumbersome and unnatural. Sometimes they project pictures of aspects of nature about which they know less than we do, and we find these pictures abstruse and unnatural. I am sure that the theoretical chemists could render an equally critical judgment of our own work—both on metallic and on nuclear structure, the two subjects which were principally in my mind just now. Yet, if we want to maintain the unity of science, if we want it to provide a living and elevating picture of the world, physicists and chemists must learn to understand one another, and must understand each other not only in a half-hearted way. The same applies even to the relation between biologists and physicists, and to unite their thinking now appears an even more arduous though *perhaps* not entirely hopeless task.

Alvin Weinberg has not only discussed the problem of the lack of communication among the sciences, he has also proposed some remedies. The first step in this direction, it seems to me, is to become fully aware of the problem and of its implications. Only if we understand the seriousness of the problem shall we be able to strive toward a solution wholeheartedly. Only then will we be willing, and perhaps even eager, to follow Alvin Weinberg's injunctions: to write our articles for a less specialized readership, to devote time and energy to the composition of reviews and to reading more of the reviews covering the results of sister sciences.

I shall add only a few words here to the foregoing. The schism about which I spoke, the communications gap to which I referred, separates not only the different natural sciences such as biology, chemistry, and physics; it yawns also between different parts of physics. Only a few dispersion theorists understand those working in axiomatic field theory; virtually no one whose principal interest is in cosmology and the general theory of relativity knows what the nuclear physicist means by an analogue or even a two-particle–one-hole state. The problem of communication is not only an external problem of physics; it threatens also to develop into an internal one. In fact, so many phenomena are being discovered and investigated, so many new ideas are advanced, that no one can fully keep track of them.

The Second External Problem: The Question of Purpose

What is the purpose of the study of physics or of the natural sciences? Every scientist must have asked himself this question, and most of us have answered it quietly, to the satisfaction of our souls. Our interest in science, and our devotion to the effort to make it deeper and more embracing, has continued as a result. Today, the same question of the purpose of engaging in research, the "relevance," as it is called, of the scientific effort, is asked in strident voices by hundreds.

It is not easy to articulate convictions of purpose, or of values, but I shall try. The first purpose is obvious to all of us: it gives us great pleasure to learn, to expand our knowledge and to add to the knowledge of the human race. I need hardly expound on this here. Some years ago, I expressed the hope that this pleasure will be shared by an increasing part of all humanity.

The second purpose is closely related to the first: it is a noble enterprise to see how far the human mind can go in "understanding" nature, that is, in uncovering hidden relations between the events that we experience. Stephen Weinberg, not a relative of Alvin's, articulated the underlying desire and motive very eloquently when pleading for additional support for high-energy physics.

The third motive which fired scientists in the past was more altruistic than the first two. By expanding our knowledge and applying it to technical problems, we make it possible for more men to lead a carefree life, to devote more time to leisure and enjoyment. It is this purpose—the only truly selfless one—which is being questioned by the experiences of recent years. This is a bitter pill for all of us to swallow: for hundreds of years our spiritual ancestors and we ourselves have endeavored to make life easier for everbody. We are now wondering whether this was a valid objective, whether man can live if he does not have to struggle for tomorrow's bread. If his livelihood is assured, does man not begin to strive in too large numbers after other objectives, such as power; and will the striving of too many, too early in life, not destroy our dearest ideals? When I was young, to earn a decent living was a valid purpose for me. Today, this is taken for granted and young people are deprived of a purpose whose pursuit gave a great deal of satisfaction to me.

These are somber thoughts, deeply disturbing, and I do not want to pursue them now. Let me go over the more technical internal problems of our discipline where I can speak as a physicist.

The two problems I have mentioned—the problem of communication and that of purpose—are not the only external problems of physics. We have probably many others which I have failed to recognize; we also have the problem of the bigness, of the impersonality, of much of modern experimental research. However, these are perhaps, and I do mean perhaps, less pressing than those which I chose to emphasize particularly.

Cosmology: The Problem of the Beginning

The simplest picture of cosmology is that of the big bang: initially, according to this theory, all matter was concentrated in a very small volume, with an enormously high energy density. Matter then spread with tremendous velocity in every direction, undergoing various transformations as the energy density decreased. What we now see is a rather mild later stage of this expansion: the volume of the universe is large, the energy density is small, but the expansion continues. There is much observational evidence to support this conception of the universe and of its origin. The most recent one that I know of is the cosmic black body radiation, corresponding to a few degrees absolute in temperature—the temperature which the big bang theory predicted. Its existence was first confirmed by colleagues of mine in Princeton, by R. H. Dicke's group.

The question then arises: what was the state of the universe before the big bang? This is neither a new nor a popular question. I understand that St. Augustine was once asked: what did the Lord do before He created the world? St. Augustine's answer was, I was once told but perhaps incorrectly, that He created Hell for people who ask such questions. Nevertheless, as far as the big bang theory, in its primitive form, is concerned, the question can be well asked.

According to the primitive form of the big bang theory, there was, originally, a singularity in the metric space and this dissolved at time zero. Few people are inclined to accept this fully. It is more reasonable to assume, instead, that the "original" density was very high but not infinite. One is led to believe, then, that it was the result of a contraction, that is, a collapse, of matter as a result of the gravitational attraction. Similarly, the final state is then not one of infinitely low density: the universe, after expanding for several billion years, will contract again, eventually to a very small volume and correspondingly high density. This will produce, in many billion years, a new big bang, and a renewed expansion, and so on. One is led in this way to the picture of a pulsating universe, alternating between states of very high and very low densities, corresponding to the very small and very large volume which matter occupies.

The picture of a pulsating, that is, essentially periodic, universe seems a very satisfactory one. In particular, if we make a mistake today, our successors, in some billion years, will have a chance to do better. The absolute uniqueness of all events, the absence of any fresh start, is a profoundly perturbing idea.

The trouble is that the physical chemists, early in this century, raised valid objections against the concept of a periodic universe. They pointed out that, as a result of the continuing increase in the entropy, the world will have to come to a steady state, with uniform temperature everywhere, everything in perfect equilibrium. The solutions of the gravitational equations do permit a periodic universe, with infinitely many successive pulsations. However, the periodic solution of the gravitational equations does not correspond to reality because it does not take the inhomogeneities of the actual universe into account. Similarly, one can conceive of a gas in which all atoms fly with equal velocities radially outward from the center of a spherical container, strike this container all at the same time, and are reflected back, and so on. However, actual gases do not have this structure, and the radial oscillations of actual gases in a spherical container eventually are damped out and the gas assumes the state of an equilibrium. Similarly, the final state of equilibrium, the state that the physical chemists called the "heat death" (*Wärmetod*), is an unavoidable consequence of our present concepts. This, then, leads us to the picture of oscillations of the universe with diminishing amplitudes—not as satisfactory as the picture of a truly periodic universe, but a conceivable one.

The trouble is again that the initial state cannot be conceived very well. The entropy always increased—what was it originally? As a result of the quantum conditions, we know that the entropy has an absolute minimum, zero. Was it ever zero? If so, why and when did it start to increase? Or, was it never actually zero, but infinitely close to this value in the infinitely distant past? We do not know. Both alternatives are perturbing and we simply do not have a cosmological picture "in the large." which does not contradict some simple and general law of physics. John A. Wheeler, also a colleague of mine in Princeton, has devoted a great deal of thought to questions similar to the one I raised, but no fully satisfactory answer has emerged so far.

Epistemology: The Role of the Observer

The epistemology forced on us by quantum mechanics has proved, for some of the most eminent physicists, very difficult to accept. The problem can be formulated in many different ways. Perhaps the most simple formulation is as follows. The equation of motion of quantum mechanics, such as the time-dependent Schrödinger equation, are causal in the sense that knowing the state of an isolated system now (i.e., knowing its state vector) permits one to determine its state (its state vector) for all future time. Nevertheless, there is a fundamental stochastic element in the theory: the result of an observation (also called measurement) on the system can yield several results and only the probabilities of the various results can be predicted.

This situation, then, leaves several alternatives. One can assume, as did Von Neumann, that the observation is a process which cannot be described by the equations of motion of quantum mechanics, that it involves an interaction with an individual, and such interaction is outside the domain of validity of the quantum equations. This is an entirely possible point of view but involves a duality—interactions that are and that are not subject to the quantum equations—which many dislike. It is possible to assume that the state vector of the observer is complicated, that it is never known, and that the result of the observation depends on the particular state, among the many undistinguishable states, in which the observer was before carrying out his measurement. This second alternative is mentioned only as a reasonable suggestion; it has been shown, both by d'Espagnat and by me, that it cannot account for the probability laws postulated by the quantum theory of observation. Finally, it is possible to avoid Von Neumann's dualism by eliminating the concept of the state vector from the theory, or equivalently by considering it to be only a mathematical tool, and formulating the laws to begin with as probability connections between subsequent observations. Thus, in the case of the simplest quantum mechanical system, that of a spin $\frac{1}{2}$, one can say that, if an observation yielded the result that the spin assumed a definite direction rather than the opposite one, the probability that an observation whether it assumes another direction (rather than the direction opposite to it) is $\cos^2 \frac{1}{2}\vartheta$ where ϑ is the angle between the original and the second direction. Naturally, the probabilities for the different possible behaviors of more complicated systems, when subjected to more complicated observations, are not so easy to give in terms of equally explicit formulas. However, even for complicated systems, it is possible to eliminate the concept of the state vector as a basic concept from the theory and formulate the laws of the theory directly in terms of the probabilities of the outcomes of observations—after all, these are the quantities in which one is ultimately interested.

Whether one adopts Von Neumann's terminology or the one just suggested, the observer occupies a particular position in the theory. It is *his* observations that are the subject of the theory; the predicates of the theory are the probabilities with which he observes one or another outcome of the observation.

This separation of the world into two parts, the observer and the outside world, has many unpleasant consequences. These formed, unquestionably, the reasons for the reluctance of Einstein, of Schrödinger, to accept the quantum epistemology at face value. I myself, though convinced that no other epistemology is compatible with the principles of quantum mechanics, have often pointed to the difficulties and concluded that new principles will be needed when the phenomena of life and of consciousness become subjects of the theory. Among the many difficulties, I now mention only the one which relates to the preceding discussion: since the theory is intended to foresee only results of observations, it is doubtful whether it considers cosmology to present meaningful questions. Yet we all feel that it does.

The particular problem which I want to point out now is a different one, however. It has been raised by the German physicist Zeh. He points out that if the observer is macroscopic—and in the ultimate analysis he always is—he cannot be separated from the rest of the world. The energy levels of a macroscopic object are so close together that the electric charge of a single electron, 1 mile away, can cause transitions between these or, more generally, can change their state. Hence, in the quantum sense, an isolated macroscopic body does not exist in practice and the dividing line between observer and the rest of the world is unrealistic.

Evidently, this is a new chink in the armor of quantum mechanics' epistemology. It is a chink, however, which is subject to investigation while one can still stand on the foundations of present theory. It should be solvable, though it is at present an unsolved problem.

Conclusion

I have spoken about two external and two internal problems of physics. The former give us pain, the latter pleasure. We hope the former will not proliferate; we could add many others to the latter. Will this ever end? Will we ever run out of problems? Or, what I fear much more, will we ever get tired of them? I hope and pray that we will not.

Adolf Grünbaum

IS THE COARSE-GRAINED ENTROPY OF CLASSICAL STATISTICAL MECHANICS AN ANTHROPOMORPHISM?

I am delighted that the editors of this volume are violating the injunction "*Fest* no *Schriften*," issued in a recent letter published in *Science*. As a former student of Professor Margenau, I am glad to have this opportunity to join in a tribute to him. In his course on the foundations of physics at Yale, I came to appreciate that classical statistical mechanics contains some important lessons for the philosophy of physics despite the physical inadequacies of that prequantum theory. The present paper is intended to argue for what I believe to be a philosophical moral of the status of the coarse-grained entropy of classical statistical mechanics.

To introduce the issue posed in the title of this essay, let me caution against a possible misconstrual of the classification "anthropomorphism" as a pejorative characterization of an attribute or entity postulated by a physical theory. It is clear enough that those who charge a *particular* concept employed in such a theory with being an anthropomorphism are *not* concerned to convey thereby the following truism: *qua being a theory*, any known theory whatever is not only propounded and devised by man (or by a humanoid computer), but is also unavoidably *fallible* because of the

Adolf Grünbaum • Department of Philosophy, University of Pittsburgh.

postulational (inductive) risks inherent in the very logic of its intellectual construction. Thus it is surely *not* the latter truism that E. T. Jaynes was interested in espousing when he wrote: "entropy is an anthropomorphic concept, not only in the well-known statistical sense that it measures the extent of human ignorance as to the microstate. *Even at the purely phenomenological level, entropy is an anthropomorphic concept.* For it is a property, not of the physical system, but of the particular experiments you or I choose to perform on it" (Jaynes[(1)]).* What may be overlooked, however, is that those who level the charge of anthropomorphism at a particular, specified ingredient of a physical theory ought *not* to be saddled with the following claim: for any and every item in such a theory, an unambiguous identification can always be made as to whether or not the item is an anthropomorphism within the postulational framework of the given theory. Nor should such a dichotomous claim be imputed to those who *deny* a *specific* charge of anthropomorphism in physics! This is not to say that the notion of anthropomorphism has not been employed in an avowedly dichotomous fashion, as, for example, in L. Kronecker's Pythagorean dictum "God made the integers; all the rest is the work of man" (*cf.* Bell[(2)]).

When considering the respective logical constitutions of the phenomenological and statistical entropies ascribed to a physical system, Jaynes singled out the extent of the role of human decision in the former and of human ignorance in the latter as touchstones for characterizing, or perhaps even indicting, them as anthropomorphisms. Whereas Jaynes was concerned to spell out his view of the bearing of human choice or decision on the status of the *phenomenological* entropy, I shall not deal with the latter kind of entropy at all. Instead, I shall focus on the significance of human volition in the logical constitution of the statistical coarse-grained entropy of classical statistical mechanics by addressing myself to the following question: is the import of the role of human choice and decision for the *physical* credentials of that statistical entropy such as to sustain the charge of anthropomorphism against it?

As I shall set forth below in detail, such a charge might be made plausible by reference to the following fact: the occurrence and direction of a temporal change of the entropy $S = k \log W$ assigned to a given physical system depend essentially on *our human choice* of the *size* of the finite equal cells or boxes into which we partition the six-dimensional

*Italics in the original. Jaynes mentions in his Acknowledgments that he first heard the remark "Entropy is an anthropomorphic concept" from E. P. Wigner.

position-velocity phase space ("μ-space"). In the prequantum classical statistical mechanics, there is enormous scope for the exercise of such choice in coarse-graining. For classical theory viewed action as a magnitude susceptible of having any value, and Planck's quantum of action h is unavailable in that theory to provide a *physical* basis for carving up the phase space into boxes each of which has the volume h^3. (*cf.* d'Abro[(3)]). It is precisely because there is such relatively great latitude for human decision in coarse-graining the specified phase space of *pre*quantum statistical mechanics that I wish to examine the soundness of the charge of anthropomorphism as leveled against the statistical entropy of that classical theory. It will then be a corollary of our analysis to determine whether the physical significance of that entropy can be validly impugned by regarding it as a measure of the extent of human ignorance as to the underlying microstate.

But before we can deal with the issue of anthropomorphism we must give some detailed consideration to the aforementioned dependence of the temporal change of the entropy on the human decision as to the partitioning of the phase into cells.

Entropy Change and Arbitrariness of the Partitioning of Phase Space

One of the things we must examine is the effect of *alternative* choices of a partition on the *direction* in which the entropy changes with time as the physical system develops. Let us briefly recall the meanings of some of the fundamental terminology amid explanation of the special notation I shall use to indicate for each of the various pertinent quantities to what particular partition they are being referred.

Consider one particular system of particles of given finite total energy and given finite three-dimensional volume. Call that system α. Let the portion of six-dimensional phase space whose occupation by the representative points of these particles is compatible with the given energy and physical dimensions be divided into a finite number m of equal cells $c_1, c_2, \ldots, c_m$. Also, let the number of particles which occupy, respectively, these various particular cells at a given time by $n_1, n_2, \ldots, n_m$. Such a set of numbers $n_i (i = 1, 2, \ldots, m)$ specifies the *macro*state of the system by stating for each of the cells how many particles occupy it at the time. Thus the macrostate is characterized by the numerical *distribution* of the total finite number n of particles among the various particular cells, irrespective of which individual particles these may be. Suppose, for example, cells c_1 and c_2 were occupied, respectively, by unequal numbers of particles x and y at time t_1, but then were occupied by y particles and x particles, respectively, at time t_2. Then, since $x \neq y$, the macrostates (distributions) I and II at these two times t_1 and t_2 would be *different*, even though we assume that the numbers of particles in each of the remaining cells $c_3, c_4, \ldots, c_m$ had not changed at all. For the macrostate does depend on the identity of the *cells* among which differing numbers of particles are distributed, although the *macro*state does not depend on the identity of the *particles* which are thus distributed.

The identity of the particles themselves is, of course, relevant to the *micro*state (arrangement, complexion). The latter is specified by stating for each of the n particles in which one of the m equal six-dimensional cells of the coarse-grained partition its representative point lies at a given time. By thus being specified only to within the finite cells of the chosen partition, the microstate as such is *relative to the chosen partition.* By contrast, the *precise* microstate of classical mechanics is given by the exact or punctal values of the six position-velocity coordinates and is therefore *independent* of the chosen partition. To distinguish the latter kind of microstate from the former, I shall use the qualifying adjectives "precise" or "punctal." Although the unqualified term "microstate" will always refer to the partition-dependent, *non*punctal kind of microstate, I shall occasionally emphasize that this relative kind of microstate is intended by using the adjective "nonpunctal" or "coarse-grained."

For any one chosen partition, each kind of macrostate or distribution is clearly constituted by a certain class of microstates: any occurrence of one of the members of this class of microstates constitutes an occurrence of the macrostate in question. The number W of different microstates belonging to the particular macrostate specified by the set of distribution numbers $n_1, n_2, \ldots, n_m$ is given by $W = n!/(n_1!, n_2!, \ldots, n_m!)$. (It will be convenient in the sequel to denote the product of the factorials in the denominator of this expression by "$\pi n_i!$") And the entropy S assigned to any one macrostate to which W microstates belong is $S = k \log W$. Since increases or decreases of W are tantamount, respectively, to increases or decreases of S, we can conveniently work with W below rather than with S when calculating the directions of entropy changes. Indeed, it should be borne in mind that it will be permissible for our particular purposes to talk about W and S interchangeably.

Since multiplication in the product $\pi n_i!$ is commutative, our aforementioned two *different* macrostates I and II are each assigned the same entropy S. Indeed, *any* distribution of the n particles of the given system that gives rise to the same value of the product π will be assigned the same value of the entropy. Thus one and the same entropy state S_0 of a given system can correspond to *different* distributions or macrostates *each* of which has W_0 microstates belonging to it, where the value of W_0 is given by $S_0 = k \log W_0$. Hence the number of different microstates which can underlie or belong to one and the same entropy state S_0 is given *not* by W_0 but rather by the *product* of W_0 with the number d_0 of different distributions having the same entropy S_0.*

Now let A and B be two different kinds of precise, punctal microstates in which the given physical system α may be at different times. Denote a particular partition of the corresponding phase space by the numeral 1. And let $1A$ and $1B$ be the respective coarse, *non*punctal microstates of the system *with respect to partition 1*, when it is in the respective punctal microstates A and B. We shall generally be interested in cases in which the *non*punctal microstates are different from one another in kind, no less than the punctal ones.

If we consider the characterization of the system in states A and B to within an *alternative* partition 2, then $2A$ and $2B$ are the respective nonpunctal microstates of the system with respect to partition 2. Clearly, the one precise microstate A gives rise to the two different *non*punctal microstates $1A$ and $2A$, since the latter are specified only to within the cells of their different partions. And similarly for B, $1B$, and $2B$.

Returning to partition 1, we note that $1A$ and $1B$ each uniquely specify the distribution (macrostate) to which they belong with respect to partition 1. These distributions may or may not be the same, but in either case we shall denote them by $D1A$ and $D1B$, respectively. Furthermore, we shall use W_{D1A} to represent the number of microstates belonging to $D1A$. For important reasons that we shall discuss later on, with respect to *the particular partition to which it pertains*, the quantity W_{D1A} does *triple duty* as follows: (i) as just noted, it is the number of different microstates belonging to the given distribution (macrostate) $D1A$; (ii) it is the (unnormalized) probability of occurrence of that distribution in the time ensemble of the macrostates which the system attains relatively to partition

*The fact that $W_0 \cdot d_0$ microstates rather than only W_0 of them underlie an entropy state S_0 was erroneously overlooked on p. 162 of my paper "The anisotropy of time."[(4)] Hence the statement of boundary conditions governing the so-called branch systems given there must be amended: either the random samples of the formulation must be particularized to each distribution of given entropy, as I shall do later on in this section, *or* the random samples must be taken from the *total* number $(W_1 \cdot d_1)$ of microstates which can underlie an entropy state S_1 of the given kind of system.

1, i.e., the so-called thermodynamic probability of the macrostate in question; and (iii) it is a measure of the degree of *homogeneity* or evenness, equalization, disorder, and "well-shuffledness" of the given macrostate *D*1*A* in the class of macrostates which the system attains relatively to partition 1. This *triple significance* of the quantity *W* will turn out to have an important bearing on our central issue of anthropomorphism later on. Using the above notational devices, we shall say that S_{D1A} and S_{D1B} are the entropies of macrostates *D*1*A* and *D*1*B*, respectively, i.e., the entropies of the system relative to partition 1 when it is in the respective precise microstates *A* and *B*. By the same token, the macrostates which *A* and *B* constitute relatively to partition 2 have entropies S_{D2A} and S_{D2B}, respectively.

We shall now illustrate the following important sets of facts concerning the bearing of the chosen partition on the direction of the entropy changes of the given system α:

i. There are partitions 1 and 2 of the corresponding phase space as well as precise microstates *A* and *B* such that if the system evolves from *A* to *B*,

$$S_{D1A} = S_{D1B} \qquad \text{but} \qquad S_{D2A} \neq S_{D2B}$$

i.e., although the system does *not* change its entropy relatively to partition 1 in the course of its transition from *A* to *B*, its entropy does change relatively to partition 2. Thus the system's two macrostates *D*1*A* and *D*1*B* with respect to partition 1 are equiprobable, whereas its two macrostates *D*2*A* and *D*2*B* with respect to partition 2 have different probabilities. There is no inconsistency between these various probabilities.

ii. There are partitions 3 and 4 as well as precise microstates *E* and *F* such that if the system evolves from *E* to *F*,

$$S_{D3E} < S_{D3F} \qquad \text{but} \qquad S_{D4E} > S_{D4F}$$

i.e., although the entropy of the system *increases* relatively to partition 3 in the course of its transition from *E* to *F*, it *decreases* relatively to partition 4. Since these two entropy changes have opposite signs, the *probability ranking* of *D*3*E* and *D*3*F* will be opposite to that of *D*4*E* and *D*4*F*. There is no inconsistency between these various probabilities.

To make our diagramming and arithmetic quite simple, we shall represent the six-dimensional phase space of the system α by a mere rectangle, and we shall choose four convenient numbers *m* of equal boxes (cells) to generate the four different partitions of the representative rectangle. Also, it will simplify our formulation without detriment to rigor to speak of the particles themselves rather than of their representative points as occupying certain cells in phase space.

Case (i)

We shall, of course, not label and depict the individual n particles, since we shall not diagram the precise microstates A and B but only their corresponding respective pairs of macrostates relative to partitions 1 and 2. The first partition will consist of two boxes ($m = 2$), while the number m of boxes of partition 2 will be 4.

A macrostate relative to a given partition is specified *only to within the particular cells* of that partition. Hence the shading of a given box in the diagram will signify that the positive integral number of particles appropriate to the given distribution of particles is to be found *somewhere or other* in that box. If a given box is not shaded, then it contains no particles *anywhere* within it.

Let the precise microstate A be such that the first of the two boxes of partition 1 contains half of the particles *somewhere within it*, while the second, of course, contains the remaining half of the n particles *somewhere within it*. Thus $D1A$ is specified by the sequence $n/2$, $n/2$. Let B be a precise microstate different from A but such that the macrostate $D1B$ which it generates relatively to partition 1 is the same as $D1A$. Thus we diagram:

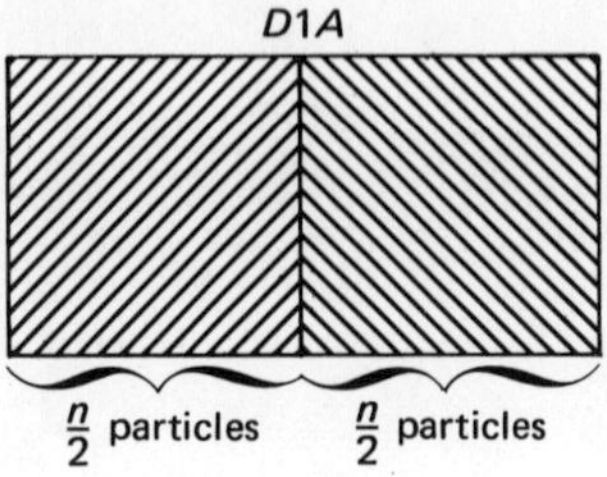

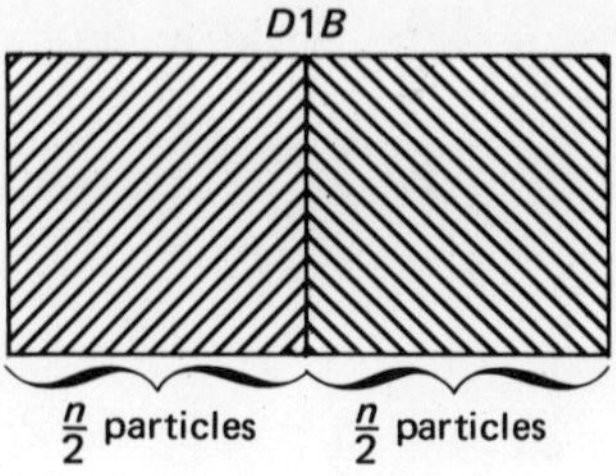

Since A and B generate identical distributions $D1A$ and $D1B$ relatively to partition 1, their thermodynamic probabilities relative to that partition will have the same value, *viz.*, $n!/[(n/2)!]^2$. Hence relatively to partition 1, the system will have the same entropy when it is in punctal microstate A as when it is in the different punctal microstate B, and we can write

$$S_{D1A} = S_{D1B}$$

as claimed initially.

Turning to partition 2 ($m = 4$), it is clear that A and B and the integral number n of particles can also be such as to generate the following quite different distributions relatively to partition 2: in macrostate $D2A$, boxes 1, 2, 3, and 4, respectively, contain $n/2, 0, n/2$, and 0 particles, whereas $D2B$ involves the uniform presence of $n/4$ particles in each of the four boxes. A and B can and must be so chosen that the requirement of energy conservation and other relevant constraints on the closed system α are indeed satisfied.

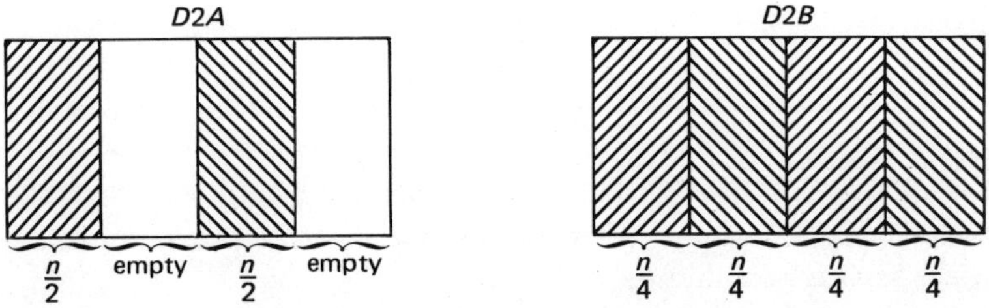

Hence we have

$$W_{D2A} = \frac{n!}{\frac{n}{2}!0!\frac{n}{2}!0!} = \frac{n!}{\left(\frac{n}{2}!\right)^2}$$

a value which, incidentally, is the same as that of W_{D1A} and W_{D1B} above. But

$$W_{D2B} = \frac{n!}{\left(\frac{n}{4}!\right)^4}$$

Since

$$\left(\frac{n}{4}!\right) < \left(\frac{n}{2}!\right)^2$$

the quantities W pertaining to the different macrostates $D2A$ and $D2B$ satisfy the inequality $W_{D2A} < W_{D2B}$, so that we can write

$$S_{D2A} \neq S_{D2B}$$

as claimed initially.

We have illustrated the joint validity of the relations

$$S_{D1A} = S_{D1B} \qquad \text{and} \qquad S_{D2A} \neq S_{D2B}$$

by reference to a case in which $m = 2$ for partition 1 while $m = 4$ for partition 2. Thus in our particular illustration the entropy of the system is *constant* for the transition from A to B relatively to the partition possessing the *smaller* number of cells. Thus our particular illustration is one in which the entropy *changes* relatively to the partition having the *larger* number of cells. But it is easy to provide another illustration of the joint validity of the above entropy relations in which the *constancy* of the entropy obtains relatively to that partition which has the *larger* number of cells. Hence consider the *general* case in which a transition of the system from A to B involves no entropy change relatively to one of two partitions while producing an entropy change relatively to the other. Then we can say that the *change* in entropy is *not* generally correlated with the partition containing the larger number of cells. In other words, the success or failure of one of two partitions in effecting an entropic differentiation between A and B is *not* generally a matter of the relative coarseness of the partition.

It is patent that there is no inconsistency between (a) the assertion $W_{D1A} = W_{D1B}$, when construed as claiming the *equiprobability* of the two macrostates $D1A$ and $D1B$, and (b) the assertion $W_{D2A} \neq W_{D2B}$, when interpreted as affirming that the distributions $D2A$ and $D2B$ have *different* probabilities of occurrence in time. There are two different time ensembles of macrostates which our physical system α exhibits relatively to the two different partitions 1 and 2. In one of these two time ensembles, the two *different* distributions $D2A$ and $D2B$ generated, respectively, by A and B relatively to partition 2 can readily have *unequal* probabilities of occurrence, while the identical distributions $D1A$ and $D1B$ which A and B generate relatively to partition 1 will, of course, have one and the same probability. For the different probability *rankings*, i.e., the respective unequal and equal probabilities corresponding to these two partitions, are *not* ascribed to the pair of precise microstates A and B themselves but only to the two different pairs of macrostates $D1A$, $D1B$ and $D2A$, $D2B$ which are associated with the one pair A and B relatively to the two differing partitions. The consistency of the two different probability rankings corresponding to the two partitions we have diagrammed above is made palpable by these diagrams as follows: it would be a blatant physical falsehood to claim that the different macrostates $D2A$ and $D2B$ depicted in our second diagram are equiprobable, but it is plain that the two identical distributions of our first diagram are equiprobable.

Case (ii)

A further important facet of the bearing of the chosen partition on the entropy changes of the system α will now emerge from illustrating the *oppositely directed* entropy changes

$$S_{D3E} < S_{D3F} \qquad \text{and} \qquad S_{D4E} > S_{D4F}$$

stated as (ii) above.

I am indebted to Allen Janis for the particular illustration which is about to follow and for a very helpful discussion of its significance for the entropy statistics of ensembles of systems.

Let partition 3 consist of nine boxes, and let the precise microstates E and F as well as the number n of particles be such that the following two distributions are generated in the course of the system's transition from E to F: $D3E$ is constituted by the presence of *all* of the n particles in the first box of partition 3, while $D3F$ is a uniform distribution of the particles among the nine boxes, so that each of the nine boxes of partition 3 contains $n/9$ particles. Hence we have

$$W_{D3E} = \frac{n!}{n!} = 1 \qquad \text{and} \qquad W_{D3F} = \frac{n!}{\left(\frac{n}{9}!\right)^9}$$

Without computing the value of the second of these two probabilities, it is evident that many more than one nonpunctal microstate belong to $D3F$, while only one such microstate belongs to $D3E$. Therefore, $W_{D3E} < W_{D3F}$, and $S_{D3E} < S_{D3F}$, as claimed initially.

Without specifying the actual number m of cells in partition 4, let that number (as well as the states E and F) be such that $D4E$ and $D4F$ have the following respective properties: in macrostate $D4E$, each box of the comparatively fine partition 4 contains *at most one* particle, whereas in macrostate $D4F$ each cell of this partition contains either two particles or none. Note that in the case of $D4F$, the n particles are so distributed that there are $n/2$ cells, each of which contains exactly two particles while the remaining cells are each empty. Hence we have

$$W_{D4E} = \frac{n!}{1} = n! \qquad \text{and} \qquad W_{D4F} = \frac{n!}{(2!)^{n/2}}$$

Evidently $W_{D4E} > W_{D4F}$, so that $S_{D4E} > S_{D4F}$, as claimed initially.

Therefore, during the transition of the system α from E to F, the entropy relative to partition 3 *in*creases whereas the entropy with respect to partition 4 *de*creases. For reasons of the kind discussed under case (i) above, the probability ranking $W_{D3E} < W_{D3F}$ is entirely *consistent* with the *opposite* probability ranking $W_{D4E} > W_{D4F}$.

Joseph Camp has given me an illustration of the existence of partitions 5 and 6, and of a temporal sequence of precise microstates A, B, and C such that both

$$W_{D5A} < W_{D5B} < W_{D5C} \quad \text{and} \quad W_{D6B} < W_{D6C} < W_{D6A}$$

We have set forth the *relativity to the chosen partition* of the very occurrence *and* direction of an entropy change in *one* particular physical system α during its transition from one punctal microstate to another. This relativity to the chosen partition for a single system α raises the important question of whether the *statistics* of temporal entropy change in large but finite *ensembles* of like closed systems are similarly partition-dependent and might thereby be claimed to depend indirectly on human choices.

To deal with this question, let us be more precise as to the nature of the ensembles of systems to which it pertains. The systems in each and every ensemble will be alike: each system will consist of, say, the same kind of gas confined to regions of three-dimensional space having the same volume and possessing the same total energy. In *one* of these ensembles, say K_1, to which our one system α above belongs, each system is in an initial state which is the same kind of distribution (macrostate) as $D3E$ relatively to partition 3, and the same kind of distribution as $D4E$ relatively to partition 4. But the precise microstates which underlie these same initial distributions in the various members of K_1 are emphatically *different* in kind from E in all except α; i.e., the precise microstates all differ from one another in kind. Moreover, the respective nonpunctal microstates which generate the initial macrostate $D4E$ in the various systems of K_1 will *each* be a *random sample* of the FINITE set of W_{D4E} different microstates compatible with $D4E$ in the pertinent member of K_1. I recognize that the concept of a random sample or of a random selection is not wholly unproblematic as characterized in the following standard kind of definition: "A sample obtained by a selection of items from the population is a *random sample* if each item in the population has an equal chance of being drawn. *Random* describes a method of drawing a sample, rather than some resulting property of the sample discoverable after the observance of the sample."(5) But I assume that such difficulties as may beset the concept of a random sample in its application to *finite* sets—which is the application I have made of it here—can be resolved as follows: the clarified concept of random sample accords with the physical correctness of the statistical claim that I shall make below concerning the entropic behavior of the ensembles of systems which we are engaged in discussing. In the case of the initial distribution *D3E*, there is only one underlying microstate for each system since $W_{D3E} = 1$, and hence this random sample requirement is trivially satisfied. All the members of K_1 have, of course, the same initial entropy S_{D3E} relatively to partition 3, and also the same initial entropy S_{D4E} with respect to partition 4.

A *second* ensemble, K_2, resembles K_1 in that the initial macrostate of its members relative to partition 3 has the same entropy S_{D3E}, and similarly for the initial macrostate of entropy S_{D4E} relative to partition 4. But the members of K_2 differ from those of K_1 in that their initial macrostates relative to our two partitions are constituted by distributions which differ in kind from *D3E* and *D4E*. Except for the latter difference in the characteristics of K_1 and K_2, the other statements I made about the initial conditions governing the members of K_1 apply, *mutatis mutandis*, to those of K_2. By the same token, we can consider further ensembles of systems $K_3, K_4, \ldots, K_n$, where n is a finite integer, in *each* of which the systems start out in entropy state S_{D3E} and in entropy state S_{D4E} but with the following difference between the various ensembles: relative to a *given* partition, the kind of distribution which constitutes the initial entropy state of the systems in *one* of the ensembles *differs* from the kind of distribution constituting the self-same initial entropy state of the systems in any *other* ensemble. Furthermore, the cardinality of each ensemble is finite though very large, say $n!$, where n is the number of particles in each system and is of the order of 6×10^{23}. Therefore, the union U of our ensembles will contain only finitely many systems. And hence it will be meaningful to make an assertion about a *majority* of the members of U.

We have been characterizing our ensembles with respect to two particular partitions, 3 and 4. But clearly we can introduce at least a denumerable infinity of other partitions and can characterize our ensembles relatively to them, *mutatis mutandis*, in the same terms as we did with respect to 3 and 4. Hereafter we shall therefore not, in general, be restricted to any particular finite number of partitions.

The precise microstates E and F of our primary single system α were separated by some particular time interval Δt. We saw that during α's transition from E to F in the time Δt the direction of its entropy changes was partition-dependent: the entropy of α increased relatively to partition

3 while decreasing with respect to partition 4. Let us now consider each of the at least temporarily closed member-systems of our various ensembles $K_1, K_2, \ldots, K_n$ at two different times t and $t + \Delta t$ in order to point out a fact of basic importance to our concerns: there is complete compatibility between the partition *dependence* of the direction of the entropy changes undergone by α, on the one hand, and, on the other hand, the following partition *invariance* of the temporal statistics of entropy change in the union U of our ensembles of systems: for any and every partition, either the common initial entropy of all the systems in U will have increased *in a vast majority* of the systems by the time $t + \Delta t$, or it will have remained the same.* Note that we here used the indefinite article "a" rather than the definite article "the" when speaking of "*a* vast majority of systems."

For it is crucial to emphasize that the partition *invariance* of an entropy increase or of entropy constancy is being asserted only in the sense that, for every partition, *some majority or other* of the systems in U will exhibit the specified entropic behavior. Thus the partition invariance of the stated *entropic behavior* of a majority M of the systems in U does *not* at all require that the *membership* of the set constituting M relatively to a particular partition be itself partition invariant! Evidently, the stated partition invariance of the statistics of entropy change fully allows our finding above that during its transition from E to F the single system α underwent specified oppositely directed entropy changes relatively to partitions 3 and 4. The latter finding does illustrate that the *membership* of M is NOT partition invariant as follows: relatively to partition 3, the particular system α does belong to M, whereas, relatively to partition 4, α belongs to the *minority* of systems which decrease their entropy during Δt.

The claim that α thus belongs to what is only a *minority* of systems undergoing an entropy decrease relatively to partition 4 appears to be quite justified. The reason is that even within the single ensemble K_1, which is included in U, in all likelihood *most* systems that started out in macrostate $D4E$ with entropy S_{D4E} will *not* have evolved, after an interval Δt, into any distribution which is such that each cell of partition 4 contains either exactly two particles or zero particles; nor into any other distribution possessing the lower entropy S_{D4F} of the latter kind of distribution; nor yet into any distribution whose entropy is lower than S_{D4E}. Relatively to

*For a statement of the physical reasons underlying this statistical claim as such, see pp. 162–163 of my "The anisotropy of time"[(4)] cited in a prior footnote, as amended in that footnote.

partition 4, only a certain minority of systems decrease their initial entropy S_{D4E} during Δt, and α does belong to that minority with respect to this particular partition. But relatively to partition 3 a *different* minority of systems—though still only a minority!—decrease their initial entropy S_{D3E} during Δt, and α does *not* belong to that minority relatively to the latter partition; instead, with respect to partition 3, α increases its entropy. Thus our partition-dependent findings concerning the system α accord entirely with the partition invariance of the entropy statistics for the class U of systems. *Mutatis mutandis*, the same partition invariance holds for the finitist entropy statistics in the respective unions of ensembles of *other* kinds of *like* closed systems.

Hence we draw the following very important conclusion: although the direction of the entropy change exhibited by a single system such as α is indeed relative to the chosen partition, *no such relativity to the chosen partition characterizes the statistics of entropy change in the specified ensembles of systems*. For there is an invariance in the entropic behavior of a majority of the systems in U amid the relativity to the partition of the membership of that majority. I have called attention to the partition invariance of the latter statistics of entropy change in order to discuss the bearing of the *physical significance* of this invariance on our central issue of anthropomorphism. Before we can turn to giving a statement of that bearing, we must deal more explicitly than above with the following triple status of the entropy: as we recall, it is the logarithm of a number W which is a measure of the degree of homogeneity as well as of the probability of occurrence, and not just the number of underlying microstates.

What Is the Physical Significance of the Triple Role of the Entropy for the Entropy Statistics in the Class U?

For the time being, our attention will be devoted to an individual physical system and only later will we turn to our class U of systems.

Entropy as a Measure of the Degree of Homogeneity

Perhaps it is possible to develop a notion of the degree of homogeneity or equalization of a distribution which is *not* relativized to a partition of the phase space. At least in regard to the speeds of the particles, the possibility of such a measure is suggested by the fact that the Maxwell distribution of molecular velocities for the *equilibrium* state of uniform temperature gives a *partition-independent maximum spread* among the speeds of the particles, the spread being maximum within the confines of the fixed total energy of the system. And perhaps some function depending on the spatial distances or spatial spread among the n particles as well as on their velocity spread could be introduced to provide a partition-independent measure of homogeneity for the various macrostates of the system. To my knowledge, no such partition-*independent* measure of homogeneity has been worked out for six-dimensional position–velocity space. As noted on p. 16, homogeneity is often called "disorder."

In any case, let us employ a concept of degree of homogeneity which is relativized to a partition. Hence we ask whether the intuitively suggested, avowedly partition-dependent attribute of degree of homogeneity of a distribution D is given numerically by the quantity $W = n!/\pi n_i!$, *initially* construed as merely the number of different microstates belonging to D. After mentioning some obvious reasons for an affirmative answer to this question, I shall call attention to cases in which our intuition of *qualitative* relative homogeneity gives no clear-cut verdicts, in order to point out that I see no conflict between intuition and the homogeneity rankings furnished by the function W for these cases.

As for the obvious reasons, we observe that W is indeed maximum when the n particles are evenly allocated to the various cells, as compared to more or less *un*even allocations among the cells of the given partition. Furthermore, two distributions that differ only in that the sequence of particular, generally unequal numbers $n_1, n_2, \ldots, n_m$ which specifies one of them is obtained by some permutation of the particular numbers specifying the other, clearly have the same degree of homogeneity on intuitive grounds. For, intuitively, the degree of homogeneity does *not* depend on the *identity* of the particular cells in which given numbers of particles are distributed. And hence by assigning the same number to two distributions which differ in only this way, the function W implements our intuitive idea of relative degree of homogeneity in such cases.

But we can see that the function W also assigns the same number to certain distributions which differ far more strongly from one another. To see this, note first that for a sufficiently large number of cells ($m \geq 4$), any given value of the product $\pi n_i!$ ($i = 1, 2, \ldots, m$) resulting from one particular distribution is capable of various decompositions into factorials, such that $\sum_{i=1}^{i=m} n_i = n$ *and* such that the sequences of factorials in the different decompositions are *not* obtainable from one another by permutation. An arithmetically simple example of this kind of nonunique decomposition into *different* sets of m factorials is furnished by the physically uninteresting case $n = 7$. Note that $24 = 4!1!1!1! = 3!2!2!0!$, while the condition $\sum_1^m n_i = n$ is satisfied, since $4 + 1 + 1 + 1 = 3 + 2 + 2 + 0 = 7$. Thus the function W assigns the same degree of homogeneity even to those distributions which differ such that they are *not* obtainable from one another by moving the various *fixed* numbers of particles into other cells of the partition. Yet it seems to me that our qualitative intuition of degree of homogeneity does not interdict the verdicts of equihomogeneity furnished by W for distributions that differ from one another as just specified.

Thus we can say that the numerical measures furnished by the function W render our *qualitative* intuitive idea that the relative degree of homogeneity of a distribution with respect to a given partition P is determined by *how evenly* the particles are distributed among the cells of P, the degree of evenness being a matter of how many different cells contain more or less nearly equal numbers of particles. And since we can speak of W and S interchangeably for our purposes, we can say that the entropy S provides a measure of the degree of homogeneity.

Entropy as a Measure of the Probability of Occurrence of a Distribution

Our concept of degree of homogeneity is relativized to a partition, and that degree is indeed measured by the entropy. Hence to say that a macrostate is a state of high entropy does entail that it is a highly homogeneous distribution relatively to the partition with respect to which the given macrostate is both defined and of high entropy. But the assertion that high entropy states are states of great homogeneity does *not* thereby become a tautology. For to say that a macrostate is a distribution D of high entropy relatively to a partition P is to say much *more*, physically speaking, than that D is highly homogeneous relatively to P: it is to say *as well* that D has a high probability of occurrence in the time ensemble of macrostates which the given system attains relatively to the partition P. Hence the entropy is a property of D which has at least the following objective macroscopic physical significance. It links or correlates the probability of the occurrence of a macrostate in time with its degree of homogeneity. I do not see how the physical significance of this important connection can be gainsaid, just because the entropy can *also* be regarded as a measure of the extent of human ignorance as to the actual microstate in the following sense: in any given occurrence of a certain macrostate D, only *one* of the W microstates that belong to D is actually being physically realized, but we don't know which one it is.

Hence I deny that W, as ascribed to the macrostate of the system at a given time, has a significance which is confined to the fact that, for all we know, *any one* of W different microstates is actually being realized at the time.

Even in its role as the mere number of different microstates belonging to a given kind of distribution *D*, the quantity *W* plays a physically relevant role and seems to me to be misleadingly belittled by being dubbed a measure of human ignorance. My reason for this claim will become clear after first noting the following: As is vouchsafed by the Birkhoff–von Neumann quasi-ergodic hypothesis, each one of the m^n microstates of which the permanently closed system is capable actually occurs *with the same frequency in time* or has the same probability $1/m^n$. And this ascription of equiprobability to all microstates is therefore *not* an assertion expressive of our lack of knowledge as to the actual (relative) frequency of the different microstates. In other words having been justified by the quasi-ergodic hypothesis, this claim of equiprobability of microstates does *not* rest on an argument from *in*sufficient reason, akin to the Laplacian's invocation of the principle of equi-ignorance (indifference) in defense of his *a priori* probability metric for dice. Let me therefore take *this* physical and epistemological status of the equiprobability of all microstates for granted. Then I can justify my contention above that even *qua* just being the number of different microstates belonging to a given distribution *D* the quantity *W* is physically relevant, although only one of the *W* microstates belonging to *D* is being realized in any given occurrence of *D*. For note that since all microstates of which the closed system is capable actually occur equally often, the number *W* of microstates belonging to a given kind of distribution *D makes for* and determines the actual frequency of occurrence of that macrostate *D* in the time ensemble of the system's macrostates.

So much for the physical and epistemological credentials of the entropy as an attribute of the macrostates of an individual physical system. Let us now turn to our issue of anthropomorphism by using the results of our analysis to assess the physical significance of the statistics of entropy change in our class *U*. As will be recalled, the focus of our interest is whether the role of partition dependence and thereby of human choice of the partition is such as to vitiate the physical significance of these statistics of entropy change and to relegate these statistics to being expressive of an anthropomorphism.

Do the Roles of Human Decision and Ignorance Impugn the Physical Significance of the Entropy Statistics for the Class U?

We noted the fundamental presumed fact that in the class *U* of systems, there is the following kind of partition *invariance* of entropy increase or entropy constancy: for any and every partition, the common initial entropy of all the systems in U will either have increased in a vast majority M of the systems by the time $t + \Delta t$, or it will have remained the same. The substitution of one partition for another affects *only the membership of M* while leaving the specified partition invariance of the entropy statistics intact. On this basis, we can now proceed to characterize several states of affairs as objective macroscopic physical facts, i.e., as *neither generated by a human choice of a partition nor expressive of human ignorance*. It is to be understood, of course, that the impending ascriptions of factual physical status are relative to the metrics of space and time ingredient in any of the partitionings of the six-dimensional phase space and in the total volume and energy of each system.

It is a presumed fact that for any and every partition P, most systems in U which start out in a macrostate that is *comparatively inhomogeneous* relatively to P, after the time interval Δt will be in macrostates which are more homogeneous with respect to P as measured by W or S. The relativity of homogeneity itself to a partition P does not make it any less an objective physical fact that, by the time $t + \Delta t$, most members of the specified subclass of U will be in macrostates which *are* more homogeneous relatively to P. Moreover, these more or less homogeneous macrostates have the property of occurring more or less frequently in the respective time ensembles of macrostates of their respective systems. And the fact that the lower and higher frequencies (probabilities) of occurrence pertain to distributions which are specified relatively to a partition does not detract from the objectivity of these frequencies. Nor are these various physical facts demoted to the status of anthropomorphisms just because they pertain to macrostates, and, as such, do not comprise the *still richer* factual physical content envisioned by *punctal* mechanics or fine graining.

Thus the statistics of entropy change in U codify partition-invariant macroscopic physical facts and cannot be held to be merely expressive of human ignorance of the underlying microprocesses. Human scientists do single out one or another particular partition relatively to which they characterize the macroscopic behavior of one or more systems entropically. But the partition dependence of the direction of an entropy change in any one system no more makes the change of macrostate in that system anthropomorphic than the dependence on the inertial frame of the spatial distance between two events in Minkowski space–time renders that spatial distance anthropomorphic.

The gravamen of the charge of anthropomorphism which I have considered was that such partition dependence as does obtain in the entropic descriptions of classical statistical mechanics renders these descriptions anthropomorphic. I have tried to argue that this charge is unfounded and that, at least to this extent, the use of the entropy concept is not illustrative of what J. L. Synge has aptly called the "Pygmalion syndrome."[(6)]

Acknowledgments

It is a pleasure to acknowledge the benefit of discussions with Allen Janis, Robert B. Griffiths, Joseph Camp, and Richard Creath. The author is also indebted to the National Science Foundation for their support of the research.

References

1. E. T. Jaynes. Gibbs vs. Boltzmann entropies. *Am. J. Phys. 33* (1965), 398.
2. E. T. Bell. *The Development of Mathematics*, 2nd ed. New York: McGraw-Hill (1945), p. 170.
3. A. d'Abro. *The Rise of the New Physics*. New York: Dover (1951), pp. 908–909.
4. A. Grünbaum. The anisotropy of time. In T. Gold (ed.). *The Nature of Time*. Ithaca, N.Y.: Cornell University Press (1966). [This paper should be supplemented by A. Grünbaum, *Philosophical Problems of Space and Time*, 2nd enlarged ed. Boston: D. Reidel (1973), Appendix §§22–26, pp. 829–831, and Chapter 22, §4, pp. 788–800.]
5. James and James. *Mathematics Dictionary*, 3rd ed. Princeton, N.J.: Nostrand (1968), p. 300.
6. J. L. Synge. *Talking About Relativity*. Amsterdam: North-Holland (1970), p. 8.

3

Håkan Törnebohm

ASPECTS OF THE SPECIAL THEORY OF RELATIVITY

The special theory of relativity should not be regarded as a theory about clocks and meter sticks. Here are some reasons for being critical against such a view.

a. The theory is not only valid for a macroworld inhabited by instruments of measurement but also valid for the world of fundamental particles, where there are no rods or mechanical clocks.
b. One should expect from a theory about instruments that the mechanisms and the functionings of them should be mapped in the theory. But the special theory of relativity does not include assumptions about clock mechanisms and other laboratory appliances.
c. Or, rather, such assumptions, if they occur, do not belong to the maps proper but to contexts of empirical "interpretations," serving another function than to spell out what the map has to say about the world. (That is why I put "interpretations" within quotation marks.)
d. One must distinguish very carefully between maps over territories that are not made by man on the one hand and maps over instruments used in making and testing such maps on the other hand. One must distinguish between questions such as "what informa-

Håkan Törnebohm • Institute for Theory of Science, University of Gothenburg.

tion about the world is contained in this body of systematized knowledge?" and questions such as "how are the knowledge claims made on behalf of the maps to be justified?" Meaning is far from the same as testability. A theory can be tested first when it has been made clear what it asserts. Meaning precedes testability.

A map over a territory X being a part of or an aspect of the real world may be called an X-ologial map and maps over instruments used in gathering information about X in the form of measurement data may be called X-ometric maps. The special theory of relativity is primarily an atlas of X-ological maps, where X stands for spatiotemporal features of events and sequences of events.

Infrastructure and Superstructure of STR

The special theory of relativity contains one part which is a successor of Newtonian kinematics, *viz.* the space-time theory within STR. A superstructure is built on this infrastructure.

The superstructure maps interactions among parts of physical systems and links interactions to processes. There is a relativistic mechanics and there is a relativistic electrodynamics, a synthesis of Maxwell's electromagnetics and relativistic dynamics.

The superstructure is causal. The infrastructure is not. The superstructure is linked to the infratheory by means of the special principle of relativity. This principle is on a metalevel with respect to causal principles satisfying it.

There is a perfect functional analogy between this metaprinciple and the Galilean principle of relativity in prerelativity physics. Both principles impose formal requirements on principles of interactions, serving to make them compatible with a basic infratheory which is not causal. The formal requirements affect the content of information of principles satisfying them.

My concern in this paper will be with the infratheory.

I think that a principle of relativity should not be used in building such a theory. Such a metaprinciple should rather be stated in a context in which it can be justified.

Should the Galilean principle of relativity be retained or should it be replaced by another homologous principle (i.e., one which has the same function)?

The answer to this question hinges on how this quite different question is answered: "Is Einstein's space–time theory better than Newton's kinematics?"

What is the Infratheory in STR About?

I think it makes good sense to answer this question as follows: the infratheory in STR is a general theory about spatial and temporal characteristics of events and sequences of events. Time does not exist independently of happenings and space not without extended objects spatially related to each other.

A Look at the Territory Mapped in STR

A material object is accompanied by a sequence of complex events. A sequence composed of events selected out from the complexes associated with a material object A will be called an M-sequence. Two kinds of M-sequences deserve to be singled out.

a. The events belonging to an M-sequence are either emissions or reflections or absorptions of light flashes. Such M-sequences will be called M-sequences of type L (L for "light").
b. The events are the "tick and tock" of a clock associated with an object A. M-sequences composed of such clock events will be called M-sequences of type C (C for "clock").

Events belonging to M-sequences associated with the same material object A have a definite time order. Let p and q be two events, locally associated with an object A. Then either

$$\begin{array}{lll} p \text{ is simultaneous with } q, & p \sim q, & \text{or} \\ p \text{ is earlier than } q, & p < q, & \text{or} \\ p \text{ is later than } q, & p > q. & \end{array}$$

These relations are absolute. No clock is needed to determine which one of them holds between two events p and q associated with the material object A. Clocks are acceptable as such only if they mirror such temporal relations with fidelity.

Consider two M-sequences A and B, associated with two material objects at a distance from each other. Let p be an event in A and q an event in B. We consider three cases:

a. There exists (or rather may exist) a material object (possibly composed of several parts always in the same neighborhood) such that p and q belong to an M-sequence C associated with it. Then either

$$p < q \qquad \text{or} \qquad p > q$$

p and q have a definite time order in this case.

b. p is the event of emission (reception) and q the event of reception (emission) of a light flash traveling the distance between A and B. It will be assumed that it is possible to distinguish between these two kinds of events. In this case, it will be ruled that $p < q$ or $p > q$ depending on whether p is the emission event or the reception event, respectively, in a couple of events (p, q) composed of one emission and one reception event.

c. p and q cannot belong to the same material object or to the same light flash. In this case, the time order among p and q is indefinite, unless there exists a propagation faster than that of light with respect to which it is possible in principle to decide whether p or q is an event of arrival or departure. Before any such propagation is discovered (it has not been discovered yet), the time order of p and q is indefinite.

Two Rival Views on Light in the Seventeenth Century

Newton held that light is a swarm of tiny bodies shot out from sources of light as bullets from a gun. I will refer to this view as the view N. Huygens, on the contrary, regarded light as a disturbance in a medium, the so-called luminiferous ether. I will refer to this view as the view H.

A Discriminative Thought Experiment

Let a light flash be emitted from a body A, event 1, and travel to a distant object C. It passes another object B, event 2, before it reaches C. At the event of passage, another light flash is emitted from B toward C. The two light flashes arrive at C, events r_1 and r_2, respectively, and are immediately reemitted toward B and A.

When the flashes pass B, events $2'_1$ and $2'_2$, new flashes of light are emitted toward A. The events of arrival of the four flashes are $1'_1$, $1'_2$, $1'_3$, and $1'_4$, respectively.

It will be assumed that the body *B* is in relative motion with respect both to *A* and to *C*.

The rate of propagation of light is dependent on the velocity of the source according to N but not according to H.

Hence r_1 and r_2 should be simultaneous according to H but separated in time according to N. What has been said about r_1 and r_2 is true also for the couple of events $2'_1$ and $2'_2$ and for the quadruple $1'_1, 1'_2, 1'_3$, and $1'_4$.

According to N, there is no essential difference between *M*-sequences and sequences of events belonging to the propagation of light. In both cases, the sequences are capable of branching off in the manner in which light according to N would branch off at 2 into the sequences $2r_1$ and $2r_2$ in our thought experiment. But if H is correct there is an essential difference between *M*-sequences which are capable of branching off in the manner described and sequences of events associated with light flashes. They will be called *L*-sequences. The latter ones can branch off only at reflection events in which they change their direction. It will be assumed throughout that light travels through a vacuum.

Representation of L-Sequences

If H (but not if N) is correct, it is permissible to represent the propagation of light as follows:

Let the *L*-sequence passing through the events 1, 2, and *r* be represented as follows:

$$v = v(1) = v(2) = v(r) = \text{const.}$$

Let the *L*-sequence through *r*, 2′, and 1′ be represented by

$$w = w(r) = w(2') = w(1') = \text{const.}$$

The lines $v = $ const. and $w = $ const. may be regarded as coordinate lines in a system of coordinates $\mathcal{O}$. The point-event *r* has coordinates $(v(r), w(r))$. *r* is also a reflection event. Events are in general represented as genuine or virtual reflection events in the system $\mathcal{O}$. Processes in which material objects are involved are represented as *M*-sequences of type *L* in the system $\mathcal{O}$.

A Convention

Let *p* and *q* be events in an *M*-sequence representing a material object.

We stipulate that

$$v(p) > v(q) \quad \text{and} \quad w(p) > w(q)$$

if and only if $p > q$ (p is later than q).

Introduction of a Metric and Geodesics in a Space of Virtual (or Genuine) Reflection Events

The (squared) distance between a pair of (virtual or genuine) reflection events p and q is defined as follows:

$$\Delta(p, q)^2 = (v(p) - v(q))(w(p) - w(q))$$

$\Delta(p, q)^2$ can be positive, zero, or negative.

i. $\Delta(p, q)^2 > 0$ if $v(p) - v(q)$ has the same sign as $w(p) - w(q)$. This is the case if p and q belong to or could belong to an M-sequence. It holds that $\Delta^2 > 0$ for any two nonsimultaneous events in an M-sequence.

ii. $\Delta(p, q)^2 = 0$ if either $v(p) = v(q)$ or $w(p) = w(q)$. p and q belong to or could belong to an L-sequence in either case. It holds that $\Delta^2 = 0$ for any pair of events on an L-sequence.

iii. $\Delta(p, q)^2 < 0$ if $v(p) - v(q)$ and $w(p) - w(q)$ have different signs. This is the case only if it is impossible for p and q to belong either to an M-sequence or to an L-sequence. The time order of p and q is indefinite if $\Delta(p, q)^2 < 0$.

Let C be a line segment with the end points 1 and 2. A measure M on C is defined as follows:

$$M(C) = \int_1^2 \sqrt{\dot{v}\dot{w}}\, dp$$

where p is a parameter fixing points on C and $\dot{v} = dv/dp$ and $\dot{w} = dw/dp$.

C is a geodesic line segment if the variational condition $\delta M(C) = 0$ is satisfied.

The corresponding Euler–Lagrange equations

$$\left.\begin{aligned} \frac{d}{dp}\frac{\partial}{\partial \dot{v}} \sqrt{\dot{v}\dot{w}} = 0 \\ \frac{d}{dp}\frac{\partial}{\partial \dot{w}} \sqrt{\dot{v}\dot{w}} = 0 \end{aligned}\right\}$$

have the solution

$$av + bw = c$$

The geodesics are thus straight lines.

There are three kinds of geodesics:

i. $a \neq 0 \neq b$ and $b/a < 0$. In this case, it holds for any two points on that line that

$$a\Delta v + b\Delta w = 0 \rightarrow \Delta v \Delta w = \Delta(p, q)^2 = -\frac{b}{a}\Delta^2 w > 0$$

Δ^2 is thus positive. The points have a definite time order; $av + bw = c$ could represent a M-sequence associated with a material object at rest in some inertial frame of reference. Geodesics of this kind will therefore be called inertial* lines.

ii. $a = 0 \neq b$ or $a \neq 0 = b$. $\Delta v \Delta w = 0$ for any two points on the geodesic in either case. Such geodesics represent possible L-sequences. They will accordingly be called optical lines.

iii. $a \neq 0 \neq b$ and $b/a > 0$. The (squared) distance between any two points on the geodesic is

$$\Delta(p, q)^2 = -\frac{b}{a}\Delta^2 w < 0$$

Hence the time order among p and q is indefinite. p and q can belong neither to an M-sequence nor to an L-sequence. Geodesics of this kind will be called separation lines.

If C is a geodesic line segment, then $M(C) = \Delta(1, 2)$, where 1 and 2 are the end points of the segment.

Proof

a. The geodesic is an optical line. Then $M(C) = \int_1^2 \sqrt{\dot{v}\dot{w}}\, dp = 0$ because $\dot{v}\dot{w} = 0$ for every point on the line, and $\Delta(1, 2) = 0$ as 1 and 2 belong to an optical line.

b. The geodesic is not an optical line. Then $a\dot{v} + b\dot{w} = 0$. Hence $\dot{v} = -(b/a)\dot{w}$ (because $a \neq 0$) so that

$$M(C) = \sqrt{-\frac{b}{a}} \int_1^2 \dot{w}\, dp = \sqrt{-\frac{b}{a}}\,(w(2) - w(1))$$

*This term as well as the terms "optical line" and "separation line" to be introduced in the following are borrowed from Robb.[16]

On the other hand, it holds that

$$a(v(2) - v(1)) + b(w(2) - w(1)) = 0$$

so that

$$\sqrt{-\frac{b}{a}} = \sqrt{\frac{v(2) - v(1)}{w(2) - w(1)}}$$

It follows that

$$M(C) = \sqrt{(v(2) - v(1))(w(2) - w(1))} = \Delta(1, 2)$$

Duration and Distance in Time

What is the relation between $M(C)$ and $\Delta(1, 2)$ if C is not a geodesic line segment?

It will be shown when more concepts have been introduced that

$$M(C) < \Delta(1, 2)$$

The implication of this inequality is that a distinction should be made between durations of a process and distances in time from their beginnings to their ends.

Consider an M-sequence C with a beginning event 1 and an ending event 2. The duration of C is the measure $M(C)$ and the distance in time between its beginning and end is $\Delta(1, 2)$.

Duration is an additive measure; i.e., the whole is the sum of its parts with regard to M. This follows immediately from the definition of the magnitude M. In fact,

$$M(C_1 + C_2) = M(C_1) + M(C_2)$$

Δ in contrast to M is not additive in general.

Let C be composed of two inertial line segments C_1 and C_2 ($C = C_1 + C_2$) such that

a. C_1 begins in an event 1 which is also a beginning event of C and ends with an event p.
b. C_2 begins in the event p and ends in the event 2, which is also an ending event of C.
c. C_1 and C_2 do not belong to the same geodesic.

Under these conditions, it holds that

i. $M(C) < \Delta(1, 2)$, because C is a nongeodesic line segment.

ii. $M(C_1) = \Delta(1, p)$ and $M(C_2) = \Delta(p, 2)$, because C_1 and C_2 are geodesic line segments.

It follows from (i) and (ii) that

$$\Delta(1, 2) > \Delta(1, p) + \Delta(p, 2)$$

What magnitude is measured by means of clocks? This question can be interpreted in two senses. It may be a factual question: given such and such instruments called clocks, what do they measure? It may also be interpreted as a what-to-do question: are we to build an instrument for measuring durations or are we to build an instrument for measuring distances in time?

I take it that instruments for measuring durations are more useful to have than instruments for measuring distances in time, because duration-meters can also be used as distance-in-time-meters, but not *vice versa.* Besides, additive measures are preferable to nonadditive ones if a choice is possible.

I believe also that ordinary-precision clocks do measure durations rather than distances in time.

Introduction of a New Type of Coordinates

We have so far only used optical coordinates. In this section, another type of coordinates will be introduced in terms of which it will be possible to define a set of useful concepts.

Inertial Reference Lines

An inertial line, passing through the origin 0 of the optical system $\mathcal{O}$, will be called an inertial reference line.

It will be convenient to represent such a line by means of the equation

$$ve^{u} - we^{-u} = 0$$

where u is a constant.

Let p be a point on this line. A time coordinate of p will be defined as follows:

$$\tau(p) = \Delta(0, p)$$

Combining this equation with the equation of the inertial reference line, we obtain the relations

$$\left.\begin{aligned} v &= \tau e^{-u} \\ w &= \tau e^{u} \end{aligned}\right\}$$

These relations can be regarded as transformation from the optical coordinate system $\mathcal{O}$ to another coordinate system $\mathcal{A}$, in which a point has the coordinates (τ, u). τ is the distance in time from the origin 0 to the point p or equivalently the duration of a virtual geodesic process starting with the event 0 and ending with the event p. u is a parameter, picking out from the set of inertial reference lines through 0 that member which passes through p.

This formal interpretation of τ and u carries through only if

$$\Delta(0, p)^2 > 0$$

$\mathcal{A}(0)$ is a coordinate system composed of two families of coordinate lines:

a. A family of inertial reference lines through 0.
b. A family of lines τ = const., being hyperbolas $vw = \tau^2$ in the $\mathcal{O}$-system, representing events with the same distance in time $\Delta(0, p)$ from the origin.

Isochronism and Velocities

Consider two inertial reference lines (u): $ve^{u} - we^{-u} = 0$ and (u'): $ve^{u'} - we^{-u'} = 0$ representing two objects A and B moving away from each other with constant velocity from a common place of departure. The event of departure is the origin 0 of time.

Let a clock C be associated with A recording the coordinates τ of events on A. The clock C can also be used to record coordinates of events on B as follows: Let a light flash be emitted from A at the event 1, proceed to B where it is reflected in the event p, and return to A where it arrives in the event 2.

The L-sequences $1 \rightarrow p$ and $p \rightarrow 2$ are described as follows (*cf.* Fig. 1):

$$v(1) = \tau(1)e^{-u} = v(p) = \tau(p)e^{-u'}$$

and

$$w(p) = \tau(p)e^{u'} = w(2) = \tau(2)e^{u}$$

From these equations we obtain the relations

$$\tau(p)^2 = \tau(1)\tau(2)$$

and

$$u' - u = \ln \sqrt{\frac{\tau(2)}{\tau(1)}}$$

The first one of these relations shows how the concept of running at the same rate as, or isochronism, may be defined for a clock C' associated with the object B and a clock C associated with the object A.

The two clocks run at the same rate if they satisfy the relation $C'(p)^2 = C(1)C(2)$, for every p or B, where p is the event of reflection at B of a light flash departing from A, event 1, and returning to A, event 2.

The second relation is a measuring condition on a magnitude $u' - u$, which can be regarded as the velocity with which B moves away from A.

As a partial justification for interpreting $u' - u$ in this manner may be regarded the fact that a Newtonian type of addition theorem holds for this magnitude.

Let there be a third object C represented by means of an inertial reference line (u''): $ve^{u''} - we^{-u''} = 0$. It is assumed

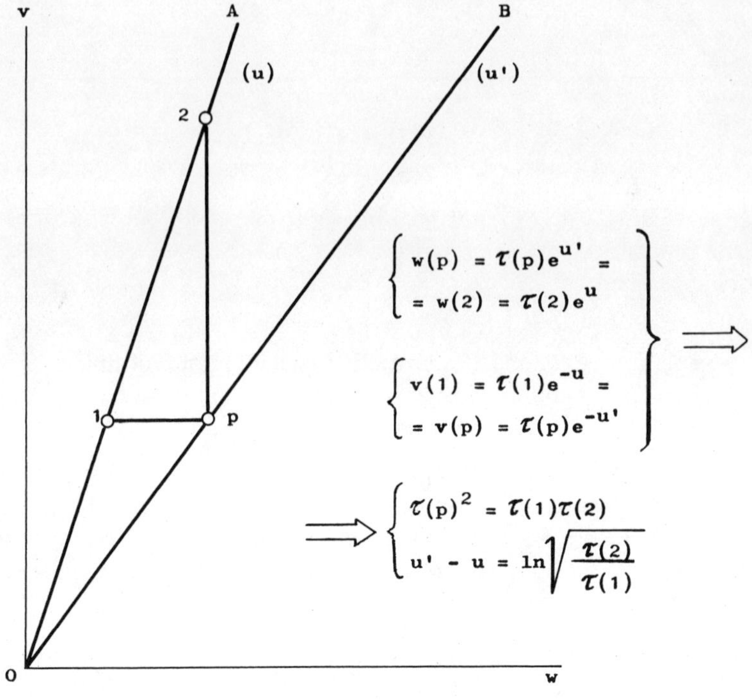

Fig. 1

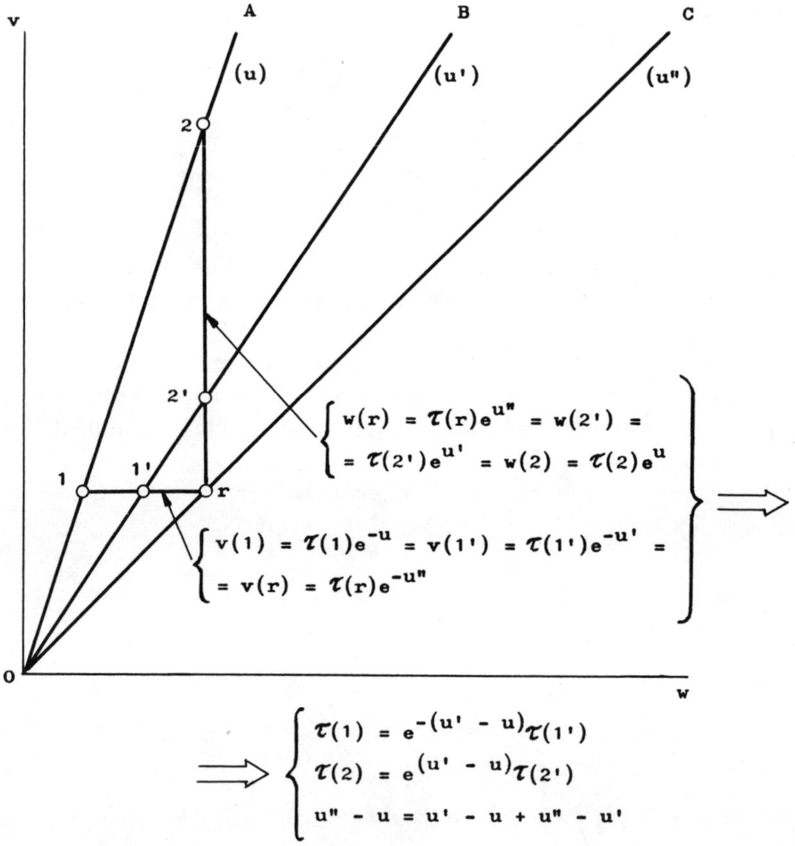

Fig. 2

a. That A, B, and C are together at the origin 0 of time.
b. That B and C move away in the same direction from A.
c. That C moves away faster than B, so that $u'' - u > u' - u$ (*cf.* Fig. 2).

Let a light flash travel from A toward B and C. It departs from A in the event 1. It passes B in the event $1'$. It is reflected from C in the event r. It returns to B and A. It passes B in the event $2'$ and arrives at A in the event 2.

The L-sequences

$$1 \to 1' \to r \qquad \text{and} \qquad r \to 2' \to 2$$

have the equations

$$v(1) = \tau(1)e^{-u} = v(1') = \tau(1')e^{-u'} = v(r) = \tau(r)e^{-u''}$$

and

$$w(r) = \tau(r)e^{u''} = w(2') = \tau(2')e^{u'} = w(2) = \tau(2)e^{u}$$

From these equations it is readily proved that

$$u'' - u = u' - u + u'' - u'$$

It is also found that

$$\tau(1) = e^{-(u'-u)}\tau(1')$$
$$\tau(2) = e^{u'-u}\tau(2')$$

These formulas are very closely related to the Lorentz equations, which will be considered in a later section.

Let a light flash emitted from A, event 1, arrive at B, event p, so that $e^{u'-u} = \tau(p)/\tau(1)$.

If B moves with a velocity approaching that of light, then $\tau(p)$ will approach $+\infty$. We conclude that the "relative" velocity of light $u' - u$ is infinite.

As $u' - u$ is a relative velocity, it is natural to regard u as an "absolute velocity." An absolute velocity cannot be measured, but it is possible to assign a value to such a velocity by convention. If an object A is singled out as a virtual station of observation, it is convenient to stipulate that its velocity is $u = 0$.

An Important Process Law

We are now in possession of a sufficient supply of concepts to prove that the duration of a process is less than or equal to the distance in time between its end points.

Let a line segment C represent a noninertial process in which a material object is involved. Let 1 and 2 be a beginning and an ending event, respectively (*cf.* Fig. 3). 1 and 2 are events on an inertial line. It will be convenient to shift the origin of time to a position on this line. It will then be an inertial reference line

$$(u): ve^{u} - we^{-u} = 0$$

where u is a constant.

A transition from 0 to 0′ is represented by the following transformations of the optical coordinates:

$$\left.\begin{aligned} v' &= v - v(0') \\ w' &= w - w(0') \end{aligned}\right\}$$

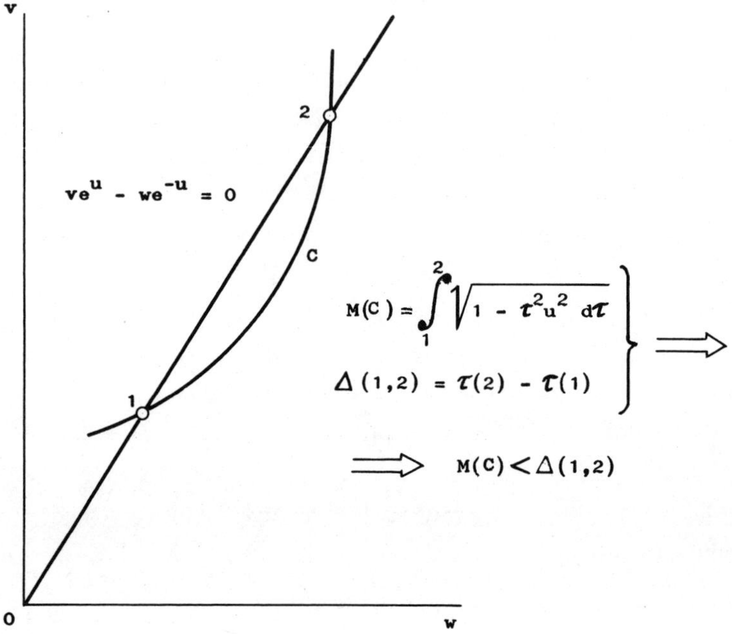

Fig. 3

It follows from them that $\Delta v' = \Delta v$ and $\Delta w' = \Delta w$ so that $\Delta(p, q)^2$ and $M(C)$ are conserved. After suitable selection of an origin (see Fig. 3), we proceed as follows:

$$M(C) = \int_1^2 \sqrt{\dot{v}\dot{w}}\, dp$$

Now

$$(v, w) = (\tau e^{-u}, \tau e^{u})$$

Hence

$$\dot{v}\dot{w} = \dot{\tau}^2 - \tau^2\dot{u}^2$$

Using τ as variable of integration, we obtain the expression

$$M(C) = \int_1^2 \sqrt{1 - \tau^2\dot{u}^2}\, d\tau$$

As C is not an inertial reference line, $\dot{u}$ is not zero everywhere on the segment. It follows that

$$M(C) < \int_1^2 d\tau = \tau(2) - \tau(1)$$

As 1 and 2 are located on the inertial reference line, it follows that

$$\tau(2) - \tau(1) = \sqrt{(v(2) - v(1))(w(2) - w(1))} = \Delta(1, 2)$$

We conclude that

$$M(C) < \Delta(1, 2)$$

This relation shows that the duration of a process in which an object is involved is less than the distance in time between its beginning and its end, if the object is accelerated.

This is an *X*-ological way of expressing a fundamental law, which is frequently described in *X*-ometric terms as follows.

A clock attached to an object *B* moving away from and then returning to an object *A* at rest runs at a slower rate than a clock attached to the resting body.

It is true that the *A*-clock would be ahead of the *B*-clock when the two clocks meet again in the event 2, if the two clocks were set to read the same time wnen they depart from each other in the event 1. This fact, however, means that the *B*-clock is slower than the *A*-clock only if the two clocks measure the same magnitude, *viz*, the distance in time from the event 1 to the event 2.

It is not mandatory to hold that they do. It is possible to hold that they measure different magnitudes, the *A*-clock measuring the duration of a process in which *A* is involved, or, which is the same thing, the distance in time from the beginning to the end of the process, and the *B*-clock the duration of a different process in which *B* is involved.

It is possible to hold that the two clocks run at the same rate. It is also possible in principle to test the assertion that they do.

We have seen how an observer attached to the resting object *A* may plot the (τ, u)-coordinates of the moving body *B* by means of "radar signals" and a clock. It is thus possible for him to calculate the integral

$$\int_1^2 \sqrt{1 - \tau^2\dot{u}^2}\, d\tau$$

using data thus obtained.

If the value obtained in this manner were to agree with the difference between the times recorded by the *B*-clock in the events 2 and 1, respectively, then the two clocks would run at the same rate. If, however, this should prove not to be the case, then there would be an occasion for asking "why do the clocks run at different rates?" An adequate explanation would have to give an account for the effects of an acceleration on the mechanism of the B-clock. Such accounts cannot belong to the infratheory of STR, which is an essentially acausal theory.

I have here talked about alternative interpretations of the process law,

$$M(C) \leq \Delta(1, 2)$$

I don't think that they are of equal merit. The X-ological interpretation is better, I believe, than the other one, because

a. The law is X-ological and should not be interpreted as an X-ometric one.
b. It is based on the contention that proper timekeepers measure additive magnitudes.
c. The alternative interpretation is an example of how a new result (the process law under discussion) is interpreted in terms of an older theory (the Newtonian space-time theory in which durations and distances in time are equivalent) which is incompatible with the new result (according to which these magnitudes are not equivalent).

Pairs of Conjugate Magnitudes

Consider two functions $f(\tau, u)$ and $g(\tau, u)$ on events in space-time. They will be said to be conjugate if they satisfy the following conditions:

$$\left.\begin{aligned} f(\tau, u) &= \frac{\partial}{\partial u} g(\tau, u) \\ g(\tau, u) &= \frac{\partial}{\partial u} f(\tau, u) \end{aligned}\right\}$$

Pairs of conjugate magnitudes have the following properties:

a. The above differential equations have the solutions

$$f(\tau, u) = A(\tau)e^{u} + B(\tau)e^{-u}$$

and

$$g(\tau, u) = A(\tau)e^{u} - B(\tau)e^{-u}$$

b. $f^2 - g^2$ is independent of u. Using 1 we find that

$$f^2 - g^2 = 4A(\tau)B(\tau)$$

c. Let an object A represented by an inertial reference line (u_0): $ve^{u_0} - we^{-u_0} = 0$ be assigned the absolute velocity 0. This convention induces a change of the coordinates to a point p from (τ, u) to $(\tau, u - u_0)$. The couple $(f(\tau, u), g(\tau, u))$ is transformed into the couple $(f(\tau, u - u_0), g(\tau, u - u_0))$ abbreviated (f_0, g_0) as follows:

$$f_0 - g_0 = 2B(\tau)e^{-(u-u_0)} = e^{u_0}(f - g)$$

and

$$f_0 + g_0 = 2A(\tau)e^{u-u_0} = e^{-u_0}(f + g)$$

The convention of setting $u = 0$ for the object A amounts to a choice of A as a station from which the properties represented by f and g of an event p are to be measured.

d. The coordinates of p are changed from $(\tau, u - u_0)$ into $(\tau, u - u_1)$ corresponding to a new choice of station from which the conjugate properties f and g of an event p are to be measured. The transformations of the conjugate magnitudes as a result of a transition from a station A represented by an inertial reference line (u_0) to a station B represented by an inertial reference line (u_1) are obtained from (c) above. They are as follows:

$$f_0 - g_0 = e^{-(u_1-u_0)}(f_1 - g_1)$$

$$f_0 + g_0 = e^{u_1-u_0}(f_1 + g_1)$$

$u_1 - u_0$ is the relative velocity of B with respect to A.

Remark

These transformation equations in the system $\mathcal{A}(0)$ correspond to the Lorentz transformations in another kind of coordinate systems to be introduced later.

This being the case, we may conclude that all pairs of conjugate magnitudes transform in accordance with the Lorentz formulas. [It is easily shown—but I will not do so here—that pairs of magnitudes that transform in accordance with the Lorentz formulas correspond to the pair of conjugate magnitudes in $\mathcal{A}(0)$].

The above properties indicate that the notion of couples of conjugate magnitudes is an important one. Examples of such pairs will be considered below.

The Notion of Conjugateness Transferred from Pairs of Magnitudes to Geodesics

A geodesic in space-time has the equation

$$av + bw = c$$

in the optical system $\mathcal{O}$. Its equation in the system $\mathfrak{a}(0)$ is

$$a\tau e^{-u} + b\tau e^{u} = c$$

Comparing this equation with the expression in terms of τ and u for a magnitude f belonging to a couple (f, g) of conjugate magnitudes, we find it plausible to regard the equation

$$-a\tau e^{-u} + b\tau e^{u} = c'$$

as that of a geodesic connected with the first one by means of a relation of conjugation.

Taking what has just been said as a clue, it is natural to define a relation of conjugation between geodesics as follows. The geodesic

$av + bw = c$ is conjugate to the geodesic

$a'v + b'w = c'$ if and only if

$ab' + ba' = 0$

It follows from this definition that any two optical lines belonging to the family $v =$ const. are conjugate. Likewise, any two optical lines belonging to the family $w =$ const. are conjugate.

A geodesic conjugate to an inertial line is a separation line and *vice versa*. This is evident from the fact that

$\frac{b}{a}$ and $\frac{b'}{a'}$ must have different signs if the geodesics

$av + bw = c$ and $a'v + b'w = c'$ are conjugate

A geodesic conjugate to an inertial reference line (u): $ve^{u} - we^{-u} = 0$ has the equation

$$ve^{u} + we^{-u} = \text{const.}$$

Let q be a point on the line (u). A tangent in q to the hyperbola $vw = \tau(q)^2$ is conjugate to (u). This assertion is proved as follows: The equation of the tangent is

$$vw(q) + v(q)w = 2\tau(q)^2$$

As

$$(v(q), w(q)) = \tau(q)(e^{-u}, e^{u})$$

the equation may be written in the form

$$ve^{u} + we^{-u} = 2\tau(q)$$

This is the equation of a separation line conjugate to the inertial line (u).

What has just been shown suggests the following extension of the concept of conjugateness as applied to curves in space-time.

A curve C is conjugate to a geodesic if the tangent in every point in which the geodesic cuts the curve C is conjugate to the geodesic.

It follows from this definition and from what has been shown above that the family of hyperbolas τ = const. are conjugate to all inertial reference lines in a system $\mathcal{A}(0)$.

A Third Kind of Coordinate Systems

If the time coordinates τ are used, then the velocity of light in a direction from an object A to an object B, C_{AB}, does not have the same value as the velocity of light, C_{BA}, from B to A. Let a light flash be emitted from A, represented by an inertial reference line (u_0) at an instant $\tau(1)$, be reflected from B at an instant $\tau(p)$ and return to A at an instant $\tau(2)$. The ratio between the velocities of light in both directions would then be

$$\frac{C_{AB}}{C_{BA}} = \frac{\tau(2) - \tau(p)}{\tau(p) - \tau(1)} = \frac{\tau(2) - \sqrt{\tau(1)\tau(2)}}{\sqrt{\tau(1)\tau(2)} - \tau(1)} = \frac{\sqrt{\tau(2)}}{\sqrt{\tau(1)}} = \frac{\sqrt{w(2)e^{-u_0}}}{\sqrt{v(1)e^{u_0}}}$$

$$= \frac{\sqrt{w(p)e^{-u_0}}}{\sqrt{v(p)e^{u_0}}} = e^{u-u_0} \neq 1$$

where u is the u-coordinate of the reflection event p.

As it is awkward to have $C_{AB} \neq C_{BA}$, it is desirable to define a new kind of time coordinates such that $C_{AB} = C_{BA}$.

The new time coordinate should satisfy this condition

$$t(p) - \tau(1) = \tau(2) - t(p)$$

i.e.,

$$t(p) = \tfrac{1}{2}(\tau(2) + \tau(1))$$

Let q be a point on (u_0) such that

$$\tau(q) = \tfrac{1}{2}(\tau(2) + \tau(1)) = t(p)$$

A geodesic passing through q and p has the following equation (*cf.* Fig. 4):

$$ve^{u_0} + we^{-u_0} = 2\tau(q)$$

Verification

It is immediately evident that q belongs to this geodesic. p belongs to the line also, because

$$v(p)e^{u_0} + w(p)e^{-u_0} = v(1)e^{u_0} + w(2)e^{-u_0} = \tau(1) + \tau(2) = 2\tau(q)$$

The geodesic

$$ve^{u_0} + we^{-u_0} = 2\tau(q)$$

is a separation line conjugate to the inertial reference line (u_0). It is also, we have seen, a tangent to the hyperbola $vw = \tau(q)^2$ in the point q on (u_0) half way between the event of departure and the event of reflection of a light flash which is reflected in the event p on B.

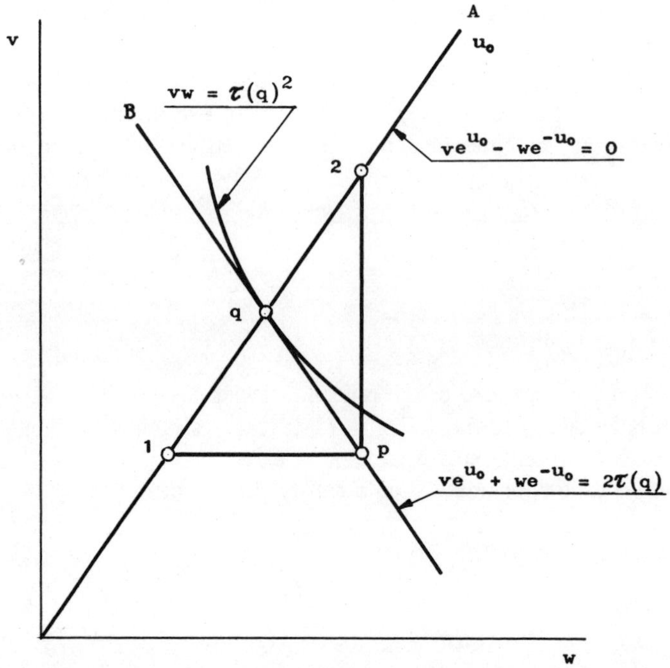

Fig. 4

The coordinate system $\mathcal{A}(0)$ is centered on the origin 0. A new kind of coordinate system $\mathcal{R}(u_0)$ centered on an inertial reference line will now be defined as follows:

a. The inertial reference line (u_0): $ve^{u_0} - we^{-u_0} = 0$ belongs to $\mathcal{R}(u_0)$.
b. Every geodesic which is conjugate to a coordinate line $\mathcal{R}(u_0)$ belongs to $\mathcal{R}(u_0)$.

It follows from this definition that $\mathcal{R}(u_0)$ is composed of two families of coordinate lines:

a. A family of parallel inertial lines $ve^{u_0} - we^{-u_0} = \text{const.}$
b. A family of parallel separation lines $ve^{u_0} + we^{-u_0} = \text{const.}$ that are conjugate to the inertial reference line in $\mathcal{R}(u_0)$.

Consider two coordinate systems of the new type $\mathcal{R}(u_0)$ and $\mathcal{R}(u_1)$. If $(u_1) \neq (u_0)$, then $\mathcal{R}(u_1)$ and $\mathcal{R}(u_0)$ do not contain any common coordinate line. $t(p)$ is expressed in $\mathcal{A}(0)$ as follows:

$$t(p) = \tfrac{1}{2}(\tau(2) + \tau(1)) = \tfrac{1}{2}(w(2)e^{-u_0} + v(1)e^{u_0}) = \tfrac{1}{2}(w(p)e^{-u_0} + v(p)e^{u_0})$$

$$= \frac{\tau(p)}{2}\,(e^{u-u_0} + e^{-(u-u_0)})$$

This equation indicates that $t(p)$ belongs to a pair of conjugate magnitudes.
Its partner $x(p)$ is by definition of conjugateness equal to

$$x(p) = \frac{\partial}{\partial u}\,t(p) = \frac{\tau(p)}{2}\,(e^{u-u_0} - e^{-(u-u_0)}) = \tfrac{1}{2}(w(p)e^{-u_0} - v(p)e^{u_0})$$

$$= \tfrac{1}{2}(\tau(2) - \tau(1))$$

This relation shows that $x(p)$ can be identified as the "radar" distance $D(u_0, p)$ of an event p from the inertial reference line (u_0) on which $\mathcal{R}(u_0)$ is based.

The coordinates of an event p in $\mathcal{R}(u_0)$ being conjugate to each other, we conclude from what was shown in the section on pairs of conjugate magnitudes

a. That $t^2 - x^2$ is independent of u [in fact, $t^2 - x^2 = \tau^2 = \Delta(0, p)^2$].
b. That the coordinates (t, x) of p in $\mathcal{R}(u_0)$ are related to the coordinates of p in another coordinate system $\mathcal{R}(u_1)$ by means of the following relations (*cf.* Fig. 5):

$$\left.\begin{aligned} t - x &= e^{-(u_1-u_0)}(t' - x') \\ t + x &= e^{u_1-u_0}(t' + x') \end{aligned}\right\}$$

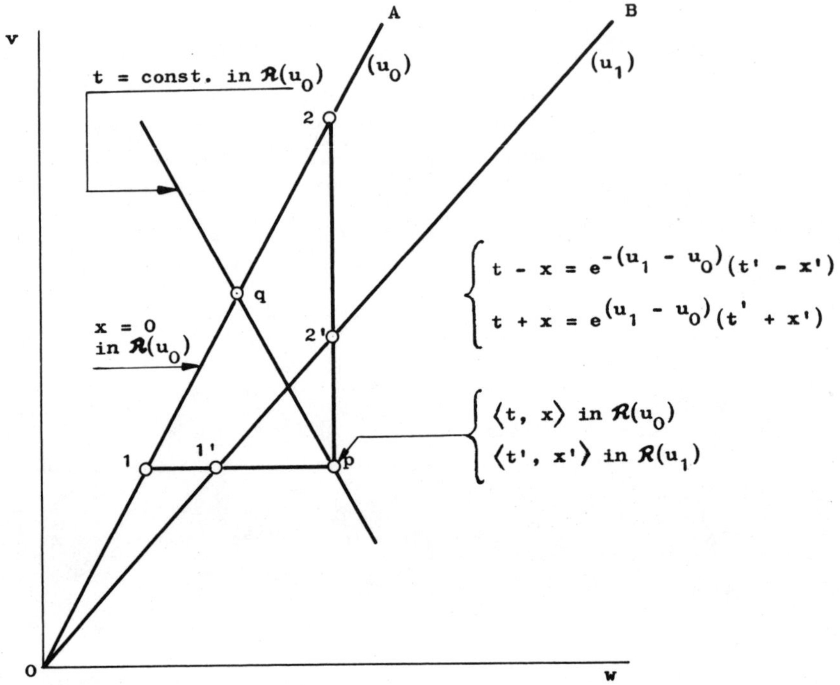

Fig. 5

Relative Velocity in Systems of Type $\mathcal{R}(u_0)$

In $\mathcal{A}(0)$ the relative velocity of an object B represented by an inertial reference line (u): $ve^u - we^{-u} = 0$ with respect to an object A represented by an inertial reference line (u_0): $ve^{u_0} - we^{-u_0} = 0$ is the magnitude $u - u_0$.

In $\mathcal{R}(u_0)$ this velocity is defined as follows: $V = x/t$, where (t, x) are the coordinates of a point on the line (u) with respect to the inertial reference line (u_0), from which $\mathcal{R}(u_0)$ is built up.

It was shown in the last section that the coordinates (t, x) of a point p in $\mathcal{R}(u_0)$ satisfy the relation

$$(t, x) = \tfrac{1}{2}\tau(p)(e^{u-u_0} + e^{-(u-u_0)}, e^{u-u_0} - e^{-(u-u_0)})$$

We conclude that

$$V = \frac{e^{u-u_0} - e^{-(u-u_0)}}{e^{u-u_0} + e^{-(u-u_0)}} = \frac{e^{2(u-u_0)} - 1}{e^{2(u-u_0)} + 1}$$

Hence

$$u - u_0 = \ln \sqrt{\frac{1 + V}{1 - V}}$$

Using this relation, we may rewrite the transformation equations from the coordinates (t, x) of a point p in $\mathcal{R}(u_0)$ to the coordinates (t', x') of p in $\mathcal{R}(u_1)$ as follows:

$$\left.\begin{aligned} t - x &= \sqrt{\frac{1 - V}{1 + V}}\,(t' - x') \\ t + x &= \sqrt{\frac{1 + V}{1 - V}}\,(t' + x') \end{aligned}\right\}$$

It follows that

$$\left.\begin{aligned} t &= \frac{1}{\sqrt{1 - V^2}}\,(t' + Vx') \\ x &= \frac{1}{\sqrt{1 - V^2}}\,(x' + Vt') \end{aligned}\right\}$$

These are the Lorentz formulas (*cf.* Fig. 5).

Another consequence of the new definition of relative velocity is Einstein's addition theorem for V-velocities.

Let A, B, and C be objects in relative constant motion as represented in Fig. 2.

Let the relative velocities be as follows:

Motion of	u-velocity	V-velocity
B with respect to A	$u' - u$	V_1
C with respect to B	$u'' - u'$	V_2
C with respect to A	$u'' - u$	V_3

As $u'' - u = u' - u + u'' - u'$, it follows from the relation between the u- and V-velocities that

$$\frac{1 + V_3}{1 - V_3} = \frac{1 + V_1}{1 - V_1} \cdot \frac{1 + V_2}{1 - V_2}$$

from which Einstein's addition theorem

$$V_3 = \frac{V_1 + V_2}{1 + V_1 V_2}$$

is obtained.

An immediate corollary of the definition of V is the existence of a limiting velocity of the motion of objects, *viz.* $V = 1$. If C moves with the velocity $V_2 = 1$ with respect to B, then (according to Einstein's addition theorem for velocities) it moves with the velocity $V_3 = 1$ with respect to A; i.e., the limiting velocity is the same in all inertial frames of reference.

Velocity of light has not been defined, but it is a good conjecture at this juncture to identify the limiting velocity $V = 1$ with the velocity of light.

A Second Example of a Pair of Conjugate Magnitudes

If the relative velocity of an object B with respect to A is very small, then the momentum of B equals $m(u - u_0)$. We will construct a pair of conjugate magnitude (X, P) of which P is the momentum of B. We will ask what kind of magnitude X must be in order to be a conjugate partner of P.

P and X must have the form

$$P = Ae^{u-u_0} - Be^{-(u-u_0)}$$

$$X = Ae^{u-u_0} + Be^{-(u-u_0)}$$

If $u - u_0$ is small, then $P \sim m(u - u_0)$. This limiting condition determines A and B as follows:

$$Ae^{u-u_0} - Be^{-(u-u_0)} \sim A - B + (A + B)(u - u_0) = m(u - u_0)$$

Hence $A = B = m/2$.

The two conjugated magnitudes are thus

$$P = \frac{m}{2}\,(e^{u-u_0} - e^{-(u-u_0)})$$

$$X = \frac{m}{2}\,(e^{u-u_0} + e^{-(u-u_0)})$$

If $u - u_0$ is small, then

$$X \sim m + \frac{m}{2}\,(u - u_0)^2$$

The second term of the right-hand side can be identified as an expression for the kinetic energy of the object. This identification makes it natural to interpret X as an expression for the energy of the body B. We will accordingly replace the letter X by the conventional letter E.

The first term on the right side of the expression for E represents a contribution to the energy of the object B harbored in its mass. We have in this easy manner arrived at Einstein's mass = energy principle. As (E, P) is a pair of conjugate magnitudes, it follows that the magnitude $E^2 - P^2$ is independent of the velocity of the object B. Using the expressions for E and P in the system $\mathcal{A}(0)$, we find that $E^2 - P^2 = m^2$.

A comparison between the pair of conjugate magnitudes

$$(t, x) = \frac{T}{2}\left(e^{u-u_0} + e^{-(u-u_0)}, e^{u-u_0} - e^{-(u-u_0)}\right)$$

and the above expression for (E, P) in the system $\mathcal{A}(0)$ leads to this result:

$$P = E\frac{x}{t} = EV$$

Combining this relation with the expression

$$m^2 = E^2 - P^2$$

we obtain the relation

$$(E, P) = \left(\frac{m}{\sqrt{1 - V^2}}, \frac{mV}{\sqrt{1 - V^2}}\right)$$

familiar from standard presentations of STR (with $c = 1$).

If an object moves with the limiting velocity $V = 1$ (= the velocity of light), it follows that $m^2 = 0$. The mass of objects moving with the speed of light is thus zero.

At this juncture, it is natural to proceed to de Broglie's relations. An object moving with the velocity of light is characterized by the wave attributes (ν, k) where $k = 1/\lambda$.

The velocity of light is equal to $\nu/k = 1$. It follows that

$$\nu^2 - k^2 = 0$$

A comparison between the pair (E, P) applied to an object moving with the velocity of light and the pair (ν, k) characterizing the propagation of light strongly suggests the following relation:

$$(E, P) = h(\nu, k)$$

where h is Planck's constant.

de Broglie extended this relation from objects moving with the velocity of light to objects moving with a velocity below that of light, thereby postulating that all objects are associated with wave motions. The beginning of an extremely fertile sythesis in the sense of complementarity between the two rival conceptions of Newton and Huygens about light was thereby brought about and extended to all kinds of objects.

Concepts of Simultaneity

τ-Simultaneity

Two events p and q are simultaneous in $\mathcal{A}(0)$ if they have the same τ-coordinates. From the definition of τ-coordinates, it follows that p and q are simultaneous only if $\Delta(0, p) = \Delta(0, q)$, i.e., p and q are at the same distance in time from the origin 0.

The locus of events with the same τ-coordinates is a hyperbola $vw = \tau^2$ in the system $\mathcal{O}$. If two points p and q are τ-simultaneous with respect to an object A represented by an inertial reference line (u), then they are also simultaneous with respect to any other object B that can be represented by means of an inertial reference line (u'). Thus τ-simultaneity is independent of how a virtual station of observation is selected from among an infinite set of virtual objects that can be represented by means of inertial lines, *provided* that they all pass through the origin 0.

But τ-simultaneity is in general not independent of the choice of origin. Consider two coordinate systems of the A type, $\mathcal{A}(0)$ centered on an origin 0 and $\mathcal{A}(0')$ centered on another origin $0'$. What has been said can be expressed in this way. If two separated events p and q are simultaneous in $\mathcal{A}(0)$, they are in general not simultaneous in $\mathcal{A}(0')$.

The condition

$$\Delta(0, p) = \Delta(0, q)$$

of τ-simultaneity in $\mathcal{A}(0)$ does not entail the condition

$$\Delta(0', p) = \Delta(0', q)$$

of τ-simultaneity in $\mathcal{A}(0')$, where $0' \neq 0$.

Conditions Under Which Two Separated Events p and q Are τ-Simultaneous in Two Systems $\mathcal{A}(0)$ and $\mathcal{A}(0')$ with 0′ Separated from 0

A transition from 0 to $0'$ is represented by the following transformation of optical coordinates:

$$\left.\begin{aligned} v' &= v - v(0') \\ w' &= w - w(0') \end{aligned}\right\}$$

p and q are τ-simultaneous in both $\mathcal{A}(0)$ and $\mathcal{A}(0')$ if and only if

$$\Delta v'w' = \Delta vw = 0$$

where Δvw is short for $v(p)w(p) - v(q)w(q)$ and $\Delta v'w'$ for $v'(p)w'(p) - v'(q)w'(q)$.

The above condition is satisfied only if

$$\Delta(vw(0') + wv(0')) = 0$$

and $\Delta vw = 0$.

Hence p and q must be points on a geodesic line

$$vw(0') + wv(0') = \text{const.}$$

This geodesic must be a separation line. The condition $\Delta vw = 0$ is not satisfied otherwise. Hence it must hold that

$$v(0')w(0') = \Delta(0, 0')^2 > 0$$

0 and 0′ must therefore be points on an inertial line. The equation of that line is

$$vw(0') - wv(0') = 0$$

The separation line to which p and q belong is thus conjugate to the inertial line passing through the two origins.

We can now state sufficient and necessary conditions for two separated events p and q to be τ-simultaneous in two systems $\mathcal{A}(0)$ and $\mathcal{A}(0')$ (*cf.* Fig. 6):

a. $\mathcal{A}(0)$ and $\mathcal{A}(0')$ must have one common coordinate line, *viz.* an inertial reference line passing through 0 and 0′. This condition is satisfied if and only if $\Delta^2(0, 0') > 0$.
b. p and q must belong to a separation line that is conjugate to the common coordinate line.
c. p and q must also belong to a hyperbola $vw = \tau^2$ which is a coordinate line in $\mathcal{A}(0)$.
d. p and q belong to a hyperbola $v'w' = \tau'^2$ which is a coordinate line in $\mathcal{A}(0')$.

If these conditions are satisfied for two points p and q, then they are not satisfied if q is replaced by another point r, separated from p and q.

Absolute Simultaneity?

The dependence of τ-simultaneity on the choice of an origin of time makes it natural to ask whether one such choice may be given a privileged rank.

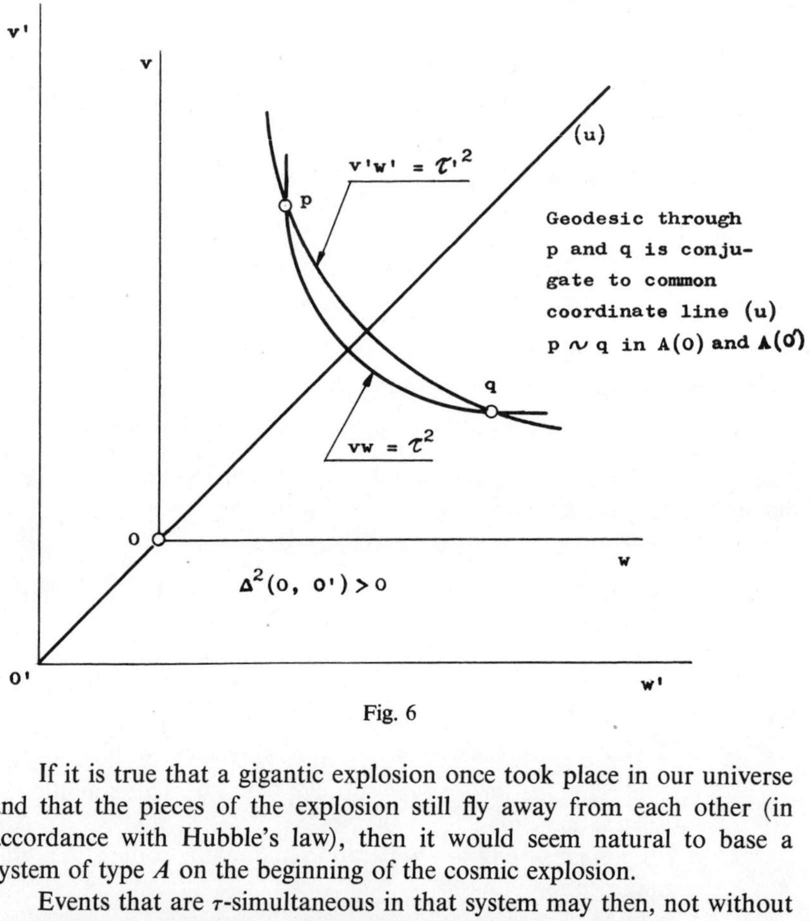

Fig. 6

If it is true that a gigantic explosion once took place in our universe and that the pieces of the explosion still fly away from each other (in accordance with Hubble's law), then it would seem natural to base a system of type *A* on the beginning of the cosmic explosion.

Events that are τ-simultaneous in that system may then, not without justification, be regarded as absolutely simultaneous.

A notion of absolute simultaneity is thus not incompatible with the special theory of relativity. Its usefulness hinges, however, on whether a controversial cosmological hypothesis can be maintained.

t-Simultaneity

Two events *p* and *q* are *t*-simultaneous in a system $\mathcal{R}(u)$ if they belong to a separation line conjugate to the inertial reference line (*u*) on which $\mathcal{R}(u)$ is centered.

Two systems $\mathcal{R}(u)$ and $\mathcal{R}(u')$ either are identical or else have no common coordinate line. It follows that two separated events that are t-simultaneous in a system $\mathcal{R}(u)$ are not t-simultaneous in any $\mathcal{R}$-system different from $\mathcal{R}(u)$.

t-simultaneity in contrast to τ-simultaneity is thus dependent on the relative motion between two stations of observation.

But t-simultaneity does not depend on the choice of an origin of time. This is another respect in which t- and τ-simultaneity differ.

Concluding Remarks

Material objects are involved in processes. The involvements have causal features. Objects act upon each other by way of transports of matter or light. The transports are viewed in the perspectives of the infratheory of STR but not the powers to emit transported agents, or the susceptibilities to be influenced by them. The infratheory is acausal.

An object is associated with a sequence of complex events. Selections of parts from those complexes give rise to what I call M-sequences. Events can also be coupled into sequences of another type than M-sequences. They can belong to propagations of light. Such sequences are called L-sequences.

Can an event be isolated? The answer is probably "no, every event belongs to an M-sequence or an L-sequence." It can belong to more than one such sequence, however, The event of emission of a light flash from an object A belongs to an M-sequence associated with A and to an L-sequence which links A to any object at which the light flash is received. STR is a general theory about L- and M-sequences.

In this paper, three types of coordinate systems have been introduced Each type is like a language. Ways of viewing the world are linked to languages. Likewise, in our case, events and processes are viewed in different ways in the three systems.

In the A-system they are more similar to the way in which Newtonians saw them.

Space-time is in both cases partitioned into space-manifolds in a manner not dependent on the state of (uniform relative) motion of virtual observers. Different virtual observers will agree on whether separated events are simultaneous or not. Velocities are composed in the same way in Newtonian kinematics and in the A-version of STR.

There are also differences. The choice of an origin will affect the partitioning of space-time in one theory but not in the other. Space-manifolds are Euclidean according to one of them but not according to the other.

Languages can be translated into each other. Ways of looking at the world linked to them may differ considerably, yet be equivalent. Translations from version *A* to version *R* of STR lead to considerable changes in the ways in which sequences of events are represented, yet the representations are equivalent.

But representations of sequences of events in Newtonian kinematics are not equivalent to the representation of them in any version of STR, although the apparent difference between Newtonian representations and *A*-representations within STR is not obviously larger than the apparent differences between the equivalent *A*- and *R*-representations of them. One should therefore be cautious in passing judgments on relations between theories on the basis of versions of them. One version of a theory may be translated into another equivalent one whereby differences may either seem to increase or to decrease.

How much does STR differ from Newtonian kinematics? The difference is larger in the version *R* than in the version *A* of STR.

Causal superstructures were built on an acausal infratheory in STR. A principle of relativity played a fundamental role in the building operations. A homologous principle was operative in Newtonian mechanics. A rational choice between the two principles of relativity depends on a verdict as to which one of the two infratheories, Newton's or Einstin's, is the better one.

a. Einstein's theory embodies Huygen's wave theory of light or rather its successor within Maxwell's electromagnetics. Many traits in which Newton's and Einstein's theories on space-time are incompatible hinge upon the view on light. Newton's view on light is well adapted to his theory of space-time (i.e., his kinematics), but Huygen's wave theory of light is not. It follows that evidence in favor of the latter view is also favorable to S T R and unfavorable to the Newtonian kinematics. An impressive body of evidence of this kind provides strong reasons to favor Einstein's theory.

b. There is an important difference between the two rival theories as they once were (I will now refer to them as N and E, respectively) in that E does whereas N does not exclude the possibility of a material body moving with a velocity surpassing that of light.

Let $\mathfrak{T}$ and $\mathfrak{T}'$ be two theories, $\mathfrak{T}$ excluding and $\mathfrak{T}'$ including the possibility of certain phenomena φ. We will disregard any other differences which $\mathfrak{J}$ and $\mathfrak{J}'$ may have. How should $\mathfrak{J}$ be evaluated in comparison with $\mathfrak{J}'$? A non-Popperian methodologist would argue that $\mathfrak{J}'$ is a better theory than $\mathfrak{J}$, because

i. Failure to discover φ will not affect the credibility of either one of the two theories but
ii. A discovery of φ would refute $\mathfrak{J}$ but not $\mathfrak{J}'$. It follows from (i) and (ii) that
iii. $\mathfrak{J}$ is more refutable (falsifiable) than $\mathfrak{J}'$. Hence
iv. $\mathfrak{J}$ is less credible than $\mathfrak{J}'$. Therefore, $\mathfrak{J}'$ is better than $\mathfrak{J}$.

An adherent of Popperian methodology like myself would arrive at the opposite verdict. $\mathfrak{J}$ is *ceteris paribus* better than $\mathfrak{J}'$ precisely because $\mathfrak{J}$ is more falsifiable than $\mathfrak{J}'$, which means that it contains more information. A theory is better the more possibilities it excludes. Einstein's theory excludes more possibilities than Newton's theory does. This is an additional reason for praising in hindsight Einstein's research program of re-forming causal theories in classical physics using his special principle of relativity as a heuristic tool. It is also a reason for believing that theories about fundamental particles should be based on Lorentz-invariant principles.

References

1. J. Aharony. *The Special Theory of Relativity*. Oxford: Clarendon Press (1959).
2. D. Bohm. *The Special Theory of Relativity*. New York: Benjamin (1965).
3. M. Bunge. *Foundations of Physics*. New York: Springer (1968).
4. A. Einstein, H. A. Lorentz, H. Weyl, and H. Minowsky. *The Principle of Relativity: A Collection of Original Memoirs on the Special and General Theory of Relativity*. London: Methuen (1923). (Reprinted New York: Dover Publications.)
5. A. Grünbaum. *Philosophical Problems of Space and Time*. New York: Knopf (1963)
6. A. Grünbaum. *Geometry and Chronometry in Philosophical Perspective*. Minneapolis: University of Minnesota Press (1968).
7. R. B. Lindsay and H. Margenau. *Foundations of Physics*. New York: Dover (1957).
8. H. Margenau. *The Nature of Physical Reality*. New York: McGraw-Hill (1950).
9. E. A. Milne. *Kinematic Relativity*. Oxford: Clarendon Press (1948).
10. C. Möller. *The Theory of Relativity*. Oxford: Clarendon Press (1951).
11. R. Nevanlinna. *Raum, Zeit und Relativität*. Basel: Birkhäuserverlag (1964).
12. E. P. Ney. *Electromagnetism and Relativity*. New York: Harper and Row (1965).
13. A. Papapetrou. *Spezielle Relativitätstheorie*. Berlin: DVW (1955).
14. S. J. Prokhovnik. *The Logic of Special Relativity*. Cambridge: Cambridge University Press (1967).

15. H. Reichenbach. *The Philosophy of Space and Time*. New York: Dover (1958).
16. A. Robb. *The Absolute Relations of Space and Time*. Cambridge: Cambridge University Press (1921).
17. H. M. Schwartz. *Introduction to Special Relativity*. New York: McGraw-Hill (1968).
18. E. F. Taylor and J. A. Wheeler. *Space-Time Physics*. San Francisco: Freeman (1966).
19. H. Törnebohm. *A Logical Analysis of the Theory of Relativity*. Stockholm: Almqvist and Wiksell (1952).
20. H. Törnebohm. *Fysik och Filosofi* (*Physics and Philosophy*). Stockholm: Almqvist and Wiksell (1957).
21. H. Törnebohm. *Concepts and Principles of Space-Time Theory Within Einstein's Special Theory of Relativity*. Stockholm: Almqvist and Wiksell (1963).
22. H. Törnebohm. A foundational study of Einstein's space-time theory. *Scientia 104*, No. 687–688 (1969).
23. H. Törnebohm. Two studies concerning the Michelson-Morley experiment. *Found. Phys. 1*, No. 1 (1970).
24. H. Törnebohm. Durations and distances in time. *Theoria*, Vol. XXXVII (1971), Part 3.

4

James L. Park

THE MEASUREMENT ACT IN QUANTUM PHYSICS

Until the advent of modern quantum theory with its profoundly statistical nature, most uncritical scientists probably espoused a doctrine of epistemological realism which presumed, rather boldly, that the mechanistic world described by the classical theories constituted quite literally a model, a facsimile, of ontological reality. With the empirical failures of mechanism and the striking successes of the irreducibly probabilistic quantum theory, the prevailing philosophical inclinations of physical theorists shifted during the 1920s toward epistemological idealism. Unfortunately, the considerable critical merits of the latter, an enlightened approach to science founded by Kant centuries before, were almost lost as physicists of the so-called Copenhagen school extrapolated idealism into solipsism until the orthodox natural philosophy of quantum physics had degenerated to a subjectivistic blend of physics and psychology wherein physical states of physical systems could be altered by the mere deliverance of information into the consciousness of the experimenter.

In dissent against this grotesque form of scientific philosophy, a minority chorus arose which included the voices of such distinguished iconoclasts as Einstein and Schrödinger, who condemned the structure of quantum physics itself as well as its Copenhagen interpretation. It has since become part of the folklore of physics that Bohr, the prophet of

---

James L. Park • Department of Physics, Washington State University.

Copenhagen, outwitted Einstein by adroit dialectical argumentation. I believe that the opposite is true: that Einstein understood the essence and the consequences of quantum theory better than Bohr ever did, that Einstein for his own metaphysical reasons would not accept those consequences, and that Bohr simply smothered Einstein's cogent analysis in that obfuscous fog called Complementarity.

Henry Margenau emerged in the 1930s as a physicist-philosopher who shared neither Bohr's abstruse views nor Einstein's reluctance to accept the new physics. Professor Margenau has since shown us that epistemological idealism does not require physical theory to exhibit subjectivist constructs, that indeed *objectivity*—physical reality—may be identified *within experience.*[(1)] In the Margenau epistemology, quantum physics may be comprehended with remarkable lucidity, yet without recourse to the ponderous verbiage of wave-particle duality, complementarity, and subjectivism.

In this essay I wish to honor Professor Margenau by offering an analysis* of quantal measurement inspired by his latency theory of observables. In particular, several types of measurement procedures will be described in order to illuminate some of the philosophical issues associated with the quantal measurement act.

The Primitive Measurement M_1

Classical and quantum physics alike focus attention on constructs called physical systems, which serve as the fundamental constituents of the theoretical models in terms of which experience is catalogued. All physical systems are regarded as carriers, in some sense, of observables.

A philosphical analysis which elegantly pinpoints the essential difference between the fundamental models employed in classical and quantum physics has been given by Margenau in his latency theory of observables.[(1–3)] This theory classifies the conceptual relations which link systems with their observables under two headings, possession and latency. Possessed observables are the hallmark of classical physics, where observables were regarded as labels attached to a system, i.e., as properties of the system. Within such a framework, it did not seem unreasonable to

*The following section of this paper is an abridged version of parts of several articles in *Scientia: 103* (1968), 569; *105* (1970), 269; *105* (1970), 320.

interpret a system with its observables as a pictorialization of the *Ding an Sich* responsible for, but beyond the grasp of, experience.

The characteristic quantal relation of latency is subtler than that of possession. The organized data of physics consist of numbers which emerge from experiments in which the acts called measurement are performed on physical systems. Each observable associated with a system is equipped in principle with operational definitions which give methods for the measurement of that observable. When a measurement of a given observable is performed, the numerical measurement result which emerges is recorded as the value of that observable at the instant defined by the onset of the measurement act. The difference between latency and possession lies in the theoretical interpretation of such numerical data. If possession is the assumed relation between system and observable, the measurement result is taken to be a revelation of which value of the observable the system possessed just prior to the measurement act.

If latency is the assumed relation between system and observable, the measurement result is not given this transcendent interpretation; the system is not regarded as an object bearing definite numerical values for all its observables either before or after measurement. Instead, a report of experimental results is confined to a minimal statement of how the systems were prepared for study, which observables were measured, and what numbers these measurements yielded. Similarly, a theoretical prediction of experimental results can have only the dispositional form "if system S is prepared in the manner Π and measurements of observable A are performed on S, then the numerical result a will emerge with probability $W(a; \Pi)$."

I shall call this minimal account of natural phenomena the *preparation–measurement format* for physical problems to distinguish it from the classical ideal of describing with Cartesian clarity the continuous evolution of an external world. Epistemologically, the latency viewpoint is entirely adequate for the scientific confrontation of experiential data and abstract concepts; but it stops short of identifying the numerical data of science with prepossessed properties of systems. Instead, numerical values of latent observables are considered to emerge from systems only when educed by an act of measurement.

Margenau has stressed for decades that the important concepts of preparation and measurement are *not equivalent*.[3] The term "preparation" signifies a reproducible physical act involving the system of interest, whereas the term "measurement" refers to a physical procedure on the prepared system which yields a numerical datum. To gather the statistical information from which the probabilities of quantum physics may be extracted, it is necessary to reprepare in an identical manner the system of interest (or another of the same kind) many times, and to perform a measurement subsequent to each repreparation. The probability $W(a; \Pi)$ associated with the measurement result a of an observable A for a system prepared in the manner Π is operationally defined as the relative frequency with which a emerges from such *ensembles* of identically prepared systems.

Unlike the classical approach to physical phenomena, the preparation-measurement format embodies a certain finality in the measurement act. The impact of measurement on a physical system can range from negligible (macrosystems) to catastrophic (microsystems); sometimes systems are even annihilated. Accordingly, quantum physics cannot include any *universal* proposition concerning the postmeasurement condition of physical systems.

The mathematical structure of quantum theory may be displayed in three simple axioms which employ the language of the preparation-measurement format:

A1: For every physical system S there is a Hilbert space $\mathcal{H}$. If S_1 and S_2 have Hilbert spaces $\mathcal{H}_1$ and $\mathcal{H}_2$, the Hilbert space of the composite system $S_1 + S_2$ is the tensor product space $\mathcal{H}_1 \otimes \mathcal{H}_2$.

A2: The linear Hermitean operators $A, B, \ldots$ on $\mathcal{H}$ correspond to physical *observables* of the system S. When a *measurement* $M_1(A)$ of observable A is performed on S, a real number, an A-datum, emerges.

A3: Every mode of *preparation* Π of a system S is represented by a statistical operator ρ. The arithmetic mean of A-data obtained from measurements $M_1(A)$ on an ensemble of systems S each prepared in the manner Π is given by $\langle A \rangle = Tr(\rho A)$.

Now, when as in classical theoretical physics, system and observable are connected by the primitive relation of possession, the preparation-measurement format seems an overly cumbersome expression of a simpler idea, *viz.*, that the act of measurement is merely a careful observation of the properties borne by the system. If a system can be thought of as carrying numerical labels, one for each observable, measurement becomes nothing more than a procedure for discovering at some instant of interest the values of one or more of these labels.

However, in the mathematical framework set forth above, it is impossible even to conceive theoretically of a preparation Π which would result in certain predictable values for all observables on measurement M_1. This is the essence of von Neumann's proof that there are no dispersionless ensembles, of Heisenberg's uncertainty relations, and it is the logical background for Margenau's latency interpretation.

In quantum physics, experiments remain describable in the preparation-measurement format; but, because of the inevitable quantal uncertainty, no consistent labeling scheme is possible. Consequently, the primitive relation *possession* must be abandoned, and measurement can no longer be interpreted as yielding revelations of preexisting values for observables. The preparation-measurement format itself is in quantum physics a primitive concept. More precisely, quantum theory employs the notion of measurement M_1 in a manner that must be regarded as logically primitive. Thus when a quantum theorist asserts that if a system is prepared in a specified manner subsequent measurements will yield particular numbers with stated probabilities, no deeper, hidden meaning is intended. In particular, no prediction is made about properties of the system (since it possesses none); the prediction is literally (and physically!) about measurement results subsequent to a specified preparation. Hence that which is predicted cannot even be stated without using the notion of measurement; therefore, measurement M_1 is a primitive term.

M_1 is, in fact, the primitive relation that plays the same role in quantum physics that possession did in classical physics. Wherever in classical theory the statement "system S possesses observable A with value a" would occur, the analogous place in quantum theory will feature the dispositional statement "if observable A is measured on S, the number a will emerge (with some specified probability)." Thus a quantum observable is associated with the system but no numerical label is attached to the system; however, when elicited by measurement, a numerical datum will emerge for that latent observable.

Complementarity is not without its contemporary apologists. Among these is Feyerabend,[(4)] who several years ago declared that Margenau's latency theory consisted merely of Bohr's views expressed in "fancy terminology." In rejoinder to this rather snide allegation, I should like to place the quantum philosophies of Bohr and Margenau in historical perspective.

Consider Copernicus, precursor to Kepler, Galileo, and Newton, whose works culminated in the decline and fall of Aristotelian science and its attendant philosophies. Aristotelian physics classified all phenomena into the categories of terrestial and celestial. For terrestial bodies linearity of motion was considered natural, but for celestial bodies circularity of motion was the dogmatic requirement. Thus in his *De Revolutionibus* Copernicus,[5] following Aristotle, asserts that

> . . . the motion of heavenly bodies is circular. Rotation is natural to a sphere and by that very act is its shape expressed.

He next proceeds to review the complexities of planetary movements, the departures from uniformity which would seem to argue against the necessity of circular perfection in the heavens. But then, having exposed the very data which inspired his bold advocacy of heliocentric astronomy, which would later motivate total abandonment of the circularity dogma by others, Copernicus made a remarkably conservative statement:

> Nevertheless, despite these irregularities, we must conclude that the motions of these bodies are ever circular or compounded of circles.

In the present century, the dismal failure of classical physics to account for atomic phenomena has resulted in what might be called the Copernican stage of a developing quantum revolution. At the microscopic level, the classical approach to physical problems has essentially been renounced. Moreover, it is generally acknowledged that quantum physics is in principle applicable to all physical phenomena, large and small. Nevertheless, the prevailing intuition with which even many modern physicists approach macroscopic phenomena is deeply embedded in the empirically discredited fabric of classical physics.

This situation is somewhat reminiscent of the medieval celestial-terrestrial dichotomy. In fact, some of the pronouncements of pioneer quantum theorists exhibit a striking resemblance to the perorations of Copernicus in defense of circularity. For example, Bohr[6] seems to insist on the necessity of the classical outlook for macroscopic experiences:

> . . . however far the phenomena transcend the scope of classical physical explanation, the account of all evidence must be expressed in classical terms.

Similarly, Heisenberg,[7] in explaining the so-called Copenhagen interpretation of quantum theory, ascribes to classical physics a rather sacrosanct position in scientific methodology:

> Any experiment in physics, whether it refers to the phenomena of daily life or to atomic events, is to be described in the terms of classical physics . . . the application of these [classical] concepts is limited by the relations of uncertainty. We must keep in mind this limited range of applicability of the classical concepts while using them, but we cannot and should not try to improve them.

It is my contention that the Copenhagen interpretation of quantum physics, a key point of which is indicated by the preceding quotations, is not the last word, but that it bears the same relationship to the current quantum revolution that the sixteenth-century work of Copernicus bore to the great seventeenth-century scientific revolution. The Copenhagen bifurcation of physical intuition into microscopic and macroscopic components is like the old celestial-terrestrial dichotomy in that it attempts to conventionalize a paradoxical melange of mutually incompatible concepts in order to conserve, in part, the no longer adequate Newtonian *Weltansicht*.

In the latency interpretation of quantum physics, and in Margenau's epistemology generally, no such absoluteness, no inevitability, is attributed to classical constructs; hence there is no "wave–particle duality." And that is a significant distinction between complementarity and latency.

The Operational Measurement M_2

If taken literally, the term "observable" as used in theoretical physics is quite misleading; for if observation denotes immediate apprehension by the senses, then most physical "observables," classical or quantal, are certainly unobservable! Consider, for example, the linear momentum P_x of a system S. Classical mechanics assumes that S has a P_x-value at any time t; quantum mechanics assumes that a P_x-value will emerge if $M_1(P_x)$ is performed at any time t. Yet in neither case is P_x directly observable. To obtain a P_x-value, a procedure is required which is considerably more elaborate than mere observation. Somehow an operation must be devised which establishes a correlation between the sought-after but unobservable P_x-value and another observable D for which the act $M_1(D)$ is a matter of direct apperception.

The creation of such correlative procedures is the art of experimental physics, which entails complex operations of data extraction commonly called measurements. Like any other physical process, measurements in this sense can be described quantally only in terms of the primitive measurements M_1. For example, an empirical measurement scheme for P_x is legitimized theoretically by proving that the procedure relates $M_1(P_x)$ and $M_1(D)$ in such a manner that the D-data actually gathered are in some sense equivalent to, or translatable to, the desired but not directly accessible P_x-data. To distinguish the term "measurement" as a physical operation which correlates observables from the *primitive measurement* construct M_1, let the former be designated M_2, the *operational measurement*.

A familiar illustration of an $M_2(P_x)$ is the time-of-flight method, wherein it is assumed that for the position coordinate X, $M_1(X)$ is either directly performable or an $M_2(X)$ is readily available. If x is the X-datum obtained from $M_1(X)$ at time t and m the mass of the system, then the number mx/t is taken to be equal to the P_x-datum that $M_1(P_x)$ would have yielded at $t = 0$, provided that three conditions are fulfilled: (a) The preparation at $t = 0$ must be one for which X-data at that initial time would display a probability distribution of compact support about $x = 0$; i.e., the probability density for X-data at $t = 0$ must vanish outside a finite neighborhood of $x = 0$. (b) The system must evolve *freely* in the interval from $t = 0$ until the terminal act $M_1(X)$ which yields the datum mx/t. (c) The precision of this "P_x-meter" increases as t increases; thus t must be as large as possible.

It is instructive to compare the classical and quantal measurement (M_2) theories which justify the claim that the foregoing procedure measures P_x. In the classical case it need only be noted that a particle always *possesses* some value for P_x and that in free dynamic evolution that value is constant; hence the final $M_1(X)$ simply enables the experimenter to ascertain which P_x-value the particle had all along its flight.

In the quantal version, on the other hand, it is physically meaningless to assert that the system had any P_x-value at any time during the process. If the initial preparation had been characterized by a statistical operator which assigned unit probability to a particular P_x-value, then to be sure P_x would retain its spiked probability distribution as long as the system remained dynamically free. However, due to the uncertainty relation $\Delta X \Delta P_x \geq \hbar/2$, the above condition (a) for the efficacy of the time-of-flight method would then be violated. Hence preparations for which the time-of-flight $M_2(P_x)$ is valid do not admit of the interpretation that the system possesses an unknown P_x-value.

To be specific, the detailed quantum mechanical analysis[8] of this $M_2(P_x)$ can and does affirm only this proposition: the probability that $M_1(X)$ at time t will yield the number mx/t equals the probability that $M_1(P_x)$ at $t = 0$ would have yielded the same number provided that conditions (a), (b), and (c) are satisfied. The answer, if there be one, to the somewhat ontological question as to whether in a single run of the experiment the value mx/t is truly the same as the datum that $M_1(P_x)$ would have yielded at $t = 0$ is beyond the scope of the probabilistic quantum theory.

Although the time-of-flight $M_2(P_x)$ is only one type among a diverse variety of conceivable operational measurements, our careful if somewhat belabored scrutiny has revealed one structural feature paradigmatic of all *quantum* theories of M_2: a procedure is quantally certified as an $M_2(A)$ if it can be proved that the theoretical probability distribution for A-data *matches* the probability distribution for D-data, where $M_1(D)$ is a directly performable act.[9]

From a classical viewpoint, simple *probability matching* would seem inadequate to justify identification of a given operation as a measurement scheme; nevertheless, it is the best justification that quantum mechanics, with its latent observables and irreducible probabilities, can possibly offer.

Most textbooks discuss the Stern-Gerlach "spin-meter" as the prototype operational measurement. At first glance, the correlations it establishes may seem stronger than simple probability matching. Indeed, after the system has interacted with the inhomogeneous magnetic field, there is a *correlated joint probability distribution* in the variables position and spin such that from a position measurement datum one may infer the value that a simultaneous primitive spin measurement would have yielded. However, the function of the "spin-meter," of which the field is merely a component, is to measure the spin of the system prior to its interaction with the meter. The only sense in which this is achieved is that the probability distribution for spin is unchanged by the interaction, a *probability match* which does not offer any extraphysical guarantee that a preinteraction primitive spin measurement would have yielded the same result as a postinteraction primitive spin measurement in the same run of the experiment. But no such assurance should be expected from an indeterministic theory which predicts only probability distributions, a theory in which observables are fundamentally latent.

To summarize: an empirical scheme is certified to be an operational measurement $M_2(A)$ if the quantum theoretical probability distribution of $M_1(A)$ data matches that for $M_1(D)$ data. Each number on the "D-scale" which emerges from the act $M_1(D)$ may be transformed by means of the $M_2(A)$ theory to that number on the "A-scale" having the same theoretical probability, and as a matter of convention that A-value may be defined as the datum that the unperformed primitive measurement $M_1(A)$ would have yielded. However, it is only the collective of such A-data obtained from an ensemble of many runs that is scientifically important.

Misconceptions Concerning Quantal Measurement

There have been many attempts to characterize the measurement act in quantum mechanics more fully, more explicitly, than does the foregoing discussion of Π, M_1, M_2, and probability matching. Consequently the literature of physics abounds with references to a number of alleged after-effects of measurement, among these the following propositions:

a. Measurement determines the unknown quantum state vector.
b. Measurement destroys coherence.
c. Measurement entails uncontrollable disturbance of the measured system.

We shall demonstrate below that none of these is a universal trait of the operational measurement M_2, hence that all are physical overspecifications of the measurement act which, if regarded as axiomatic, would unreasonably constrict the quantal meaning of the term "measurement."

To focus attention on structure rather than detail and to avoid fruitless tedium, we consider only the simplest nontrivial physical systems. *viz.*, those whose individual Hilbert spaces $\mathcal{H}$ are only two-dimensional. The observables for each such system comprise all Hermitean linear combinations of the identity operator 1 and the Pauli operators σ_x, σ_y, σ_z on $\mathcal{H}$. Let α, β be the normalized eigenvectors of σ_z belonging to eigenvalues $+1$, -1, respectively. If two systems S_1, S_2 are combined to form a single composite system, the Hilbert space $\mathcal{H}_1 \otimes \mathcal{H}_2$ for the latter will be four-dimensional and may be spanned by the orthonormal set $\alpha\alpha$, $\alpha\beta$, $\beta\alpha$, $\beta\beta$, where, for instance, $\beta\alpha$ denotes the tensor product of the β in $\mathcal{H}_1$ with the α in $\mathcal{H}_2$. This notation may be extended in an obvious way to embrace composite systems having N constituent subsystems. Similarly, let the direct product of operators be represented by ordered juxtaposition; for example, $\sigma_x 1$ signifies the direct product of the σ_x on $\mathcal{H}_1$ and the 1 on $\mathcal{H}_2$, where $\sigma_x 1$ itself is of course an operator on $\mathcal{H}_1 \otimes \mathcal{H}_2$.

Does Measurement Determine the State Vector?

Let ρ be the statistical operator for a given preparation of a system S and P_α the projector onto α. All presentations of quantum theory attribute special significance to the number $\mathrm{Tr}(\rho P_\alpha)$, but there are two distinct versions:

i. $\mathrm{Tr}(\rho P_\alpha)$ is the probability that $M_1(\sigma_z)$ or $M_2(\sigma_z)$ will yield the datum $+1$.
ii. $\mathrm{Tr}(\rho P_\alpha)$ is the probability that S will be found in the state α.

The language of the rule (ii) is based on the projection, or wave packet reduction, postulate according to which a measurement, in addition to yielding a number, *prepares* S so that an immediate remeasurement will yield the same number. Thus if σ_z is measured with result $+1$, S then will be in, or "have been found in," the eigenstate α. The irrationality of the projection postulate was exposed long ago by Margenau on both physical and philosophical grounds.[3] Indeed, it is somewhat vexing to find that wave packet reduction has yet to be exorcised from contemporary physics literature.

The pragmatic experimenter, who gathers and records in his log book *numerical* data—not Hilbert vectors—would probably regard (ii) as the theoretician's quaint rendition of (i); thus it might seem that the only difference between the two versions is a semantic one.

To see that the difference is quite physical and that (ii) is in fact erroneous, consider the following M_2 scheme for measuring observables of a system S_1. Let S_1 interact with an auxiliary system S_2 (the "apparatus") for a time interval in which the causal evolution of the composite system is described by the unitary operator T whose matrix representation in the basis $\alpha\alpha$, $\alpha\beta$, $\beta\alpha$, $\beta\beta$ is given by

$$(T) = \begin{pmatrix} 1 & 0 & 0 & 0 \\ 0 & 0 & 1 & 0 \\ 0 & 1 & 0 & 0 \\ 0 & 0 & 0 & 1 \end{pmatrix} \tag{1}$$

We have shown elsewhere[10] that such an evolution operator may be generated by a Hamiltonian proportional to $\vec{\sigma}\cdot\vec{\sigma}$; but this kind of detail is unimportant in the present analysis. It is always possible in principle to find a Hamiltonian H and a time interval τ such that any given unitary evolution operator T may be expressed as $T = \exp(-i\tau H)$.

If S_1 is prepared such that $\rho_1 = P_\psi$, $\psi = a\alpha + b\beta$, and S_2 such that $\rho_2 = P_\alpha$, then at the onset of the measurement interaction $S_1 + S_2$ will be characterized by the global state vector $\psi\alpha = a\alpha\alpha + b\beta\alpha$.

Since

$$\begin{Bmatrix} T\alpha\alpha = \alpha\alpha, & T\alpha\beta = \beta\alpha \\ T\beta\alpha = \alpha\beta, & T\beta\beta = \beta\beta \end{Bmatrix} \tag{2}$$

the global state vector after the interaction will be $a\alpha\alpha + b\alpha\beta = \alpha\psi$. Thus the probability distributions for all S_2 observables after the interaction are identical to the distributions for corresponding S_1 observables before the interaction, a *probability match*; hence we have an operational measurement M_2. If one now asks for (i) the probability that $M_2(\sigma_z)$ will yield $+1$, the answer is certainly

$$\mathrm{Tr}[(P_{\alpha\psi})(1P_\alpha)] = \mathrm{Tr}[(P_{\psi\alpha})(P_\alpha 1)] = \mathrm{Tr}_1(P_\psi P_\alpha), = |\langle\alpha \mid \psi\rangle|^2 \tag{3}$$

where the first equality is a formal statement of the probability match on which the M_2 is based. On the other hand, if one asks which state S_1 is in after the M_2, the answer is that S_1 is certainly "in the state α," and this physical fact is completely independent of whether any individual run of the $M_2(\sigma_z)$ procedure yielded $+1$ or -1. Therefore, (ii) is not universally true, and it is not equivalent to (i).

Although a single measurement cannot determine an unknown quantum state, there is a procedure for doing so. It must be understood, however, that the role of *state* in quantum theory is played by the statistical operator ρ, which characterizes modes of *preparation*. Hence the problem of state determination must be posed as follows: given a preparation Π, how many data must be collected in order to find the statistical operator ρ which describes Π? One answer, developed by Band and me,[(11)] is that ρ may be determined from the mean values of a set of observables called the *quorum*. The size of a quorum depends on the dimensionality of the Hilbert space. In general, neither a single measurement result nor even an entire statistical collective of data from a single observable is sufficient information from which to deduce the quantum state. If the statistical operator turns out to be of the form $\rho = P_\psi$, then as a convention we may use this somewhat misleading but established jargon: Π prepares systems "in the state ψ."

Does Measurement Destroy Coherence?

This is not the place to enter again[(9)] into the interminable philosophical controversy that rages over the uniquely quantal phenomenon of interference of probabilities. There are a number of philosophers of science who may never accept the primitivity of the preparation-measurement format and who will therefore always see recondite paradoxes in the fabric of quantum physics.

This metaphysical debate has encroached upon theoretical physics itself to this extent: in terms of our simple two-dimensional example, it is often postulated that a measurement of σ_z upon S initially prepared in the pure state $\rho = P_\psi$, $\psi = a\alpha + b\beta$, *must* convert that state to the mixture $\rho_M = |a|^2 P_\alpha + |b|^2 P_\beta$. The rationale for this claim has much in common with the belief, discredited above, that each measurement act reveals which state vector S possesses. Thus, according to this view, the ensemble of measured systems should be a mixture, divisible into two pure subensembles characterized by the state vectors α, β uncovered by the measurements.

Since ρ incorporates phase relations between a and b not present in ρ_M, a measurement act which converts ρ to ρ_M is said to "destroy coherence."

It is indeed possible to envisage operational measurements M_2 which do destroy coherence. Among these is the famous two-slit gedankenexperiment with electrons, wherein the placement of counters to detect the electrons at the slits destroys the interference pattern that develops on the screen when the counters are absent.

This phenomenon is often given an extravagant subjectivistic interpretation according to which the interference is lost because the experimenter, in placing the counters, comes to know too much about the electrons; thus since quantum theory allegedly limits knowledge (complementarity) the experimenter must be penalized by forfeiture of his interference pattern. Fortunately there is, in terms of the preparation-measurement format, a more elegant and objective interpretation. Data can be altered in two ways: by changing the preparation or by measuring a different observable. In the two-slit experiment, with or without counters, the screen measures the same observable, *viz.*, electron position at some instant following preparation. Hence the reason that the screen exhibits different data in the presence and absence of counters at the slits must be that two distinct preparations are involved. Electron-source-plus-double-slit and electron-source-plus-double-slit-plus-counters are not identical preparations; therefore, measurements do not yield the same data for the two cases.

It is merely an *accidental* feature of the two-slit experiment that the difference in data with and without counters may be described as destruction of coherence due to measurement by the counters. Indeed, it is possible to conceive, as will be demonstrated below, of operational measurements M_2 which have the effect of *creating coherence*. Destruction of coherence is therefore not a universal characteristic of the measurement act.

To prove this, we consider first an M_2 which is mathematically similar to the two-slit experiment with counters but which involves only simple two-level systems. Let S_1 be a system prepared with $\rho_1 = P_\psi$, $\psi = a\alpha + b\beta$, a coherent superposition of σ_z-eigenstates. If S_1 interacts with another system S_2 (the "counters") in a manner described by the evolution matrix

$$(U) = \begin{pmatrix} 1 & 0 & 0 & 0 \\ 0 & 1 & 0 & 0 \\ 0 & 0 & 0 & 1 \\ 0 & 0 & 1 & 0 \end{pmatrix} \tag{4}$$

and S_2 is initially prepared with $\rho_2 = P_\alpha$, then since

$$\begin{Bmatrix} U\alpha\alpha = \alpha\alpha \\ U\beta\alpha = \beta\beta \end{Bmatrix} \quad (5)$$

the global state vector after the interaction will be $\Phi = a\alpha\alpha + b\beta\beta$. The reduced statistical operator for S_1 is then the partial trace over $\mathcal{H}_2$,

$$\rho_{1M} = \mathrm{Tr}_2 P_\Phi = |a|^2 P_\alpha + |b|^2 P_\beta \quad (6)$$

an incoherent superposition of σ_z-eigenstates. Thus the preparation of ρ_1 followed by the interaction with S_2 constitutes preparation of ρ_{1M}, which is quite distinct from ρ_1. Now, since there is a probability match between $M_1(\sigma_z)$ on S_2 after the interaction and $M_1(\sigma_z)$ on S_1 before the interaction, the procedure may be utilized as an $M_2(\sigma_z)$ for S_1; thus it is possible to say here that "measurement destroys coherence." But the coherence is destroyed not by the acquisition of knowledge but by the interaction, whether or not the interaction is actually applied as an operational measurement.

On the other hand, an act of measurement may occur which incidentally results in the *creation* of coherence. Suppose that S_1 is prepared initially in an incoherent state $\rho_1 = w_+ P_\alpha + w_- P_\beta$ and that S_2 is initially in a pure state $\rho_2 = P_\psi$, $\psi = a\alpha + b\beta$. The initial global state is then given by

$$\begin{aligned} \rho &= (w_+ P_\alpha + w_- P_\beta) P_\psi \\ &= w_+ P_{\alpha\psi} + w_- P_{\beta\psi} \end{aligned} \quad (7)$$

Let S_1 and S_2 interact according to evolution operator T, whose matrix was given by (1). Applying (2) to (7), we obtain for the postinteraction global statistical operator

$$\begin{aligned} \rho_M &= w_+ P_{\psi\alpha} + w_- P_{\psi\beta} \\ &= P_\psi (w_+ P_\alpha + w_- P_\beta) \end{aligned} \quad (8)$$

The state of S_1 has now been transformed *from* an incoherent superposition of σ_z-eigenstates *to* a coherent superposition of σ_z-eigenstates. Moreover, the interaction meets the test of establishing a probability match between preinteraction M_1s on S_1 and postinteraction M_1s on S_2, and may therefore be used as an M_2 for S_1. Of this measurement scheme it could truly be said that "measurement creates coherence"! The general proposition that measurement destroys coherence is therefore false.

Does Measurement Entail Uncontrollable Disturbance?

The normally cataclysmic impact of the measurement act on a microsystem is undeniable; however, characterization of the measurement disturbance as unpredictable, hence uncontrollable *in principle*, is somewhat ambiguous. Historically the uncontrollability doctrine arose in complementarist analysis of Heisenberg's microscope, wherein it was essentially noted that the final momenta in a single electron-photon collision cannot be predicted in advance. This is of course true, since classical mechanics is wrong and quantum mechanics predicts only probability distributions. Thus such discussions of uncontrollability really assert, in a colorful way, only this: quantum mechanics cannot in general make predictions concerning a single run of an experiment. In other words, quantum theory offers no classical predictions, hence no classical controllability of individual events.

However, this lack of control in the classical sense should not be construed to mean that the measurement act is quantally uncontrollable. On the contrary, it is possible to design operational measurements which do not disturb the quantum state at all or, to be precise, which leave the system after the measurement reprepared just as it was before the measurement interaction.

As an example, suppose observables of S_1 are to be measured by interaction with S_2. To devise the nondisturbing M_2, we first ascertain the quantum state of S_1 by measuring by any means available a quorum of observables. Assume for simplicity that the preparation of S_1 turns out to be pure, with state vector $\psi = a\alpha + b\beta$, a, b real. Let S_1 interact with S_2 according to the evolution matrix

$$(R) = \begin{pmatrix} a & b & 0 & 0 \\ 0 & 0 & a & b \\ b & -a & 0 & 0 \\ 0 & 0 & b & -a \end{pmatrix} \tag{9}$$

If S_2 is prepared with initial state vector α, the initial global state vector for $S_1 + S_2$ will be $(a\alpha + b\beta)\alpha = a\alpha\alpha + b\beta\alpha$. Applying (9), we find that

$$\begin{pmatrix} a & b & 0 & 0 \\ 0 & 0 & a & b \\ b & -a & 0 & 0 \\ 0 & 0 & b & -a \end{pmatrix} \begin{pmatrix} a \\ 0 \\ b \\ 0 \end{pmatrix} = \begin{pmatrix} a^2 \\ ab \\ ba \\ b^2 \end{pmatrix} \tag{10}$$

The postinteraction global state vector is therefore given by

$$a^2\alpha\alpha + ab\alpha\beta + ba\beta\alpha + b^2\beta\beta = (a\alpha + b\beta)(a\alpha + b\beta) \tag{11}$$

The structure of (11) establishes that there is a probability match between the M_1s on S_2 after the interaction and corresponding M_1s on S_1 before the interaction and that the interaction also reprepares S_1 as it had been prepared originally. Hence we have a measurement act whose effect on the system has been "controlled" so well that the system emerges as though nothing had happened.

Finally, it is interesting to note that such a "nondisturbing" M_2 may be used to generate a time ensemble. Consider a large number N of systems initially prepared so that the initial global state vector is $(a\alpha + b\beta)\alpha\alpha \ldots \alpha[N \text{ factors}]$. If S_1 interacts successively with S_2, then S_3, etc., in the manner of evolution operator R, eventually the global state vector will be $(a\alpha + b\beta)(a\alpha + b\beta) \ldots (a\alpha + b\beta)$, provided that each S_n participates in no other interactions and that its own free evolution is stationary. There is now probability matching between the original M_1s of S_1 and the analogous M_1s of each S_n after the interaction. Suppose an $M_2(\sigma_z)$ is now performed simultaneously on $S_1, \ldots, S_N$; the yield will be a collective of σ_z-data which, according to the usual conventions, could be described theoretically as the results that would have been obtained by performing N $M_1(\sigma_z)$ acts simultaneously on S_1 after its original preparation.

Because latency, not possession, is the theoretical link between S_1 and σ_z, it is perfectly rational to imagine that simultaneous $M_1(\sigma_z)$ acts might yield different results. However, this does *not* refute, for example, the well-known experimental fact that an electron is never found in two places at once. The methods by which electron position is operationally measured involve interactions described by evolution operators more akin to T than R. Consequently, the Stern-Gerlach apparatus would never yield both $+1$ and -1 as simultaneous σ_z-data for a single system. Nevertheless, the formal possibility of a nondisturbing quantal measurement act calls attention in a new way to the many interesting but relatively unexplored questions regarding simultaneous measurements, sequential measurements, and macroscopic observations which continue to stimulate research into the foundations of physics.

References

1. H. Margenau. *The Nature of Physical Reality*. New York: McGraw-Hill (1950).
2. H. Margenau. *Physics Today* 7 (1954), 6.

3. H. Margenau. *Phil. Sci. 4* (1937), 337; *Phil. Sci. 30* (1963), 6; and many other papers.
4. P. Feyerabend. *Phil. Sci. 35* (1968), 311.
5. N. Copernicus. *De Revolutionibus Orbium Caelestium* (1543). Translated by J. F. Dobson and S. Brodetsky. *Occasional Notes of the Royal Astronomical Society 2* (1947), 9.
6. N. Bohr. In P. Schilpp (ed.). *Albert Einstein: Philosopher-Physicist*. New York: Harper (1959), p. 209.
7. W. Heisenberg. *Physics and Philosophy*. New York: Harper (1958), p. 44.
8. J. Park and H. Margenau. *Int. J. Theor. Phys. 1* (1968), 211.
9. J. Park. *Phil. Sci. 35* (1968), Part I, 205, Part II, 389.
10. J. Park. *Found. Phys. 1* (1970), 23.
11. W. Band and J. Park. *Found. Phys. 1* (1970), 133. *Cf.* also *Found. Phys. 1* (1971), Part I, 211, Part II, 339.

5

R. Bruce Lindsay

ACOUSTICS–OLD AND NEW

Acoustics is the science of sound. But what is sound? In its historical sense, sound is a disturbance produced by an object or set of objects in a material medium which somehow manages to reach the ear and produces there the sensation called hearing. Just as the word *optics*, for the science of light, comes from the Green meaning sight, so *acoustics* comes from the Greek meaning hearing. These two important sensory windows on the world of our human experience (which interestingly are both of roughly the same sensitivity from the energy standpoint) have naturally for ages stimulated inquiry into how and why they work, an inquiry which has by no means reached completion even today. But just as light has come to mean much more than what we see, so sound has come to incorporate a much wider range of experience than what we hear. What we hear is plenty and indeed often too much for comfort and safety, but the sounds we do not hear (ultrasonics) provide a more far-reaching and fascinating branch of science.

From the day when the Frenchman Joseph Sauveur[(1)] first suggested (1701) that the name *acoustics* be applied to the science of sound, it has had an enormous development and now plays a role in practically every aspect of human experience, touching on all branches of science.

Acoustics and Physics

An interesting philosophical question, perhaps more appropriately termed sociological, is now often raised: is acoustics a branch of physics?

R. Bruce Lindsay • Department of Physics, Brown University.

This may seem to be a foolish question, yet it must be admitted that many modern physicists answer it in the negative in spite of the fact that the Acoustical Society of America is one of the founder members of the American Institute of Physics. Substantial courses in acoustics are now rarely taught in departments of physics in universities in the United States. The contemporary physicist, if he admits acoustics to the sacred precincts at all, is apt to dismiss it as a part of what is euphemistically termed classical physics, a relegation to a realm of things so outdated as no longer have any relevance to the progress of physical science.

To the question at issue and the implied negative attitude several responses may be made. In the first place, acoustics certainly grew up as a branch of physics. The ancients, faced with the problem of explaining how the sound of a speaker's voice could be heard by a listener some distance away, obviously had to assume that something was happening in the intervening medium. This was physics. Even Aristotle(2) in his treatises *De Anima* and *De Audibilibus* stresses his conviction that the propagation of sound has something to do with the motion of the air, and it is possible to read into his description a vague idea of wave motion. It is in any case clear that the famous Roman architectural engineer Marcus Vitruvius Pollio(2) of the first century B.C. had a clear grasp of the wave theory of sound in the analogies he drew between the transmission of sound and surface waves in water.

The idea that the transmission of sound takes place by wave motion through a material medium did not go unchallenged. The revival of the Greek atomic theory in the Renaissance, notably by the Frenchman Pierre Gassendi,(3) led to the attempt to explain the propagation of both sound and light through the motion of corpuscles. The atomic theory of sound transmission, however, did not survive very long, certainly not as long as the corpuscular theory of light in the formulation which Newton gave to it. It is an interesting commentary on the role which acoustics early played and has continued to play in the theoretical development of other branches of physics that it was the wave theory of sound which encouraged natural philosophers like Huygens to apply the same sort of thing to light. At the same time, the celebrated principle of Huygens governing the propagation of wave fronts ultimately materially advanced the understanding of certain acoustical phenomena, once again illustrating the close interconnection of different branches of physical science.

This is scarcely the appropriate place to review the whole history of acoustics, but in the light of the question of its relation with physics as a whole we must of course refer to the preoccupation of giants of physics like Galileo and Newton with acoustical problems. It was apparently Newton[4] who carried out the first derivation of the general expression for the velocity of sound in a gas. During the eighteenth century, this derivation was criticized and improved by such scholars as Euler,[5] Daniel Bernoulli,[6] and Lagrange.[7] It was Euler, indeed, who as a youth of 20 in 1727 wrote the first serious treatise on the nature of sound, the beginning of a long line of works in theoretical physics which are still in process of republication in his collected works, though he died in 1783.

The problem of the velocity of sound and its dependence on the other properties of the medium opened up a field of physical research which has continued to attract attention to the present day. It began with the mysterious failure of Newton's formula and the equivalent ones of his immediate successors to agree with experimentally measured values: the theoretical value was too low. It was Laplace[8] who suggested in 1816 that the trouble was due to the fact that the Newtonian formula implies that sound disturbances take place isothermally. Laplace reasoned that in sound the compressions and rarefaction in the medium take place so rapidly that the isothermal condition cannot be maintained. Rather, he insisted, the adiabatic condition is more realistic. This assumption worked, although Laplace's reasoning was somewhat at fault.

As a further illustration of the role of acoustics in the development of physics, particularly in its theoretical aspects, we may cite the stimulus given to the investigation of the properties of continuous elastic media under both constant and time-varying stresses. In a certain sense, we may regard art in the form of music as the initial stimulus: the desire to understand the behavior of vibrating strings as used in musical instruments. The problem of the vibrating string was a key eighteenth-century problem and engaged the attention of the major mathematicians of the period. The setting up of the appropriate partial differential equation and its solution were achieved by d'Alembert.[9] This led to the development of the practical theory of partial differential equations of the second order with incalculable consequences for the whole future course of mathematical physics. There was hardly a great applied mathematician of the eighteenth or nineteenth century who did not give attention to the motion of elastic media. All this had a definite acoustical basis.

Specific problems related to the propagation of sound waves in fluids were, of course, not neglected, as witness the studies of George Green[10] on the reflection and refraction of sound waves at interfaces, the profound treatises of Stokes[11] and Kirchhoff[12] on the attenuation of sound in fluids due to the transport properties of viscosity and heat conduction, as well as the labors of Earnshaw,[13] Rankine,[14] Riemann,[15] and others on the changes in the form of sound wave profiles due to large increase in the sound intensity.

The culmination of all this activity came with the publication in 1877 of the *Theory of Sound* by Lord Rayleigh,[16] one of the great landmarks of physical literature. Rayleigh felt that it was time for the great works on acoustical wave production, propagation, and reception to be pulled together in a unified fashion, to which he added the discussion of numerous problems whose solution was still open, e.g., acoustic resonators and filters as well as sound diffraction. This classic work has indeed been the subject of some misconstructions. The very thoroughness with which Rayleigh did his job and the many ingenious suggestions he made for the attack on outstanding acoustical problems have led many people to believe that he had said all there was to say about acoustics and that thanks to him the subject was essentially finished as a branch of physics. That this view is erroneous is made plain by Rayleigh's own involvement with acoustical research after the publication of his treatise and until his death in 1919. The record shows that he published some 75 articles on acoustics from 1877 to 1919. An unpublished paper on acoustics was found on his desk when he died.

Rayleigh's penetrating insight saw many acoustical problems in need of solution. Some he tackled himself, but others he merely mentioned for his successors to follow up. One of these provides an excellent example of the way acoustics came to be involved in present-day atomic and molecular physics. The attenuation of sound in polyatomic gases is generally considerably in excess of that predicted by viscosity and heat conduction and its dependence on the frequency is also of a different character. In a paper entitled "On the Cooling of Air by Radiation and Conduction and on the Propagation of Sound" (1899), Rayleigh[17] suggested that the excess attenuation might be attributed to the transfer of some of the sound energy represented in increased translational kinetic energy of the molecules of the gas to internal rotational and vibrational energy states. It takes a finite time for this energy to get back into the translational form and hence be restored to the energy flow in the sound wave. If this time is of the order of the period of the sound wave, considerable transmission loss will take place.

It is somewhat strange that Rayleigh did not follow up his suggestion with actual calculations. He was of course familiar with the molecular theory of sound, since he had discovered in 1892 the now-famous paper by the Scottish engineer and amateur scientist John James Waterston.[18] This paper, entitled "On the Physics of Media That Are Composed of Free and Perfectly Elastic Molecules in a State of Motion," had been originally submitted to the Royal Society for publication in the *Philosophical Transactions* in 1845, but was rejected on the basis of adverse criticism by two referees. It remained in the archives of the Society until Rayleigh exhumed it and realizing its worth had it finally published. In this paper, Waterston revived the molecular theory of gases initiated by Daniel Bernoulli in the eighteenth century and further discussed in some analytical detail in 1821 by J. Herapath, though with unfortunate errors due to his inadequate grasp of the principles of mechanics. Waterston was apparently the first to attempt a serious derivation of the laws of gases on the basis of the molecular theory. He succeeded in deriving the laws of Boyle, Charles, and Gay-Lussac, though with some questionable assumptions and some numerical errors. In the process he of course obtained an expression for the root-mean-square molecular velocity in terms of gas pressure p and density ρ and in the first instance found the now conventional expression for an ideal monatomic gas, *viz.* $v_m = \sqrt{3p/\rho}$. He later felt that this should be modified by replacing the 3 by 6.

From the acoustical point of view, Waterston's paper is particularly interesting since he introduced a molecular theory of sound propagation, considering sound pulses to be transmitted by means of molecular collisions. From his theory, he concluded that the velocity of a sound pulse should be one-half the root-mean-square molecular velocity. Using the figure 6 in his expression for the latter velocity gives for the velocity of sound $v_s = \sqrt{1.5p/\rho}$. When applied to air, this gives a result, which though too high, was not far enough off to discourage Waterston. It must be remembered that there was at the time Waterston was writing still some controversy over Laplace's adiabatic assumption in which the factor in front of p/ρ in the above expression becomes γ, the ratio of the two specific heats of the gas. Many experimental values of γ were too low to give the correct experimental value of v_s.

Waterston was in some sense a frustrated man. Many of his papers on which he lavished much time and care were rejected for publication and even those that were accepted (e.g., by the *Philosophical Magazine*) largely fell on deaf ears, as many scientists evidently considered him a crank. It would be of some interest to compare him in this respect with Robert Mayer, the German physician in Heilbronn, who was at that very same period trying to secure a hearing for his ideas on the concept of energy and the mechanical equivalent of heat. Both men kept working in spite of lack of recognition. Mayer received his during his lifetime, while Waterston's came only after his death. Mayer was determined to establish and claim his priority. It does not appear that Waterston had any interest in the problem of priority. He merely wanted people to know what the molecular theory (which he also applied at some length to liquids) could accomplish.

Our purpose in discussing Waterston's work is to emphasize that the molecular theory of sound was available to Rayleigh and his contemporaries. It therefore remains somewhat strange that Rayleigh did not go into the problem seriously himself. He may have felt that it was a difficult problem in statistical mechanics, about which there was much controversy in the closing decade of the nineteenth century. It remains curious, however, that the molecular theory of sound made so little progress after Rayleigh's suggestions. J. H. Jeans(19) did indeed make an attempt in his *The Dynamical Theory of Gases* but with little real success.

That the fundamental concept of a relaxation phenomenon, whereby some of the energy of a sound wave would get trapped temporarily in a molecular energy-receiving process and then after a measurable time would be released again out of phase with the incident flow, was bound to attract the attention of physicists of high competence is shown by the fact that Einstein(20) in 1920 first showed that dissociation relaxation in a gas could lead to acoustic attenuation and dispersion. This made it possible to understand the pioneer work of G. W. Pierce(21) in the mid-1920s on absorption and dispersion in carbon dioxide. The theoretical attack on the kind of relaxation which Rayleigh had in mind (called nowadays thermal relaxation) was taken up in earnest by H. O. Kneser(22) in the early 1930s with brilliant experimental confirmation provided by the work of V. O. Knudsen(23) on acoustic attenuation in air due to water vapor. This was the start of a new branch of the science of sound, called, appropriately enough, molecular acoustics. Rayleigh's forecast was completely confirmed. Since the relaxation times involved in the experimental studies were found to correspond to ultrasonic frequencies, this stimulated the development of transducers for the production of sound waves of very high frequency (up to 10^{10} Hz) and of very high intensity.

Ultrasonic attenuation and dispersion measurements coupled with relaxation time theory have proved a valuable tool in the study of molecular interactions in gases. Thus transition probabilities between internal energy states of molecules can be estimated. Similar theoretical and experimental techniques have also proved very fruitful in application to liquids and solids.

No branch of radiation physics is respectably complete without its particle. Electromagnetic radiation has its photon or quantized energy entity, whose quantum significance increases with the frequency. Similarly it has proved desirable to quantize the energy of the very high frequency waves due to the vibration of crystal lattices. Following Landau[24] these acoustic energy quanta have been called *phonons*. The energy of a phonon is equal to the momentum associated with the sound wave multiplied by the velocity of sound. Here again phonons are important only if the sound frequency is in the far ultrasonic range. But they have proved useful in discussing relaxation processes in the interaction between electrons and crystal lattices in superconductivity. Landau has also employed them in the study of the peculiar propagation of sound in liquid helium.

For the very highest acoustic frequencies possible (and molecular acoustics shows that there must be an upper limit), one has to forego producing them by a conventional transducer and fall back on the sound waves produced by the thermal agitation of the atoms and molecules themselves. This radiation, now known as hypersonic or pretersonic (the former term has been stolen by the aerodynamicists for fluid flow velocities greater than Mach 5), makes itself evident by the way in which it scatters light. The domain of so-called Brillouin scattering[25] has become an important asset in the study of the liquid state.

We could go on at much greater length to exemplify the relevance of modern acoustics as a physical science. But enough has been presented to convince even the skeptical that the adjective "classical" is utterly misleading as applied to acoustical physics.

The More General View: Acoustics as a Science

It is possible to dismiss the problem of the relation between acoustics and physics as irrelevant and assert categorically that acoustics is a science in its own right. This view is a wholly tenable one in the light of the enormous role which sound in all its aspects plays in human experience. Acoustics is indeed a science in the fundamental sense that it seeks to describe a portion of human experience by finding or imposing regularity or pattern upon it, seeks to create new experience through the process of experimentation, and finally seeks to understand by building theories in the attempt to explain the observed regularities. As an example, consider the velocity of sound in a gas. The dependence of this velocity on the temperature and its independence of pressure is a regularity or pattern in our acoustical experience expressible in terms of the appropriate physical law. Acoustics creates experience by performing experiments such as measuring the velocity of sound in different media or measuring the reverberation time in a room. Finally, it invents theories such as the molecular theory of sound already referred to. But acoustics is more than a science; it is a technology and an art as well. It is a technology in the sense that it involves as part of its sphere of activity the manipulation of the environment for the satisfaction of real or imagined human needs. An example would be the development through the use of electroacoustics of means for the high-fidelity recording and reproduction of sound. The tremendous influence of this on human activity can scarcely be overestimated. But acoustics is also essentially an art since it concerns itself with speech and music and therefore with the humanistic tradition.

This association of the science of acoustics with technological and artistic applications is of course not unique with acoustics. But the degree of involvement of acoustics in all aspects of human experience is unusual. One simple sociological exemplification of this is the very large number of common words in the English language (or in any Western European language, for that matter) which have to do with sound in its various manifestations. The reader might like to jot these down sometime for his amusement. In English he should have no trouble at all in thinking of at least 50. With a little more sophistication many more can doubtless be identified.

The index classification of the *Journal of the Acoustical Society of America* serves well to illustrate the wide scope of the subject. Beginning with room or architectural acoustics, it proceeds through physiological and psychological acoustics (hearing) to acoustic instrumentation, standards, and signal processing. Then follow in turn musical acoustics, noise and its control, speech communication, ultrasonics, and basic physical acoustics including the fundamental properties of sound radiation in any material medium. But this does not end the tale, for there remain vibrations of material structures of all kinds and the effect of shock on them, underwater sound, macrosonics (sounds of very high intensity), and biological acoustics in the sense of the effect of sound on living things as well as the production of sound by animals of all kinds. Even a casual glance at this list, each item of which of course contains a large number of subtopics, should convince anyone of the breadth of acoustics as science, technology, and art. A few examples are instructive of both the old and the new in this discipline.

Though Vitruvius (already mentioned earlier) had some appropriate ideas on open air theater acoustics, the essential problems involved in room acoustics were not fully appreciated until studied in 1856 by Joseph Henry,[26] at that time Secretary of the Smithsonian Institution. Henry made no detailed numerical calculations. Such had to wait for the pioneering work of Wallace Clement Sabine,[27] who in 1900 hit upon the law connecting reverberation time in a room with the volume and absorbing material. This revolutionized the subject of auditorium acoustics, but was by no means the last word. The importance of the diffusion of sound in a room and the role of the normal modes produced by the relative dimensions of the room have come in for recent intensive study. It has been realized that psychological factors involving, for example, the binaural effect, have great influence on hearing in rooms. On the side of design, the use of the computer has immensely increased the possibility of modeling in advance of construction of concert halls. Noise reduction in buildings has become a major problem, and all the results of physical acoustical research on transmission through walls and vibration damping of structures have been brought to bear on its solution. This is a good illustration of the way the various branches of acoustics have cooperated and emphasizes its essential interdisciplinary character.

Other examples are legion. Consider, for instance, the biological side of the subject, with its emphasis on the fact that acoustics has much to do with the human organism as such. The ear is a fascinating instrument, and with the combined efforts of physiologists, psychologists, and physicists more is being learned about it all the time. It is safe to say, however, that though there are numerous theories of hearing, each successful in its own field, we have not arrived at a single comprehensive view that satisfies all requirements. These requirements naturally increase with the creation of new experience, a phenomenon characteristic of all science. On what may be called the psychological side, take the so-called binaural effect. It was at one time believed that the main significance of the fact that we have two ears is the ability to detect the direction of sound reaching the observer. More recent work has shown, however, that binaural listening has the even more important consequence of enabling the observer to distinguish wanted signals from masking noise. Here is a new development calling for more intensive exploration.

Biological acoustics is not confined to hearing. It also includes speech communication and the use of acoustical radiation in the study of biological material and living things. The acoustic analysis of speech has now attained a high degree of precision and has led to the possibility of speech synthesis, of great practical importance in speech coding. Much emphasis is being placed in current research on the machine recognition of speech. A practical automatic typerwriter may emerge from this effort. Speech communication of course bears on phonetics and hence ultimately on linguistics, which brings the study of language into the acoustical field. All this relates of course to audible sound. But ultrasonics has also proved of importance in bioacoustics. The disruptive effect of high-intensity ultrasound on living cells is being studied, as well as the mutation possibilities inherent in the absorption of ultrasonic radiation. The diagnostic use of ultrasonic beams is being exploited in medicine, with increasing study of their possible use in therapy. Here again we encounter interdisciplinary activity of a high order. Closely allied is the problem of acoustic visualization. It is often convenient to be able to see with our eyes what the sound wave finds as it explores a given piece of material, living or nonliving. This has led to the development of a new branch of technology, so useful in nondestructive testing, whose culmination is the acoustic holograph, still, to be sure, in its exploratory stages. Acoustic holography has the great advantage over optical holography in that ultrasound is able to penetrate material which completely absorbs visible light and for whose examination penetrating electromagnetic radiation is inappropriate.

The whole story of acoustics—old and new—is, of course, too long for anything short of book length presentation, but we cannot conclude this brief review without some mention of an acoustical problem which is of vital concern to every city dweller in our modern society, namely the impact of noise and what we can do to control it. Of all forms of environmental pollution, noise is in many respects the most subtle. The direct effect of high-intensity sound on the human nervous system is hard to measure precisely, although much research in this area is now going on. The effect on the hearing mechanism itself is well known and the long-term deleterious influence can be pretty accurately predicted. Hence control is obviously necessary if we are to live in healthy comfort in our surroundings. The ironic fact is that enough about the nature of sound is known so that appropriate noise reduction methods are available for every practical situation, save possibly for the sonic boom. But these methods are often expensive and the inducement to use them not sufficiently compelling in many cases. Fortunately in many very high noise level environments, such as factories, ear plugs can help prevent hearing loss. But the overall problem remains without sufficient attention. Worse than that, many young people seem to be determined to deliberately ruin their hearing by exposure to so-called rock music.

Acoustics as a science, a technology, and an art is an open-ended activity. Its challenge to the imaginative researcher is increasing. As a science it is continually turning up new experience which it is prepared to describe and explain. As a technology it is determined to utilize this science for the ultimate benefit of mankind. As an art it strives to encourage the production and dissemination of sounds of the highest esthetic quality. No wonder the practitioners of this activity throughout the world are enthusiastic about its past accomplishments and its glorious future.

References

1. J. Sauveur. *Systeme General des Intervalles des Sons.* Paris: L'Academie Royale des Sciences (1701), p. 297*ff.*
2. M. R. Cohen and I. E. Drabkin. *A Source Book in Greek Science.* New York: McGraw-Hill (1943), pp. 286–310.
3. R. B. Lindsay. Pierre Gassendi and the revival of atomism in the Renaissance. *Am. J. Phys. 13* (1945), 235–242.
4. Sir Isaac Newton. *Philosophiae Naturalis Principia Mathematica.* Joseph Streater, for the Royal Society (1687) London; Cambridge: 2nd ed. (1713); London: 3rd ed., (1726). The usually accepted standard English version is the revision of F. Cajori of the translation by A. Motte (1729) (Berkeley: University of California Press, 1946).
5. L. Euler. Dissertatio physica de sono (Basel, 1727). In Leonhardi Euleri. *Opera Omnia* (III). Leipzig: B. G. Teubner (1926), Vol. 1, p. 182*ff.* L. Euler. De la propagation du son. *Mem. Acad. Sci. Berlin 15* (1759), 185–209. Or *cf.* Leonhardi Euleri. *Opera Omnia* (III). Leipzig: B. G. Teubner (1926), Vol. 1, pp. 428–451.
6. D. Bernoulli. Reflexions et éclaircissements sur les nouvelles vibrations des cordes, exposées dans les memoires de l'Academie (1747 and 1748). Berlin: Royal Academy (1755), p. 147*ff.*
7. J. L. Lagrange. Recherches sur la nature et la propagation du son. In Oeuvres de Lagrange. Paris: Gauthier-Villars (1867), Vol. 1, p. 39*ff.*
8. P. S. Laplace. *Ann. Chim. Phys. 3* (1816), 238*ff.*
9. J. Le R. d'Alembert. Recherches sur le courbe que forme une corde tendue mise en vibration. Berlin: Royal Academy (1747), p. 214*ff.*
10. G. Green. *Trans. Cambridge Phil. Soc. 6* (1838), 403*ff.*
11. G. G. Stokes. *Trans. Cambridge Phil. Soc. 8* (1845), 287.
12. G. R. Kirchhoff. *Poggendorf Ann. Phys. 134* (1868), 177.
13. S. Earnshaw. On the mathematical theory of sound. *Phil. Trans. Roy. Soc. London 150* (1858), 133*ff.*
14. W. J. M. Rankine. On the thermodynamic theory of waves of finite longitudinal disturbance. *Phil. Trans. Roy. Soc. 160* (1870), 277–288.
15. G. F. B. Riemann. Ueber die Fortpflanzung ebener Luftwellen von endlichen Schwingungsweite. *Göttingen Abh. 8* (1858–1859), 43*ff.*
16. J. W. Strutt Lord Rayleigh. *Theory of Sound.* London: Macmillan (1877; 2nd ed.

revised and enlarged 1894, reprinted 1926, 1929). First American ed. with a historical introduction by R. B. Lindsay (New York: Dover Publications, 1945).
17. J. W. Strutt Lord Rayleigh. On the cooling of air by radiation and conduction and on the propagation of sound. *Phil. Mag. 47* (1899), 308*ff.*
18. J. J. Waterson. *Phil. Trans. Roy. Soc. A 183* (1892), 1. See also J. S. Haldane (ed.). *The Collected Scientific Papers of John James Waterston.* Edinburgh: Oliver and Boyd (1928).
19. J. H. Jeans. *The Dynamical Theory of Gases.* Cambridge: Cambridge University Press (1904), p. 303*ff.*
20. A. Einstein. *Ber. Preus. Akad. Wiss.* (1920), 380–385.
21. G. W. Pierce. *Proc. Am. Acad. Boston 60* (1925), 271–302.
22. H. O. Kneser. *Ann. Phys. 11* (1931), 761–776, 777–801.
23. V. O. Knudsen. *J. Acoust. Soc. Am. 6* (1935), 199–204.
24. D. ter Haar. *Men of Physics*, L. D. Landau. Oxford: Pergamon Press (1965), Vol. 1, p. 62.
25. Brillouin memorial issue. *J. Acoust. Soc. Am. 49* (1971), 937–1068.
26. J. Henry. Acoustics applied to public buildings. In *Annual Report of the Smithsonian Institution.* Washington, D.C.: Smithsonian Institution (1856), p. 221*ff.*
27. W. C. Sabine. *Collected Papers on Acoustics.* Cambridge, Mass.: Harvard University Press (1927.) (Reprinted by New York: Dover Publications, 1964, with a new introduction by F. V. Hunt.)

II

VISTAS IN PHILOSOPHY

Wolfgang Yourgrau

SOME REFLECTIONS ON PHILOSOPHY OF SCIENCE

During the last 50 years or so, European and American universities (and those in Japan) have introduced two new disciplines that did not exist before. Granted, however, there have always been academicians interested in these fields: the history and philosophy of science. In some universities and colleges the history of science was sharply distinguished from the so-called philosophy of science, and today almost all academic institutions in this country and abroad offer courses in both subjects. These courses either are organized in autonomous, independent departments or are very often attached to existing scientific or humanistic departments.

The history of science has become a highly specialized item of the curriculum. We are inundated with a plethora of books and papers in which the histories of the respective natural sciences, such as physics, chemistry, geology, biology, and geography, and mathematics, too, are offered to the student and mature scholar alike.

The philosophy of science—and that is the present trend—is treated as an autonomous academic discipline or is very often attached to departments of philosophy, especially to logic. A few universities in this country and overseas have created research centers dedicated to the philosophy of science, and the number of books and journals treating of this subject has grown steadily, even in small countries which can ill

Wolfgang Yourgrau • History of Science and Theoretical Physics, University of Denver.

afford to support either such autonomous departments or institute-like organizations.

In the following, I shall commit the inexcusable crime of presenting my own views on this subject matter, which I have been teaching for almost 40 years and in which I received a comparatively severe training at one of the leading European universities.

After having taught the history as well as philosophy of science for so many years in several countries and lectured all over Europe on certain aspects of the philosophy of science, I have arrived at certain conclusions which may appear to many of my respected colleagues as heretical and hardly reflecting the modern trends as far as philosophy of science is concerned.

Let me emphasize that many of the specialists in this domain of knowledge have also been productive in one of the respective above-mentioned sciences. In my own case, I have never ceased to do research in theoretical physics, while many of my colleagues who teach philosophy of science continue to "dirty their hands" by publishing either textbooks or the results of their researches in their own specific discipline. Thus although the so-called pure philosopher of science has often received his entire training in the humanities—and done only occasional and haphazard study in one or several of the sciences in order to get at least a superficial insight into the workshop of the working scientist—on the whole and at best, he is a scientist fully trained in one of the many disciplines or possesses a survey knowledge of several of the natural and physical sciences and mathematics.

During the last few years, it has become more and more clear to me that it is premature and incorrect to talk of *a* philosophy of science, and the prevailing literature in English, French, and German—and some Russian publications—has convinced me that philosophy of science is something that arrogates to itself a position it is not fit to claim—the same established status as one of the individual sciences.

It seems to me that, as with so many other concepts developed by the Greeks, we have adopted a somewhat cavalier attitude toward philosophy of science that can easily be traced back to Plato and Aristotle and even to some of the pre-Socratic thinkers.

The tremendous progress made in the respective sciences during the last 50 or 100 years with their ever-advancing specialization reveals more and more that we are not yet able to make valid statements about the sciences as such and put them all under the umbrella term "science," as if it were a homogeneous academic enterprise. It is factually misleading not to treat each science and its so-called philosophy separately; the conceptual foundations of the sciences differ by far more than is customarily assumed.

Of course, we still encounter adherents of traditional reductivism who are convinced that in the near future all sciences will be reducible to physics, which therefore enjoys a privileged position among its sister sciences.

So I have become convinced, perhaps owing to recent results in all sciences, that it is virtually impossible to talk about *the* philosophy of science. The very fact that many of the sciences—for instance, biology—are becoming more dependent on physics and chemistry must not make us forget that biology, in spite of its undeniable relationship with these other sciences, is distinguished from them to a degree that the radical reductionist, or reductivist, has never considered as an established state of affairs.

Each of the sciences does share certain fundamental notions with all the other sciences, but this circumstance or truism does not permit us to think in terms of *a* philosophy of science—in the singular.

I think there is a reason for this belief in an all-embracing, omnipotent philosophy of science. It is a generally accepted fact that physics has advanced during the last 100 years or so more spectacularly than any other science and has increasingly pervaded all the others. Now it is well known among scientific scholars that many of the leading creative minds in physics proper have often at a late period of their lives "philosophized." And when we follow their endeavors in that field where they chose to talk *about* physics rather than *do* physics, we find that they have tried to generalize many concepts of physics proper and have reasoned as if the conceptual foundations of physics are of such universal nature that they automatically also apply to all the other sciences, with the exception of mathematics, which is supposed to possess its own specific, individual philosophy that it does not share with the natural or physical sciences. This claim can easily be corroborated by perusal of some of the writings of men such as Helmholtz, Bridgman, Bohr, Planck, Einstein, Schrödinger, Margenau, Heisenberg, and Peierls. For Popper, too, the conceptual foundations of physics seem to be the paradigm for his philosophy of science, although in his case the influence of a mind trained in the concepts of pure mathematics is manifest beyond any doubt. Likewise, the work of Peirce reveals that his conception of philosophy of science was markedly indebted to his reasoning in contemplating formal logic in its manifold aspects.

Let me repeat at the risk of becoming pedantic: philosophy of science was for many years the intellectual playground of many outstanding physicists who devoted the last years of their lives to inquiries into the conceptual foundations of physics proper. In addition, a few thinkers who dealt with philosophy of science were proven workers in the realms of logic and/or pure mathematics.

At present, we suffer from a veritable spate of literature, especially so-called textbooks in the philosophy of science, which show clearly that the author is trained in one of the natural sciences, logic, or mathematics; most, however, claim to treat the philosophy of science in a methodical, philosophical, and comprehensive manner.

It seems to be the ambition of contemporary scholars in the domain of philosophy of science to arrive at models or formalisms which hold for all the sciences; there is an undeniable tendency to construct an *Ueberbau*. The underlying, although rarely explicit, supposition is that the philosophy of science is capable of disclosing crucial properties of a philosophic nature which are shared by *all* the sciences. It seems to be thought that we have arrived at a stage where we can demonstrate by means of models or formalisms that the philosophic concepts presented by the ever-growing number of workers in this field are something like the fundamental, profound philosophic foundations of the natural and physical sciences, and sometimes even those of mathematics and logic.

Having succumbed to this illusion, I, like most of my fellow authors, accepted this current tendency as if it were a true revelation and a tremendous advance. But I must confess that I began to realize more and more how wrong and totally unjustified it is to think in these terms.

I should like to plead for a pluralistic viewpoint with respect to the subject under discussion and to indicate how this topic ought to be treated if we do not want to be drawn into a facile, superficial, and uncritical attitude that is bound to intensify a fundamental error. It is always difficult to be a heretic and a risky enterprise to question the justification or validity of a fashionable movement or school of thought, etc. However, having taught and researched in the history as well as in the philosophy of science for almost 40 years, I suddenly discovered that a universally valid philosophy of science cannot possibly be achieved by more or less abstract models or sophisticated formalisms.

It is common knowledge that—apart from pure mathematics—any formalism is vacuous as long as we cannot find a satisfactory interpretation, which has to contain, among other properties, at least verifiability or falsifiability. We take it for granted that basic requirements such as intelligibility, communicability, consistency, and even coherence are fulfilled. Now, this question of interpretation of formalisms or models or any other conceptual constructs leads me to the insight that we have not arrived yet or may never arrive at *a* philosophy of science. This means, in my opinion, that the only justified approach to our subject is to try to discover, discuss, and analyze the philosophic concepts underlying each of the physical and natural sciences. That such a pluralistic—personally I do not care very much for that term but it may serve our purposes—conception of philosophy of science is not just a crankish disagreement with the current method employed can easily be shown. For instance, there are many literary attempts to establish autonomous philosophic foundations in biology, chemistry, and cosmology, and recently also in such concrete and tough sciences as geology and geography. (Whenever we refer here to biology, it is understood that botany too is included among the biological sciences.) I do not know whether it makes any sense to talk about a theory of medicine in general, but I do know that the writings on theoretical chemistry or theoretical biology disclose concepts of a decidedly specific, singular nature. It also becomes obvious that the philosophic foundations of physics proper do, in very essential respects, differ rather dramatically from the philosophic concepts necessarily associated with the foundations of chemistry, biology, botany, or any other of the natural or physical sciences.

This pluralistic and perhaps pessimistic standpoint amounts to no more than a reorientation as far as the various philosophies of science are concerned. When a Planck or a Schrödinger applies, without any qualification, philosophic concepts characteristic of physics not only to the other sciences but also often to the world at large, his ideas and philosophic meditations do not lose their value, because they may act as a stimulating factor and contain some very wise, though often too premature, hypotheses concerning a philosophy of science. I realize now what escaped me during those very many years of writing on topics of theoretical physics or on logico-mathematical problems—that I also unconsciously attempted to look at the other natural and physical sciences through the tinted glasses of the philosophy of physics.

Take, for example, the concept of time. I was accustomed, by training as well as by inclination, to dismiss Bergson's two types of time, *temps longueur* and the so-called *durée*. All the various conceptions of time which differed from the t of the physicist I dismissed *a priori* as invalid or even as crankish aberrations of scholars lacking in rigor who did not understand the generally accepted conception of time dictated by t, one of the most frequent tools of the physicist or applied mathematician. However, even a superficial glance at the way in which the concept of time operates in the various sciences—and I ignore deliberately psychology—will lead one to the view that the physicists' t is in no way the only possible attitude toward the notion of time.

I know of no uniform, unique conception of the notion of time. The fact that the various sciences and many philosophers of science employ the same or similar units for the measurement of time must not deceive us. Of course, Bergson's rather naive distinction of the two categories of time is, at best, obsolete and a philosophic or psychological interpretation which can hardly be defended in that it is the product of an age for which almost every one of the individual sciences and the philosophy of science, even in terms of its most modest claims, can easily be shown to have been either naive or plainly wrong. And yet t is open to manifold interpretations, provided that one accepts the autonomy of the various natural and physical sciences. Despite the fact that any analysis of the concept of time may reveal that the notion as it appears in the various sciences and philosophic discussions shares some properties, this must not create the illusion that the diverse sciences have treated time as if it were a singular, monolithic concept that allows the same rigid interpretation for all those sciences as well as for mathematics. I think, moreover, that it could be shown that in some aspects of logic we encounter also the notion of time, although in a manner which profoundly differs from any interpretation of time in the sciences.

Let us select a few examples to show that we are not yet in the position to talk about philosophy of science as if it were a homogeneous branch of philosophy or of the sciences *in toto*. There exists in mathematics a seemingly simple problem, namely,

$$x^n + y^n \neq z^n \qquad (n > 2;\, n \neq 0)$$

This equation cannot be proved rigorously for any $n > 2$. Yet we will hardly encounter a pure mathematician who will not accept the above expression as valid. The time-dependent Schrödinger equation for waves which correspond to free particles is:

$$-\frac{h}{2\pi i}\frac{\partial\psi}{\partial t} = \frac{h^2}{2m(2\pi i)^2}\left(\frac{\partial^2\psi}{\partial q_1{}^2} + \frac{\partial^2\psi}{\partial q_2{}^2} + \frac{\partial^2\psi}{\partial q_3{}^2}\right)$$

Further, let us select from biology an arbitrary constant differential growth ratio; for instance, the fiddler crab grows according to the expression

$$y = bx \qquad \text{or} \qquad \log y = \log b + k \log x$$

Finally, electrodynamics in Hamiltonian form can be expressed by a variational principle, no less than in wave mechanics and in essential parts of relativity theory. Even if we have succeeded in producing the generalized expression for all those examples chosen from different fields, it seems to me impossible that we should be able to treat the implications and consequences of each example by means of *a* philosophy of science. And here I am referring to the current flirtation with the construction of a universal philosophy of science applicable to any discipline in the natural and physical sciences and, according to some authors, even to mathematics.

In contrast to this trend currently in vogue, I consider it more modest and defendable for the time being to deal with a philosophy of mathematics, a philosophy of physics, a philosophy of chemistry, a philosophy of biology, a philosophy of cosmology, etc. A well-known example will illustrate what I have in mind without any unnecessary complications or any pseudoprofundity. The concept of causality has been flogged to death throughout the ages. Thinkers like Max Born and many others did turn acausalist and their arguments have convinced many physicists.

An analogous situation prevails with respect to the concept of determinism. It is easy to understand from a psychological point of view why men like Planck and Einstein were never convinced that causality and determinism had forever disappeared from physics or even from the sciences in general. And yet at the same time Born maintained, like many other outstanding physicists, that all laws of nature can claim only statistical value. The belief that scientific laws obey statistics rather than strict determinism or classical causality is incompatible with Planck's thesis and Einstein's hope that one day laws of nature will acquire again a pure dynamic character—in other words, that causality and determinism will reenter into physics and perhaps also into the other sister sciences. But as soon as we treat, e.g., the law of gravitation, in any of its many presentations, we would hesitate to commit ourselves to whether or not some laws of nature in general or rather without exception are statistical and not dynamic and strictly causal.

A so-called pluralistic approach to philosophy of science has the great advantage that we avoid the professional disease from which every pure mathematician suffers, that of accepting a proof as valid only when it represents the most generalized form possible. One could even go so far as to say that without the attainment of such generalization mathematics would certainly lose its alleged position as a queen of the sciences. Nonetheless, the history of mathematics contains many famous examples where we are convinced of the general validity of a certain proposition but have been unable to discover the universal proof, which is the only thing that the mathematician seeks in order to claim that his conclusive proposition holds for all cases in question that can be subsumed under a particular problem.

I can only repeat that the individual natural and physical sciences, and in part even mathematics, share several common properties, in the practical as well as in the theoretical domain. However, the very moment we try to create a philosophy of science which holds for *all* the sciences without exception, I think we are guilty of superficial reasoning as a result of the temptation to assign to models, analogies, and some historical parallels a role that by far transcends our knowledge as far as the philosophic or conceptual foundations of each of these individual natural or physical sciences are concerned.

My friend and former coeditor of the *Foundations of Physics* will, I am sure, appreciate that I ultimately arrived at this unorthodox and unfashionable viewpoint. In his *The Nature of Physical Reality* as well as in Popper's *The Logic of Scientific Discovery*, we are confronted with inquiries into one particular philosophy of science, namely, the philosophy of physics. Granted, Popper tried to arrive at some universal characteristics as far as methodology and scientific method in general are concerned, but a careful study of the first edition, as well as of the much later English edition (1959), demonstrates indubitably that he approaches philosophy of science via philosophy of physics. It is with notions emerging from physics proper that he wrestles, and in this respect there is a definite affinity with Margenau's work. Perhaps both authors had the pluralistic approach to philosophy of science in mind without making their viewpoint explicit.

In a way, any philosophy of science is a metatheory even if this fact is not stated. These many metatheories or assumed philosophies of science, in contrast to *the* philosophy of science, will, of course, display *some* similar or identical notions: it is to be expected that we shall find several fundamental concepts the various sciences have in common. But I vehemently oppose any claim that because of those unquestionable conceptual similarities we are already in a position to combine all the diverse metatheories into one single philosophic theory or conception. This whole discussion is so very intractable because it does not depend on the discoveries or theoretical advances in one of the natural or physical sciences, however spectacular and revolutionary. Our demand for a pluralistic attitude *vis-à-vis* philosophy of science transcends any merely classificatory distinction. *It derives from an examination of the fundamental philosophic concepts which form the foundation of each of the individual sciences.*

Anybody who has taught and done research in more than one particular scientific discipline must, I think, accept the fact that we are either not ripe yet to think in terms of *a* philosophy of science or that it is highly improbable that we shall ever be capable of giving up the here advocated pluralistic assertion that there are only *philosophies of the sciences*. It may be quite a while before this pluralistic approach becomes palatable to those philosophers of science who are trained in the presupposition that their work is valid for all the sciences, with the possible exception of pure mathematics, whose close relationship with the foundations of logic has been established beyond any doubt.

Needless to say, the pluralistic approach to philosophy of science will not be accepted without sharp polemics from dogmatically committed orthodox thinkers. After having worked for so many years in this academic area, I have no illusions about the reaction of my fellow workers in this realm of systematic inquiry. But I am quite confident that gradually the pluralistic approach will be recognized as the only attitude that can correctly determine our future researches whenever we feel inclined to examine the conceptual foundations of any science.

There is, however, a strong argument against the pluralistic attitude to philosophy of science that I have endeavored to plead for in this essay: that the physical and natural sciences have become more and more reducible to or dependent on some branch of physics. As a matter of fact, during the last few years a group of mathematicians have not only further developed the relation between mathematics and physics but have also discovered that certain aspects of biology have *enriched* mathematics, whereas in the past advances in biology were only too often heavily dependent on some area of mathematics. At least we have here a modern example where, in contrast to the past history of the sciences, biology was able to exert an interesting and promising influence on its former "mentor," mathematics. Hence we are facing the following unexpected situation for which we have no parallel in the development of the sciences.

Scientists in our age, and even in the previous 300 years, became more and more accustomed to the idea that sooner or later all the natural and physical sciences would become definitely dependent on physics, in a reductivist manner. In particular, the experimental aspects of the respective sciences became categorically and demonstrably connected with some branch of physics. Parallel with this development of an empirical manner, all the sciences were alleged to acquire academic respectability and scientific reliability only when their mathematization could be performed. One need not go back to Pythagoras in order to support this claim: Kant made it a dogmatic condition that only the progressive mathematization of a science permits us to grant it the unqualified status of a proven part of objective and reliable knowledge.

We find a similar development in such subjects as psychology, sociology, and economics, where mathematics has become more or less the main criterion for their respectability. We are forced to arrive therefore at the following points of view: As far as philosophy of science is concerned, the individual physical and natural sciences will develop autonomous, independent philosophies of science so that each of them will become anchored in its own specific conceptual foundations. As far as the experimental process is concerned, it appears today that physics will play an ever-growing role in the individual development of empirical sciences. At the same time, mathematics in its classical form (e.g., differential and partial differential equations) may soon prove to be not the most efficient tool for dealing with certain physical disciplines. (This idea, incidentally, was already in the mind of Schrödinger, who mentioned it in one of his writings.) The result is and will be either that mathematicians will develop entirely new means that will satisfy the radical progress in physics and other sciences, or they will create novel developments in some algebras and especially in topology.

All this has been stressed only in order to demonstrate why it is my opinion that the so-called pluralistic approach should be reserved to philosophy of science, whereas the experimental development of the individual sciences may continue to become dependent on physics. At the same time, disciplines of knowledge such as psychology, sociology, and economics may advance in a rigorous scientific manner by becoming more and more mathematized. I do not feel qualified to talk about any philosophic foundations or even the concepts associated with these three disciplines of knowledge, which cannot be considered natural or physical sciences.

These reflections in this short article are intended to remind the reader that philosophy of science in this country and in Britain is still, although latently, the outcome of positivistic reasoning. The term "positivism" or "logical positivism" has been more or less dropped in current literature concerning philosophy of science—to the extent that many authors have openly disavowed even indirect dependence on the messages of the Vienna Circle and have rather approached the subject with people like Popper, Margenau, and Bridgman in mind. However, with a little patience and knowledge of the relevant literature in English and in German, it can clearly be shown that contemporary philosophy of science is still in some way or other affected by positivistic reasoning. It is really not important that Schlick himself and many of his followers seldom or never resorted to formalisms of a logical nature. The basic philosophic attitude of most Anglo-Saxon and American philosophers of science shows the influence of that circle.

We need only mention the work of Reichenbach, Philipp Frank, Hempel, Carnap, and Feigl—to cite only some of the more popular and influential thinkers. Popper, Margenau, and I myself have not been associated with the Vienna Circle. I attended, at Berlin University, several courses and one seminar by Reichenbach, and there was a time when I dreamt of philosophy of science being plausible only if it were approached from a positivistic viewpoint. But even before I left the University I became skeptical toward the dogmatic attitude of the so-called logical positivists, although I never went so far as G. Bergmann, who maintained that the logical positivists are infected by metaphysical reasoning no less than their opponents who admit in some form or other their dependence on such philosophizing.

It is certain that the pluralistic attitude to philosophy of science which leads to philosophies of the sciences, as far as the conceptual foundations are concerned, will meet with sharp and intransigent opposition, because we seem to have lost the ability to argue an issue without becoming personally involved and are resorting to more or less self-righteous and arrogant viewpoints or methods. It is not easy in our country to advocate a heterodox opinion, because we have reached a level where one is either "in" or "out." The best example is perhaps the feud which Popper himself was subjected to with the so-called leading authorities in the domain of philosophy of science in this country. It seems to me that Margenau, too, has not enjoyed sincere readiness to follow his views in philosophy of science, although he has always been taken very seriously by conscientious theoretical physicists.

Having been active in physics as well as in philosophy of science and even in history of science for so long, I feel that I have acquired the right to plead, however sketchily, for a pluralistic conception of philosophy of science. At this stage of our knowledge, we can consider ourselves very fortunate if we succeed in arriving at diverse philosophies rather than a monolithic philosophy of science. In this manner we may save ourselves from facile self deception and premature theorizing.

In conclusion: Let us try to work in the respective sciences not only in a purely scientific manner but also to establish, as a preliminary step, sound conceptual foundations for each. The outcome may lead to an entirely new conception of philosophy of science, and the false and deceptive dream of *a* philosophy of science or *the* philosophy of science as the last word may forever disappear, although we should never forget the many diverse phases of this subject in history and in past philosophy of science.

Ervin Laszlo

NONEMPIRICAL CRITERIA IN THE DEVELOPMENT OF SCIENCE

One of the disturbing consequences of contemporary theory of science is the haze which surrounds its notion of *progress.* As long as science was held to converge on "truth" or "objective reality" in its path of evolution, successive theories could be defined as constituting progress by being more true, or more conformant to reality, than their predecessors. When the essentially constructed character of all knowledge came to be realized, however, it became clear that we do not see reality independently of our conceptions of it. Observation, to quote Hanson,(1) is theory laden. We do not see "what is there" otherwise than by interpreting the stimuli emerging into our consciousness as sensation in terms of some preestablished conceptual or gestalt scheme.(2) Our gestalts and concepts guide our perception of reality, and there are no ways whereby we could compare our conceptions of reality with reality itself. There are no checks on scientific veridicality through direct inspection of the world "out there." And if we cannot compare our scientific theories to objective reality, we cannot tell which of them is truly conformant to it. Hence we cannot give meaning to a concept of progress in reference to reality.

Ervin Laszlo • Department of Philosophy, State University of New York at Geneseo.

Progress must then be defined in terms of theories which do not come in opposition with experience. However, a large (but possibly finite) number of theories satisfy this definition. The wider the generality of the theory, the more it avoids disconfirmation by experience. Theologies, for example, as well as myths, are practically impervious to disconfirmation. If confirmation by anything and everything we encounter were to be the mark of scientific progress, we would not have reached the stage of what goes under the name of science in the first place: systems of myth, religion, astrology, and the like do the job better. Science lays itself open to disconfirmation by making precise predictions testable against experience. But experience, if not sooner then later, hardly ever fails to conflict with some of these predictions. Areas of "doubtful fit" populate even new and powerful theories, and more of these anomalies come to light as the theories are explored. Hence, on the whole, science seems to qualify for truth a great deal less than do myths, astrology, and religion, and one's concept of scientific progress as convergence toward truth is shattered.

This raises an acute problem. Nature, to quote Whitehead, is patient. A great many mutually inconsistent and even incommensurable theories mesh at any given time with our experience of the world, and thus such experience cannot be the sole basis for our choice among them. We need nonempirical criteria to effect such choice. These criteria render science a partly convention-based discipline: the empiricism of Mills and Bacon now appears as but a pleasing mirage. "Conventional" does not mean "random" or "arbitrary," however. Science is an activity conducted according to rigorous rules, even if scientists cannot always explicitly name the rules they follow. Doing so marks off the activity of engaging in science from other activities concerned with the interpretation of experience. The rules in question are the topic of much current debate in philosophy of science. They are also the topic of this study, since they define not only what science is but also on what basis scientific theories are established, overthrown, and alternative theories decided.

I shall briefly review three influential views of science: the pertinent theories of Popper, Kuhn, and Margenau. I shall show that already in 1950 Margenau anticipated trends which took more definite shape only in the 1960s, and that in fact he went further in explicating the significant points than either Popper or Kuhn. I, in turn, shall attempt to go still further in explicating Margenau's conclusions and press for an answer to the question: what are the demarcation criteria of scientific theories? By showing that successive theories satisfy these criteria increasingly closely, we get a meaningful answer to the related question: what is gained or maximized in the development of science that can be laid out as constituting scientific *progress*?

I

a. Popper is undoubtedly still the most influential proponent of the "new look" in philosophy of science. His *Logic of Scientific Discovery* appeared in the 1930s, and challenged the then dominant positivist view of science. It reappeared in 1959 in an expanded second edition and was followed by *Conjectures and Refutations* in 1962. In these works (essentially collections of papers given over a period of some four decades) there emerges a view of science which first rejects any criteria of a theory in terms of its verification, and then brings some such criteria back cautiously, e.g., as "verisimilitude" and the Tarskian concept of truth (or "truth-content"). Popper's early disenchantment with concepts of what constitutes a genuinely scientific theory is evident in his reminiscences of his Vienna days. Theories then current of Marx, Freud, and Adler could explain everything that came within their scope, yet left room for alternate explanations. They could hardly be refuted, for the predictions flowing out of such theories were general and vague enough to cover practically all instances of observation. Consequently Popper[(3)] concentrated not on what verifies a theory but what falsifies it. The growth of scientific knowledge consists, according to him, in the *overthrow* of scientific theories, and their replacement by better or more satisfactory ones (p. 215). Science grows, in Popper's assessment, by satisfying three requirements. Two of these concern the falsification of a theory. One is that whatever new theory is proposed to replace an existing theory should be independently testable. The second requirement is that the new theory should not only explain all facts explained by the existing theory but should also predict new facts; and some of these predictions should be corroborated. Through the requirements of independent testability and excess empirical content, stagnation in science, due to theories of all-embracing confirmability, is avoided. New theories are highly testable and that means falsifiable. But one may object, with Bronowski,[(4)] that if requirements of falsification alone were to provide demarcation criteria of science and explanation of scientific progress we would find that science contains only two categories of theories: *unproved* theories and *disproved* theories. Of course, Popper himself sees the necessity of adding some positive requirement to the essentially negative ones of falsifiability, and speaks of our expectation that the theories also be true. However, truth can only be a relative measure of a theory's corroborated and non-corroborated predicted consequences, and not the result of a comparison of the theory with our independently acquired concept of nature. Thus the truth-content of a theory is at best a relative measure, and a theory which possesses it in greater measure than its competitors may yet be falsified in all or most of the relevant respects. On Popper's grounds it makes no sense to say that progress is constituted by an asymptotic approach of theories to the truth, since we have no way of knowing what the truth is independently of our theories. Thus we can never tell whether our theories approach it or not.

Dissatisfied with such accounts himself, Popper[(3)] offers a positive yet nonempirical criterion of scientific growth. It is the requirement that "the new theory should proceed from some *simple, new, and powerful, unifying idea* about some connection or relation (such as gravitational attraction) between hitherto unconnected things (such as planets and apples) or facts (such as inertial and gravitational mass) or new 'theoretical entities' (such as fields and particles)" (p. 241). Theories are never fully and definitely falsified by any disconfirmation of predicted effects: they can always be patched up to "save the phenomena." However, Popper is aware that there are cases of scientific development where such manner of procedure was discarded in favor of finding a new theory which explained the problematic facts by recourse to some "simple, new, and powerful, unifying idea." In fact, in many cases the theory thus superseded was not even seriously in doubt by disconfirming evidence. The theories of Galileo and Kepler, for example, were not refuted when Newton explained their findings in reference to more general postulates. Likewise Ptolemy's system was not refuted when Copernicus produced his, and Newton's theory was not refuted when Einstein offered his theory of relativity (the puzzling results of turn of the century experiments could always be patched up by Lorentz and others).

Having pointed out the importance of unifying principles in science and its progress, Popper nevertheless persists in affirming the primacy of his requirements of falsifiability. He says that we have no reason to regard the new theory better than the old one until we derive from it new predictions, unobtainable from the previous theory, and test them. What Popper fails to see is that while such "crucial experiments" can cinch the acceptance of a theory, scientists did not subscribe to theories such as Newton's, Copernicus's, and Einstein's because of their added falsifiability, but rather because of their greater "richness": additional empirical content combined with accrued elegance and simplicity. Einstein's relativity theory found general acceptance prior to the observation of the advance of the perihelion of Mercury, although that confirmation cinched its acceptance more firmly. Copernicus's theory reduced the number of epicycles involved in the explanation of planetary motions from 83 (in Ptolemy's system) to 17, and he himself regarded this simplification as the decisive aspect of his theory. Newton unified the laws of Kepler's planetary motions with those of Galileo and the acceptance of his theory was not contingent on detailed confirmations of new predictions (although these were forthcoming) but resided already in the "simple, new, and powerful, unifying idea" of universal gravitation. Popper recognizes the existence of the element which I shall call "adequate simplicity" in the growth of science, but (perhaps because he finds it a "bit vague" and "difficult to formulate very clearly") he fails to attach due importance to it. Yet we can agree with Bronowski,[(4)] who, in referring to the passage in which "simple, new, and powerful, unifying idea" occurs, says that it is admirably conceived: "In the most practical way, it leaves no doubt that there is something more in the human search for knowledge than the wish to get the facts right—basic as that is. We want to feel that the world can be understood as a unity, and that the rational mind can find ways of looking at it that are simple, new, and powerful exactly because they unify it."

b. Kuhn's recent views are stated in the two editions of *The Structure of Scientific Revolutions* (1962 and 1970) and his contributions to *Criticism and the Growth of Knowledge* (1970). These views substitute, in place of a conception of science as an enterprise designed to perfect concepts and methods for discovering and stating truth about the nature of reality, a conception of science as a rule-governed game engaged in by a community of specialists with the object of solving puzzles in rigorously defined experimental situations. "Normal science" is a phase of communal activity based on a paradigm (or on a "disciplinary matrix," as Kuhn specifies in the Postscript to the 1970 edition of his book) which "works": it solves puzzles in the relevant field of science. The success of its practitioners is limited only by their own skill and ingenuity in applying the concepts. An ailing paradigm, however, consistently fails to solve problems in certain key areas of a field; these remain anomalies for the practitioners. When a sufficient number of anomalies accumulate and persist, some practitioners (mostly younger men, with less vested interests in the paradigm than long-term practitioners) defect from the community and initiate a period of "extraordinary science." Here a seemingly random and frenzied search for workable concepts and theories is the defining state of activity. New hypotheses are advanced, tested, and rejected in quick succession. The anomalous paradigm is not discarded, however, until a new one can take its place; science cannot function without any paradigm at all. When, out of the hypotheses thrown up during a crisis, one comes to be accepted by a representative majority of scientists, a new community forms around it, and it henceforth defines a new paradigm or "disciplinary matrix." This matrix must be rich enough to permit solutions to all puzzles solved by the old paradigm, and in addition must resolve the anomalies. It usually constitutes a fundamentally new way of looking at the phenomena, so that "facts" and "laws" come to be reinterpreted in its light. (This was the case, for example, in relativity and quantum physics, genetic biology, evolution and cell theory, behavioral psychology, and so on.)

Now, what possible meaning have "truth" and "progress" in this account of scientific development? The concept of absolute truth, even as an ideal regulating the development of science, is expressly denied. It does not help, Kuhn[5] says, to imagine that there is some one objective and true account of nature and that the measure of scientific achievement is the extent to which it brings us close to that goal (p. 171). Rather, we must learn to account for both science's existence and its success in terms of evolution from the scientific community's state of knowledge at a given time. Progress in science is defined similarly to progress in Darwinian evolution theory: it is not progress *toward*, but progress *from*. What is accumulated in the process is not objectively true knowledge, but a list of problems solved and the precision of individual problem-solutions. As Kuhn[5] says, we may have to "relinquish the notion, explicit or implicit, that changes of paradigm carry scientists and those who learn from them closer and closer to the truth" (p. 170).

Kuhn's account of scientific development, like Popper's, refuses to concede an ideal limit in objective reality toward which science progresses. Science grows by self-correction. But correction in regard to what criteria? Popper found it necessary to bring in "verisimilitude" and "truth-content" to avoid relativism and be able to speak of the "growth" (rather than the mere change) of scientific knowledge. Kuhn seldom speaks of the "growth" of scientific knowledge, and his concept of progress is restricted to puzzle-solving range and precision. But, as Popper, he too, came back somewhat on his original position. Accused of "relativism" (e.g., by Shapere[6] and lately by Kockelmans[7]), in the 1970 Postscript

Kuhn[5] offers a list of criteria by which to distinguish later theories from earlier ones, on the branching tree of scientific evolution. Among the most useful criteria he lists *accuracy of prediction, the balance between esoteric and everyday subject matter*, and the *number of different problems solved.* As less useful he considers such "values" as *simplicity, scope*, and *compatibility with other specialties.* If this list can be completed, he adds, scientific development turns out to be, like the biological, a unidirectional and irreversible process (p. 206). In that case, however, "progress toward" is no longer inapplicable to it: successive theories could be rightfully said to be progress toward greater accuracy of prediction, balance between esoteric and everyday subject matter, greater number of solved problems, as well as simplicity, scope, and higher compatibility with other specialties. Thus, on this revised theory, science does not progress toward conformity with nature (i.e., toward an "ontological" match between predictions and what is "really there") but toward the ideals of theory-construction in scientific communities. These ideals, termed by Kuhn "values," function as the means of persuasion in choosing one paradigm in preference over another. Inconsistently with his "less useful" predication, Kuhn suggests simplicity, together with fruitfulness, accuracy, "and the like," as the values underlying "conversion" from one paradigm to another (p. 199). Kuhn tacitly acknowledges a *telos*, not in nature, to be sure, but in the values of the scientific communities. Thus he is no longer in opposition with Kockelmans'[7] point that, although we should not subscribe to a goal set *by* nature (because then we must account for this goal as well as for our knowledge of it, which we cannot do except by appealing to questionable "metaphysical" speculations), there is no reason why we should not subscribe to the view that goals are set by rational *investigators* of nature (p. 239).

It is only one step from here to reach the position I am pressing for, but Kuhn[5] does not take it. He persists in regarding puzzle-solving capacity as the leading criterion of scientific progress, notwithstanding his list of "standards" and "values." He concludes that "later scientific theories are better than earlier ones for solving puzzles in the often quite different environments to which they are applied" (p. 206). This, he hastens to add, is not a relativist's position, and it displays the sense in which he is a convinced believer in scientific progress.

Now, while it is true that improved puzzle-solving ability in different environments defines one sense of "scientific progress" beyond pure relativism, one wonders whether it is the one which follows even from Kuhn's own list of criteria. For these criteria (if you disregard the unexplained judgments "most useful" and "less useful") define a progression toward an increasingly technical and predictive science, encompassing an increasing number of problems with accrued simplicity and compatibility with neighboring specialties. Science, it would seem, becomes more *adequate* in that it deals, in rigorous and predictive fashion, with increasingly large chunks of nature, and it becomes more *simple* not only in its paradigm structure but also because it extends in scope and ties compatibly together different specialties. It is true, of course, that "adequate simplicity" and "efficient problem-solving" are correlative. But it is not clear which is the means and which the end. For Kuhn, puzzle-solving is the end, and a way of doing it that is both adequate in scope and precision and simple enough in structure to be manageable is the means. It could be also held, however, that to obtain adequate yet simple theories to convey understanding of observed effects is the end, and puzzle-solving abilities and techniques are the means to it.

c. Margenau's *The Nature of Physical Reality* (1950) antedates both Popper's *Conjectures and Refutations* (1963) and Kuhn's *The Structure of Scientific Revolutions* (first edition, 1962). While it postdates Popper's path-breaking *Logik der Forschung* (1935), it explicates non-empirical criteria concerning the development of science more clearly and unhesitatingly than any of the aforementioned works. Margenau recognizes that scientific theories are *constructed*, rather than empirically induced from observation. He is in agreement with Einstein, who repeatedly emphasized that scientific theories cannot be generated by observations, but must be the free creations of the intellect.[8] In Margenau's[9] philosophy of science, the constructs of scientific theories are correlated with the elements of sensory experience by postulated "rules of correspondence" (pp. 60*f*.). These rules are not dictated by the observed phenomena but are conferred by the theories. Thus the same phenomena can be correlated with different constructs in different theories; compare the "rule of correspondence" leading from the perception of a solid body to atoms with hooks in Democritean atomism with that leading to the construct of molecular structure in solid-state physics. Now, the decision as to what theoretical structures to correlate with what sensed experiences follows multiple criteria. Some of these criteria are empirical: the predictions flowing out of a set of constructs must check with actual experiences within limits of probability applying to both observation and prediction. But different theories could give rise to identical predictions in given instances (e.g., Newtonian and Einsteinian relativity equations generate identical predictions in regard to the motion of bodies at ordinary velocities on the surface of the earth). Hence criteria must be discovered which *de facto* guide the preferences of scientists when faced with alternative theories. Margenau[9] lists six such nonempirical (hence "metaphysical") criteria, which function as requirements on scientific constructs. These are *logical fertility*, *multiple connections*, *logical stability*, *extensibility*, *causality*, and *simplicity* (also termed *elegance*). They can be analyzed as follows:

(i) The "requirement of logical fertility" is that the constructs of a theory be so formulated as to permit logical manipulation. Positive examples are *mass*, *molecule*, *chromosome*; negative examples are Berkeley's concept of God and Hume's appeal to "human nature" for want of a better explanation. Without logically fertile components, scientific theories would not constitute manipulable systems, but merely registers of atomic concepts and principles. (ii) The "requirement of multiple connections" defines an analogous criterion with regard to the connectedness of constructs in a theory. Since constructs can be connected both to other constructs ("formal connections") and to observational data ("epistemic connections"), a systematically manipulable theory must be exclusively composed of constructs with at least two connections of one of two kinds: formal–formal or formal–epistemic (an entity with epistemic–epistemic connections is not a theoretical construct but an item in perceptual experience). Of course, more than two connections are permissible and, in view of requirements below, even preferable.

(iii) The "requirement of permanence and stability" makes scientific theories unambiguous theoretical structures. As long as a construct is upheld, it remains what it is; if it is changed, the new construct conserves its own identity as long as it is part of a theory, and so on. In virtue of this requirement, scientific constructs are clear and uniquely defined entities, notwithstanding the haze of uncertainty surrounding the observed events with which some of the constructs of every empirical theory are epistemically connected. (iv) The "requirement of extensibility of constructs" rests on an explicit recognition that "scientists judge the quality, and ultimately the correctness, of a given theory by its range of application, taking the generality of a system as a measure not only of its usefulness but of its credibility" (Margenau,[9] p. 90). This requirement says that effecting a merging of formerly separate constructs (such as Galilean gravitation and Kepler's motions of planets, gravitation and centripetal force in curvature of space-time, energy and mass in mass-energy, and space and time itself in space-time) is a persistent conceptual goal and motivation of science.

(v) The "requirement of causality" reformulates a classic philosophical principle in reference to contemporary theory of science: causality is a necessary ("required") property of *constructs* (or else we could not pass from one construct to another) but does not apply to items in the stream of experience (observations). Thus Hume's difficulty is avoided: empirical phenomena do not manifest causal relationships but remain "constantly conjoined." However, causality is an assumption we must make if our theories about empirical phenomena are to work. Constructs must generate causal laws, but since these laws are not derived from phenomena but are theoretically postulated, the openness of science, demanded by Einstein among others, is respected.

(vi) There remains the final requirement, that of "simplicity and elegance." "When two theories present themselves as competent explanations of a given complex of sensory experience, science decides in favor of the 'simpler' one" (pp. 96–97). Margenau cites the victory of Copernican over Ptolemaic astronomy, the electromagnetic theory of light's supersession of the hypothesis of luminiferous ether, and, in *Open Vistas*, the theory of relativity's acceptance over Newtonian physics *cum* Lorentz-Fitzgerald equations as cases in point. Margenau's[10] clear-cut conclusion: elegance and simplicity, although usually regarded as matters of aesthetics and not of science, are actually at work in modern physics, guiding thought and producing concrete results (p. 101).

In reviewing Margenau's criteria, further conclusions come to light. Keeping in mind that, as he says, his criteria are not cellular but may be overlapping aspects of a larger ideal of science, we can classify his "requirements" under the following headings.

"Logical fertility," "multiple connections," "permanence and stability," and "causality" are criteria of *systemicity*. They define the minimum requirements on any rational system of thought with empirical applications. "Logical fertility" and "permanence and stability" apply to constructs taken separately; "multiple connections" and "causality" apply to the relatedness of constructs in a theory. Jointly they demand that the theory be systemic, i.e., interconnected, manipulable, and unambiguous. The remaining two requirements fall under the heading of *adequacy* and *simplicity*, respectively. The requirement of "extensibility" correlates the validity of a theory with the generality of its application, recognizing the intrinsic preference of scientists for theories with applications in more than one field, over *ad hoc*, even if workable, theories. The requirement Margenau calls "simplicity and elegance" is that which guides theory choice in cases where other factors (such as confirmation of predicted effects) are equal. Thus Margenau recognizes the role of nonempirical criteria in the development of science: his "requirements on constructs" show that the development of science is guided by the ideal of interconnected, manipulable, and unambiguous theory, of greatest possible scope and simplicity.

The concept of "adequate simplicity," as an ideal asymptotically approximated in the development of science, is implicit in the here considered theories, yet, with the exception of Margenau, it is not explicitly recognized. I argue that even Margenau's recognition of this ideal can be further specified, and that such specification, itself conforming to the principle which is its subject matter, is a valid scientific procedure. To be sure, this assumption begs the question. But there is sufficient evidence contained in the studies of persons not explicitly espousing this principle to break the circularity; and there is further independent evidence available from psychology and history. Of course, one can always object that if science is truly a search for adequate simplicity any evidence for adequate simplicity based on science is likewise circular. But this need not bother us. After all, what more could we expect of any body of systematic thought than that it should entail itself consistently with all its other empirical consequences?

II

> We are seeking for the simplest possible system of thought which will bind together the observed facts. (Albert Einstein in *The World as I See It*)

"Adequate simplicity" will now be defined as follows. *It is the ideal*, never fully present in any existing theory but functioning as motivation and criterion in theory construction and criticism, *that the existing facts, and all the existing facts, be explained as consequences of the smallest possible number of primitive postulates.* Here "existing facts" provides the empirical base, "all the existing facts" provides for adequacy, and "smallest possible number of primitive postulates" defines simplicity. The history of science offers ample examples of the influence of the ideal of adequate simplicity. In fact, unless I am seriously mistaken, all recorded "revolutions" in science can be analyzed as radical approximations of this ideal, relative to previous phases of the science. The search for adequate simplicity is the main reason that science is not an accumulation or register of observations, or even of theoretical entities. In the history of science, the number of "facts" (however defined) accumulated by researchers would defy the scope of any intellect, and science would lose the forest for the trees. However, when such a situation is impending, there are attempts at theory reformulation to exhibit the quasi-unmanageable register of facts as consequences of a manageably small number of postulates. The search for simplicity is, in first approximation, a search for intelligibility. Current trends in science make this very clear.

The shift from the Bacon-Mills method of induction and fact-gathering to deductively formulated and indirectly tested theories is a response to the threat of unintelligibility in theory, conferred by data-gathering activities over the past 200 years. Normal science, to use Kuhn's term, is its own nemesis not merely in that it sooner or later leads to the discovery of anomalies but also in the sense that it accumulates quantities of information which obscure the intelligibility of the ruling paradigm. New data do not fit, as a rule, into any existing theoretical framework as neatly as cards in a file; adjustments and extensions of the framework are usually required. It does not take a few persistent anomalies to bring about the breakdown of that framework if the facts already explained by it call for too many *ad hoc* hypotheses. Newton's theory was not falsified by the Michelson-Morley findings—an auxiliary hypothesis was available. Nor was Kepler's celestial mechanics falsified in any way when Newton offered his more generalized theory. The attraction of the new theory resided in each case in its higher level of intelligibility in the ordering of the existing facts, as well as in its ability to order more of them. Newton's theory was more adequate yet simpler than Kepler's, Einstein's theory than Newton's, and the general theory of relativity than the special one. Similar considerations would apply to revolutions in fields of science other than physics, for example, to cell theory in biology, theory of class conflict in philosophy of history, theory of elites in sociology, and structuralism in cultural anthropology.

In virtue of the axiomatic-deductive form of theories, there is no tradeoff between simplicity and generality. Generality does not entail loss of simplicity because of the propensity of abstract mathematical models and definitions to remain invariant when confronted with different domains of experience. A basic law, such as that of Galileo, applies to all falling bodies on the earth's surface. Newton's law generalizes this law without loss of simplicity. And the relativity equations yield simple equivalences which are conserved (such as mass-energy relations, and the Einsteinian interval) where previous formulas broke down. The very attraction of systems concepts in contemporary biology, economics, ecology, psychology, political science, computer science, and management theory (and other fields) is due to the conservation of the same (or analogous) open system model under diverse transformations.[11] In a mathematical systems theory, formal aspects of open systems can be singled out and generalized (such as those that satisfy Onsager equations, negative or positive feedback, steady-state pattern maintenance, and development in accordance with formal laws); in qualitatively formulated theory, further analogous properties of systems can be stressed. Thus models can be outlined which apply to systems in general, regardless of whether they are physical, biological, social, or psychological.[12] (Structural isomorphies, revealed in the invariance of the systems model when translated from one field of application to another, also serve important functions in the understanding of biological and psychological phenomena too complex to be grasped in every detail, by analogy with simpler systems, such as electrical networks and computer simulation.)

The significance of mathematical methods in contemporary sciences lies precisely in the capacity of axiomatically formulated mathematical theories to combine simplicity with adequacy. Compare Newton's physics with that of Aristotle. Aristotle required a treatise to describe the several disjunctive motions and natures of the various elements, whereas Newton's primitive concepts and postulates can be put on one page, and the definition of his eight most important defined concepts takes up but 16 lines.[13] Yet such a simple minimum generates rigorous explanations for phenomena as diverse as the motion of the planets, the swing of the pendulum, the falling of an object to the ground, the lever, the balance, and the gyroscope. As Einstein emphasized, in contrast to Newton's own assertions, such a theory could never be won by the classification and generalization of observations alone: hypotheses must be creatively postulated and then deductively tested, with respect to their entailed consequences. There is nothing *per se* in the motion of a pendulum which makes it a special case of a theory which likewise encompasses the motion of the planets. Nevertheless, the search for adequate simplicity resulted in their integration within classical mechanics.

In science, the tradeoff is not between adequacy and simplicity, but between adequate simplicity and common sense. The more adequate and yet simple a theory is, the further removed it is from everyday concepts, as it makes increasing use of abstract mathematical concepts. "The striving for greater adequacy," says Margenau, "coupled with greater simplicity pushes us back farther and farther into the realm of constructs. And all of modern science documents this fact."[14]

The reduction of the richness of sensory experience to conceptual order is an act of simplification, whether it occurs in a systematic science or in the fancy of a mental patient. Science distinguishes itself from lunacy in that its reduction is not only simple but adequate to the full range of experiences forming the protocol set of a discipline. Consequently we can offer the following statement, defining demarcation as well as theory-selection criteria in science. *A valid scientific theory is one which explains the greatest number of facts taken into account in its field as deduced consequences of a systematically connected set of postulated constructs of greatest possible simplicity.* The parallel striving for adequacy and simplicity pushes theory farther into the realm of abstraction and richness. Simple and adequate postulates have multiple applications in the realm of experienced nature: they, to use a related expression, have a wider range of invariance. Inasmuch as science progresses toward adequate simplicity, it progresses toward greater range of invariance and richer empirical content in its constructs. Progress is defined not by conformity to "truth" or "objective reality" but by satisfying the scientist's striving for adequate simplicity.

It could be objected that we have herewith defined an entirely subjective criterion of scientific status and development. But whether adequate simplicity is an objective or a subjective criterion depends on whether or not adequate and simple scientific theories conform to the structure of objective reality. Although some scientists ultimately come to believe that they do,* clearly, this we cannot tell. But to conclude subjectivism from this is to commit the atheist's fallacy who denies God's existence when all he can consistently assert is that he cannot know it. Adequate simplicity need not be a matter of individual or collective psychology, or even a "given" of the history of science, accepted on force of tradition. It may be shown that processes such as typify the growth of science characterize the development of any complex open system, coupled to its environment.[(12)] Demonstration of this fact presupposes scientific theories of self-evolving complex systems, and such theories are products of the very striving which (in this case) they help to explain. Thus the proof for the objectivity of adequate simplicity, even if attainable, would be

*Cf. Einstein[(8)]: "Nobody who has really gone deeply into the matter will deny that in practice the world of phenomena uniquely determines the theoretical system, in spite of the fact that there is no logical bridge between phenomena and their theoretical principles" (p. 23). Yet such acts of faith can always be interpreted, as Northrop[(15)] does, to mean that "among the present possible theories, one is unique in its capacity to bring the widest possible range of empirical data under a single minimum set of assumptions" (p. 402).

circular. The fact is that the conditions of human knowledge are, of necessity, also the conditions of our knowledge of these very conditions. To anyone who is disturbed by this circularity, I have no consolation to offer. I can only point out that, if he construes this as a mark of subjectivity, objectivity becomes humanly unattainable.

References

1. N. R. Hanson. *Patterns of Discovery*. Cambridge: Cambridge University Press (1955), Chap. I: "Observation."
2. Ervin Laszlo. *System, Structure and Experience: Toward a Scientific Theory of Mind*. New York: Gordon and Breach (1970).
3. Sir Karl Popper. *Conjectures and Refutations*. New York: Basic Books (1963).
4. J. Bronowski. Humanism and the growth of knowledge. In P. Schillp (ed.). *The Philosophy of Karl Popper*. La Salle, Ill.: Open Court (1974).
5. T. S. Kuhn. *The Structure of Scientific Revolutions*. Chicago: University of Chicago Press (1970), 2nd ed.
6. D. Shapere. The structure of scientific revolutions. *Phil. Rev. 73*, 383–394 (1964).
7. J. J. Kockelmans. On the meaning of scientific revolutions. In R. Gotesky and E. Laszlo (eds.). *Evolution-Revolution: Patterns of Development in Nature, Society, Man and Knowledge*. New York: Gordon and Breach (1971).
8. A. Einstein. *The World as I See It* (a volume of collected papers). New York: Covici-Friede (1934).
9. Henry Margenau. *The Nature of Physical Reality*. New York: McGraw-Hill (1950).
10. Henry Margenau. *Open Vistas: Philosophical Perspectives of Modern Science*. New Haven: Yale University Press (1961).
11. Ervin Laszlo and Henry Margenau. The emergence of integrative concepts in contemporary science. *XIIIth International Congress of the History of Science*, Moscow (1971).
12. Ervin Laszlo. *Introduction to Systems Philosophy: Toward a New Paradigm of Contemporary Thought*. New York: Gordon and Breach (1972), Part One.
13. F. S. C. Northrop. The problem of integrating knowledge and the method of its solution. In *The Nature of Concepts, Their Inter-Relation and Role in Social Structure*. Proceedings of the Stillwater Conference, Stillwater, Okla. (1950), p. 30.
14. Henry Margenau, personal communication.
15. F. S. C. Northrop. Einstein's conception of science. In P. Schillp (ed.). *Albert Einstein: Philosopher-Scientist*. New York: Harper Torchbooks (1959).

8

Peter Caws

ON A CIRCULARITY IN OUR KNOWLEDGE OF THE PHYSICALLY REAL

In this essay I wish to raise a comparatively innocent-looking problem and explore the consequences of taking it seriously. The problem, briefly stated, is this: is there an essential circularity in our knowledge of the physical world? If so, does it matter? That is, does it have a systematically self-defeating effect on our attempts to understand that world? It will be seen as we proceed that a similar question can be raised for all claims to knowledge, but for the time being I restrict my attention to the epistemology of science. By way of an approach to the problem, consider an example that embodies it. Suppose we take some book about the physical world, for example, Henry Margenau's *The Nature of Physical Reality.*[(1)] It is an object in the physical world and has all the properties of such an object—location, cohesion, relative impenetrability, mass, motion, and the rest. Also, it consists of an ingenious and compact ordering of plane surfaces, about 1.7×10^5 cm^2 of them, that allows the display of an arrangement of some 9.5×10^5 marks, of roughly 75 basic types, by means of a technique of impregnating the surface at the appropriate points with a preparation that changes its reflective power. These marks constitute a code that can be decoded by certain other physical objects, namely human beings, which share the same basic properties—location, cohesion,

Peter Caws • Department of Philosophy, Hunter College of the City University of New York.

and the rest—but have in addition special facilities for receiving and analyzing visual signals and processing and storing information. The book that I hold in my hand is a member of a class of books (called *The Nature of Physical Reality*) which is in turn a member of a class of such classes of books (called just books). It is a characteristic of the members of the class of books called *The Nature of Physical Reality* that they are virtually indistinguishable from one another, except for accidental marks of individuation—stains, the yellowing of pages, tears, marginal notations, dedications. By contrast, it is a characteristic of the members of the class of human beings that they are essentially distinguished from one another, i.e., that each is a unique individual. This contrast may not be as fundamental as it appears, however, depending as it does largely on differences in the mode of production in the two cases: if books were still copied out by hand they would have much greater individuality, if humans underwent some standardizing or screening process that straightened out differences in genetics or education they would have much less. A more fundamental difference between books and people than their degree of individuation (and one that largely explains that difference) is the fact that books, once produced, are inert and do not change, either in themselves or with respect to one another, except by the operation of adventitious forces, whereas human beings not only start out with marked genetic differences but also continue throughout their lives to change in themselves and with respect to one another as well as being acted on by adventitious forces. Also, they are far more sensitive to the adventitious forces, being affected by changes in their environment and by incoming stimuli to which books remain completely indifferent. It is just because the sequences of internal changes and especially of adventitious forces are never the same for two individuals (although they may be closely similar in the case of identical twins), and because the changes are for the most part irreversible and the effects of the forces cumulative, that progressive physical individuation of human beings goes on continuously. I pursue perhaps to the point of absurdity this insistence on the common physical status of books and human beings because the interaction between them to which I now wish to draw attention is purely physical in nature, consisting as it does of the scanning of the pages of the book, under suitable conditions of illumination, by the eyes of the human being, and the transmission of the coded sequence of arrays of light and dark surface areas to the brain, where it is processed by means of rapid chains of electromagnetic and chemical events in and between some of the 0.5×10^{10} neurons normally found there. It is the configuration of the links and potential barriers between these neurons that determines the individuality of the *person* whose brain they constitute, as apart from the individuality of his body (which is more like the individuality of a book). The person is a mental structure. This structure is made progressively definite by what we call the person's "experience," which can be thought of as a long syntagmatic sequence of sensory inputs, some of them coming into the brain from his own body directly (as when he looks at parts of himself, hears his own voice, touches himself, etc.), but the greater proportion coming from the outside world in the form of light, sound, convection or radiation of heat, pressure, and the like. In the case of humans who have grown up in a social context, an extremely important, indeed the dominant, part of this syntagma consists of spoken or written or printed *words*, i.e., arrays of marks or sequences of sounds that occur in repeating patterns and set off, singly or in combination, essentially similar neural reactions on each occurrence. Depending on the state of the mental structure at the time, the other types of input that arrive simultaneously, etc., these reactions may in turn trigger others, and the cumulative influence of these inputs and reactions determines eventually the structure of the actions that the person carries out. Let me refer once again to the present case: my own verbal input syntagma has included two readings of Margenau's *The Nature of Physical Reality*, one almost 20 years ago, the other quite recent. Partly as a result of these experiences, and partly under the stimulus of other inputs—reading philosophy and talking about it, feeling gratitude in respect of certain events in my life and wishing to cooperate in a particular enterprise, receiving letters and telephone messages from editors, etc.—it came about at a quite definite time and in a quite definite location that the physical body with which I am associated sat down, took up a physical pen, and began to make marks on a plane surface by running an inked ball over it that left a trace, detectable because of the differential reflecting powers of the surface and the ink. "In this essay," I wrote, "I wish to raise a comparatively innocent-looking problem and explore the consequences of taking it seriously."

I have now constructed a circle in this essay, and it is easy to see how it mirrors a circle in the processes of our knowledge. For all knowledge that can be recognized and defended as such eventually finds its way into written (or at any rate spoken) form, and the physical character of the mode of its expression will locate it in the world alongside the things of which it constitutes knowledge. *The Nature of Physical Reality* is not, it is true, a physics book, so that the allusion to "things in the world," of our knowledge of which it constitutes the expression, may be needlessly confusing. I would be prepared in another context to defend the view that there is only one world, namely my own, and that everything that I can know finds its place there—and to argue further that this does not involve solipsism—but for the time being wish to concentrate on the case of physical knowledge: books about the physical properties of physical objects are themselves physical objects, and may indeed be composed of some of the physical objects—fundamental articles, atoms, molecules, etc.—about which they speak. This is clearly self-referential, but its essential circularity may not be so evident. That, however, is because we have to include ourselves in order to complete the circle. We believe that our eyes and brains, as well as the ink and paper, are made of physically real particles, but clearly we would not be able to know about physically real particles if it were not for the use of our eyes and brains, of the ink and paper.

Now it is a mistake to think of all circularity as being paradoxical or even undesirable. In fact, there is a long tradition in philosophy which suggests that circularity in argument is inevitable. An early hint of it is given in the *Euthyphro*, where every attempt at clarity seems to result in bringing the argument back to its starting point[(2)]; at the other end of the historical scale, Wittgenstein, in the preface to the *Philosophical Investigations*, speaks of traveling "over a wide field of thought criss-cross in every direction . . . . The same or almost the same points are always being approached afresh from different directions."[(3)] The fact is that the world in which philosophy operates is a closed world; *every* attempt on the part of sentient beings to understand the world in which they find themselves is bound to be circular if it is carried far enough. This inevitability arises out of the fact that, whatever the starting point of the inquiry which is to lead to understanding, sooner or later the starting point itself will become an object of the inquiry. Circularity can, it is true, always be *refused* by resort to *a priori* assumptions; the trouble with these, however, is that the resulting linear argument may not intersect with another linear argument constructed by somebody else on different assumptions. If my assumptions contradict yours, our conversation is ended before it is begun. (There are people who are not troubled by this kind of standoff, but this I suspect is either because they are too lazy to be interested in alternatives or because they are too dogmatic to entertain them.)

The question is, what is the character of this world which is thus closed in upon itself? Is there room for physical reality in it? In the first instance it is a world marked by discourse, but since there are parts of it that have not been adequately articulated discourse cannot be the whole story. The best way of characterizing it may be to say that it is a *thought* world—not a world *of thoughts*, but a world every element of which is thought by somebody: in my case, by me, in yours, by you. The arguments for closure have still not been better put than by Berkeley in the *Three Dialogues between Hylas and Philonous.*[4] All sensation, as well as all conception, all language, and so on, belong on the side of mind, as is clearly evidenced by the fact that in the absence of all mental activity no trace of the world remains. This is not to say that no world remains, it is only to remark on the thorough assimilation of whatever world may independently exist into mental form for purposes of human consumption. In the thought world there is clearly no room for material, if by material we mean something that is not thought.

But there is no need to identify the physical with the material; *physis* means, after all, nothing more than the nature of things naturally generated. And yet I think that most people who use the word mean by the "physical" something that is at any rate independent of *their* minds, and the independence of the material is no more problematic than, in Berkeley, the independence of the mind of God. Even for Hegel we are only the Absolute knowing itself, and its knowing itself does not in any way entail *our* knowing it. The problem of physical reality remains as acute as ever for the individual in his private circle, and the philosophical circle can be described in idealist or in materialist terms indifferently. I started this essay with a book as an object in the world, but I could have equally well have started it with the concept of the book—the point is that in both cases the exercise was an exercise in thought, a conclusion not to be escaped because it happened to be an exercise in discursive thought and even in written form.

The choice of the medium, then—ideas or matter—leaves the question of reality untouched. The problem of the real, as I understand it (and this is not, as we shall see, exactly the way that Margenau himself understands it), is the problem of clarifying the status of what there is, insofar as this is independent of my experience. My experience itself is of course also real, but this is not problematic, nor has it been since Descartes—that is, since the establishment of the distinction between the actuality of experience and its significance. To say that the problematic part of the real is the part of it that is independent of my experience does not mean that none of it could be in my experience, only that it would be as it is whether it were in my experience or not. But by far the larger part of it—everything that happened before I was born, everything that will happen after I die, everything that now happens elsewhere, or behind my back, or beneath the surface of things—is and remains outside my experience. Reality is infinitely richer, ontologically speaking, than my world (even though my world can contain a *conception* of the whole of reality).

I wish now to deal with two kinds of attempts that have been made to solve the problem of the relationship between the content of my experience on the one had and reality on the other. One strategy has been to try to construct the elements of reality out of the elements of experience, or at any rate out of classes of elements, typical examples of which are found in experience. The most elegant expositions of this strategy are to be found in Russell and Carnap. Russell poses the problem as "the construction of physical objects out of sense-data," and he arrives at the notion of things as classes of their aspects. We have seen the table from a few points of view, under a few conditions of lighting, etc.—imagine how it would look from all possible points of view under all possible conditions, and we have an aggregate that will exhaust without remainder all possibilities of perceptual knowledge of the table.[(5)] Carnap's program is more ambitious: from the perceptions of a single observer he constructs, in ascending order, autopsychological, physical, heteropsychological, and cultural objects.[(6)] (He uses for the purpose a fictional character A, the first half of whose life is spent in experiencing the world without analysis, the second half being spent in analyzing the data thus gathered, without further experience.) A third attempt along somewhat similar lines—although all three have radical differences from one another—is that of Whitehead, who by his method of extensive abstraction sought to isolate real space-time points as the termini of converging series of experienced space–time regions.[(7)]

The second strategy is that of Margenau himself. He approaches the question from a diametrically opposite point: assuming, in effect, that if the real shows up in our experience at all it will bear the marks of its independence (its coherence, its connectedness, and so on), he limits the ascription of reality to those parts of experience that have been successfully systematized by means of scientific constructs. "To us," he says in *The Nature of Physical Reality*,[(1)] "reality is not the cause but a specifiable part of experience" (p. 289). Now both these strategies—Russell's and Carnap's on the one hand, Margenau's on the other—seem to me unsatisfactory: in the former case because of ontological extravagance, in the later case because of ontological poverty. Russell thinks nothing of introducing infinite sets into the real, and he does so with an abandon that would have made Ockham shudder. Margenau[1] on the other hand excludes from the real undeniable facts of experience, if these have not been "normally standardized into scientific knowledge." "As yet," he says, "they have not been united into an organized pattern comparable with the structure of physical reality, and it would be pardonable for the scientist to suggest that the name *reality* be at present denied to them" (*ibid.*). If we subscribe to the Cartesian principle referred to above, this puts him in the odd position of saying that reality is only a part of a part of itself. Of course, the motivation for introducing a limited and specific meaning for the expression "physical reality" is quite clear in Margenau's work, and if we view it as an epistemological rather than as an ontological limitation his strategy becomes perfectly acceptable. And yet from the ontological point of view it does seem overcautious.

My own inclination is to try a completely different approach, by asking whether there could be anything with respect to which we could know that it lay outside our experience as well as being independent of it. If it were just a question of an individual confronting the physical world, there would be no possible way in which he could *know* that anything lay outside his experience, and anyone who wished stubbornly to maintain that view throughout the following argument could perfectly well do so. There is, however, a class of events of which we can come to have knowledge that would strike all but the confirmed solipsist as satisfying my criteria: I mean the contents of other people's experience. We come to know these by being told about them, and we have to admit that, for the most part at any rate, our existence makes no difference to them and their immediacy is out of our reach. Other people report to us what happened in the past, and what happened elsewhere, and what happened when we were not looking, and so we extend the basis of our knowledge of the real. I cannot here go into the details of the process according to which scientific knowledge is generated out of this collective resource; suffice it to say that the result is not a monolithic body of scientific theory or even of knowledge of the everyday world, but a series of free-standing conceptual schemes each associated with a single knower, yet overlapping in such a way that groups of people can in various times and places be taken to embody such theories and such knowledge.

Every conceptual scheme is closed upon itself just in the way in which experience was earlier said to be—in fact, every such scheme is precisely the consequence of a certain experience—and so is the aggregate of conceptual schemes, incoherent as this is bound to be even if limited to the domain of a particular science. Yet, when the scheme of an individual knower has been critically rethought by him, when out of his intuitive concepts he has formed scientific constructs, when some part of the scheme has been converted into a *system*, he finds that the overlapping with other people's schemes is more precise than before, so that among the practitioners of a given science it becomes reasonable to speak of an isomorphism between conceptual systems. And it is this that leads us, in the end, to the idea of an external and independent physical reality. "It is forced upon us by the constant pattern of the isomoprhic systems. But this in itself does not necessitate the postulation of an external: perhaps the isomorphic systems just hang together that way. All of them, however, except the postulated natural reality, are in mind; they constitute a universe of thought. The reason why it seems to me sensible to postulate a universe of being also is that in principle the universe of thought could be entirely capricious, and it is not. The simplest explanation of this constancy of form between the isomorphic systems is the existence of another system isomorphic with all of them—rather such that they are all imperfect isomorphs of it—underlying them and proceeding independently of them."(8)

The standard objection to this formulation is that it makes physical reality hypothetical, and that this hypothetical status then extends to every element of it, even those we take to be reflected in our experience. This leads to the paradoxical consequence that the real components of our experience are made up of hypothetical parts. But this is just a set of confusions. The real elements of our experience are not *made up* of anything, unless of parts really experienced at the same time (there is nothing paradoxical about an experienced person's having experienced arms and legs). What is hypothetical is the object as a whole under its real, as opposed to its experienced, modality, and there is nothing odd about a hypothetical object's having hypothetical parts. What science enables us to see is that there are more hypothetical objects than experienced ones: electrons lie in the hypothetical real world, and not in the real world of experience—but the formulation of that sentence is meant to underline the proposition that we are not talking about two worlds, only about one, most of which, however, remains outside our experiential range.

The upshot of this analysis is that the cycle of our knowledge, from perceptions to hypotheses and back to perceptions, while it is itself inscribed in the real, does not enclose the reality it claims to know. Reality lies outside the circle, rather in the way that tracks lie outside the train. (Imagine a train running on a single circular track, and a passenger able only to look horizontally out of a side window—he could tell by the repetition of the landscape that he was on a closed track even if he could never actually see it.) The idealist temptation is to identify reality with the circle, but this seems unnecessarily limiting as an ontological principle, besides being immodestly self-centered. Assuming a reality infinitely vaster than our knowledge of it, we can still make at least one assertion about it with complete assurance, albeit a negative assertion: namely, that the real has not been such as to require other experiences than those we have in fact had. In closing I offer two formulations of a somewhat more positive kind, each of which sums up the result of the argument, and the choice between which can be left to philosophical temperament. The first is skeptical, and yet conditionally affirmative: *we do not know the real at all, but what we assert about it would be knowledge if there were any way of getting out of the circle.* The second, which I prefer, has obvious affinities to Margenau's position, but avoids some unfortunately relativistic aspects of that position: *the real is the objective correlate of those aspects of conceptual structure that most men have in common or in such form that is translatable into a common structure.* This assumes that the development of science is on the right track, not necessarily that it has arrived. The possibility remains, of course, that we might all be mistaken together.

I conclude, then, that there is an essential circularity in our knowledge of the physical world—that every line of demonstration sooner or later turns back upon itself. But I conclude also that this is not a self-defeating circularity, and that two things about it even give grounds for philosophical satisfaction. One is that, as always happens in philosophy, the state of the *knower*, if not of his knowledge, changes as he goes round the circle, so that he understands each proposition anew in the light of the propositions that have intervened since his former understanding of it. The other is that if the relative sequence of the propositions remains unchanged —if they are always there, as it were, when the argument comes round to them—that in itself may be evidence of an underlying stability in the real, of which, after all, the propositions and the argument and for that matter we ourselves are an integral part.

References

1. Henry Margenau. *The Nature of Physical Reality*. New York: McGraw-Hill (1950).
2. Plato. *Euthyphro*. 15b.
3. Ludwig Wittgenstein. *Philosophical Investigations*. Tr. G. E. M. Anscombe. Oxford: Basil Blackwell (1953), p. ix^{e}.

4. George Berkeley. *Three Dialogues Between Hylas and Philonous.* New York: The Liberal Arts Press (1954) (Originally published in 1713).
5. Bertrand Russell. The relation of sense-data to physics. In *Mysticism and Logic.* Harmondsworth: Penguin Book, (1953), p. 140.
6. Rudolf Carnap. *The Logical Structure of the World.* Tr. Rolf A. George. London: Routledge and Kegan Paul (1967).
7. A. N. Whitehead. *The Organisation of Thought.* London: Williams and Norgate (1917), Chapter 7.
8. Peter Caws. The functions of definition in modern physical science. Ph.D. dissertation, Yale University (1956), p. 225. (The sponsor of this dissertation was Henry Margenau.)

9

Siegfried Müller-Markus

THE STRUCTURE OF CREATIVITY IN PHYSICS

This article is dedicated to Professor Henry Margenau as an expression of my deep esteem. I owe him in large measure the stimulus for my own scientific thinking. His manner of combining exact physics with equally exact philosophy is in itself an eminent contribution to the concept of logical fertility which he has introduced. The considerations which follow set themselves the task of establishing a connection between Margenau's idea of the *C*-field and the depth psychology of C. G. Jung. I am honored to be permitted in this way to return thanks to my highly respected friend and mentor for the encouragement and furtherance he has given to my own investigations.

Statement of the Problem

For a long time, science has discussed everything but itself. Today we possess a large store of analyses of its logical structures and its historical development. But the creative activity of the scholar remains a mystery. Here we shall make an attempt to expose the roots of theory formation. We assert: they lie hidden within the depths of the unconscious. But this unconscious is itself not without order. It, too, has a structure. Our second assertion is therefore the following: even the unconscious involved in the formation of physical theory is capable of being described in a

---

Siegfried Müller-Markus • Television Commentator, formerly University of Fribourg.

rational manner, because it itself obeys certain laws. I therefore like to call it the field of *protoreason*. All operations which take place within this field I call *protorational*.

My third assertion therefore is this: the protoreason generates the *C*-field of theoretical constructs of Margenau. It is the field of creativity of the physical subject. In investigating its structures, we make the decisive turn from physics as a science of physical objects to *protophysics*, a science of the subjects of physical theory formulation.

In a sense, Margenau's *C*-field forms the peak of an iceberg. It floats on the ocean of human creativity which we call the field of *protoreason*. The theory of the *C*-field therefore presupposes implicitly a theory of protoreason, namely protophysics. In the following we shall endeavor

a. To clarify the genesis of Margenau's constructs, i.e., to describe the theoretical creativity ("the emergence of the iceberg").
b. To formulate the conditions of this emergence.
c. To establish relations between theoretical creativity and the physical creativity which is implied by it, i.e., the creativity of physical objects.

Micro-, Macro-, and Megacreativity

A division of the creative acts of physicists into three forms is indicated. I shall designate as *microcreativity* an innovation with the following characteristics: (a) it is difficult for the outsider to recognize it; (b) it is accompanied by a slight rise in the curve of emotions, both of the author and of the outsider; (c) it takes only a small step in the direction from the unknown to the known.

Such steps are, for example, the repetition of an experiment with refined means, the derivation of a known theorem by a new method, or the discovery of a new star of a class which is already known.

I denote as *macrocreativity* an innovation with the following characteristics: it is at once recognizable, both to the author and to the outsider; with both, the "seismograph" of emotions deflects strongly; an important step from the unknown to the known takes place.

As an example, we may cite the discovery of four-dimensional geometry by Minkowski. It became the methodological basis of subsequent physics insofar as it was to be formulated in relativistically invariant fashion. This is especially true for cosmology and quantum field theory. A second example is the representation of quantum mechanical states by vectors in Hilbert-space as introduced by John von Neumann. In both instances the physicist was enabled by an ingenious mathematical process to understand known physical connections in a better way. Postulates of the theory became simpler and more transparent.

In the realm of *megacreativity*, however, we witness the opening of a new world. It offers more than a new view of already known relations. There occurs a violent deflection of the emotions, both upward and downward. The upward deflection takes the form of a euphoric agreement by contemporaries, the downward one a passionate rejection. The resistence against a new discovery becomes almost a measure of its importance. The world protests very little against small innovations, more strongly against larger ones, and against a radical revolution it raises the entire psychological mechanism of mental inertia. For all thinking has the tendency to be left to itself and to go its straight, undisturbed course. External forces produce resistance. The outsider feels threatened by a revolution within his present state of mind and defends himself against it. Nothing is more difficult than to alter one's thinking. To this even the genius of religion testifies when he preaches a metanoia, an inner reversal.

On the other hand, the glory of the "heureka!" which affects the author as well as his favorably disposed contemporaries shows that a new path has been found for theoretical thinking. Indeed, at the moment of innovation there occurs an emotional discharge: it turns out that the scientific public has already been prepared for this discovery, without knowing it. The dammed stream suddenly breaks its bounds and floods valleys which were hitherto infertile. This creative activity of the theorist breaks forth both within the author and within the recipients from the depths of the mind, because its barriers have been blasted away.

To cite examples in our century, we note the discovery of relativity by Einstein, the explanation of gravitation as an expression of Riemannian geometry, also by Einstein, the discovery of the quantum theory by Planck, of the wave theory of matter by de Broglie, and of the uncertainty relation by Heisenberg. In every one of these instances a whole continent of new insights is discovered, which their contemporaries and their successors can settle with their own theories and experiments. The authors of megacreativity are therefore justly compared with Columbus, the archetype of scientific creativity with Copernicus.

On the Incidence of Theory Formation

Micro-, macro-, and megacreativity contain the following common elements:

a. Production of a new insight or a new method.
b. Spatial and temporal localization in the history of science.
c. Personal localization as achievement of a person or group of persons within society.
d. The assessment of the achievement by contemporaries in accordance with the degree of innovation it brings about.

We shall denote as "theoretical event" the substratum of creativity. Its elements can be localized in terms of the "coordinates": history, persons, and values. Figuratively, we can thus envision a kind of world geometry, within which the acts of the physicist can be localized as events by the preceding three coordinates.

This model at once shows the existence of a dual problem. The first relates to the form, the second to content. Formally, we may regard Minkowski's geometry as a model of theoretical creativity. The reason is easy to see: both physical creativity and theoretical creativity occur within time. Both represent a process. I shall call it self-realization. If event A produces along a world line the sequence of events $B, C, D, \ldots$, then we may consider this as a process in which A becomes "what it really is," namely, the sequence of events $B, C, D, \ldots$. Thus one might identify in a very abstract way the position of the hand x of a clock with the event A, and then say: the positions $x' = B$, $x'' = C$, $x''' = D$, etc., represent the self-realization of A. A moves to B, C, D, because within the subsequent events it realizes itself, because with it, it "becomes whatever it is" (Nietzsche). This, of course, is a psychological interpretation of time which is very uncommon among physicists. For it forms the exact opposite to the general tendency of our time, which is to physicalize, so to speak, nonphysical processes and to mathematize them. Possibly, however, it may also lead to fruitful results, once we try successfully to establish relations between psychical and physical processes. Here, this is being done in a purely formal way without claiming any relevance for the actual character of physical events.

The logical bracketing of the psychical events of theoretical creativity on the one hand with the physical events in the Minkowski world is suggested by the homonymous character of the word *event*: the first meaning is characterized by a physical occurrence which is localizable in space and time, the second means a psychical process which can be, at least within some approximation, localized in space and time. Thus, for example, the discovery of the quantum of action is a psychical event having the coordinates Berlin, 14 December 1900, when Planck at a meeting of the German Physical Society published his new law of radiation.

Nevertheless, it is at once clear that the causes for physical and psychical events are of wholly different kinds. The collision of particles is occasioned by the previous motion of the particles and its characteristics. For the discovery of Planck's quantum, however, it is the "motion" of the theory of black body cavity radiation and all the characteristics of Planck and his scientific milieu, i.e., its theoretical condition or state. Hence we find it expedient to use a special symbol c_P for the physical events as abbreviation for "physical creativity"; we shall use for psychical events of theory formation the symbol c_H as abbreviation for "human creativity." Common to both is the possible localizability in space and time and the historical sequence involving such relations as earlier than, simultaneous, and later than.

The second problem concerns content. Whence this homonymity? Why does theory formation also occur in time? Does not mankind possess from the very beginning a perfect knowledge? These questions are answered by the simple observation that science is a learning process. We are always on the way toward knowledge, but we never exhaust completely all possible insights regarding the facts in the world. It is very questionable whether this will ever be the case. This is true because man's theoretical creativity is somehow tied to a sequence of physical events in the cerebral cortex. The subject of human creativity is itself an object of extrahuman, purely physical creativity, which is subject to physical interactions. Therefore, events c_H can at least in principle be recorded as events c_P within the Minkowski world. This record is the history of physics, which endeavors to assign precise spatiotemporal dates to theoretical events.

But here the previously mentioned homonymity begins to fail. For is is impossible to find for the events c_H Minkowskian world lines of physical events as their causes. Even a very precise record of all "movements" in the cerebral cortex of Planck before 14 December 1900 would never completely depict the theoretical initial condition which led to the discovery of the quantum. However, it is possible to conceive of a kind of logical "world lines" which led in the form of equations, experiments, hypotheses, and axioms almost but not quite uniquely to Planck's discovery (see Jammer,[1] pp. 10–20). We see, therefore, that so far as content is concerned the discovery of the quantum of action and the discovery of

Planck's law of radiation differ indeed, but formally they can be localized at least in principle in time and space. We shall not discuss here the relation between events c_H and c_P as far as their contents are concerned, since this would lead to hitherto unsolved problems of psychology.

Recognizing these difficulties, we shall nevertheless attempt to establish a formal analogy between the two. We construct a purely spatial field of physical events c_P with the axes x, y, z. Analogously, we construct a purely "spatial" field* of the c_H. This is Margenau's C-field, with the restriction that Margenau's constructs represent the results of theoretical events c_H, not the events themselves. With this understanding, all theoretical events (experiments, theories) form the "space" of physics. Thus at the time 1 January 1972 the "space" of physics is characterized by the manifold of all theoretical events taking place on that day.

We now draw within physical space a time axis, and thus make it analogous to a Minkowski world. In the same way, we draw a time axis within the theoretical "space," and this is the same as the time axis of physical space. We thus take note of the fact that the c_H too are subject to a temporal sequence, indeed a sequence measurable by means of the same clocks which mark the history of elementary particles and their aggregates.

"Past" thus represents all events c_H prior to a certain point of time, e.g., 1 January 1972. They form the store of data in the journals of physics. They can be mobilized under certain conditions by the memory of physicists and can be learned. Thus the "past" of physics is subject to a learning process.

In an analogous way, we shall speak of the "future" of physics as the manifold of all future "events" c_H, independently of whether they actually occur or are only given as possibilities. Thus, in contrast to the physical "past," the "future" is not subject to being learned, cannot be stored in memory, and is in principle infinite in extent. A set of facts which can be learned, however, is in principle finite inasmuch as science does not have an infinite past.

Thus we note a fundamental asymmetry between the "future" and the "past" of physics.

We now draw a temporal cut within theoretical "space," e.g., at $T = 1$ January 1972, and we thus recover the purely spatial part of all events occurring on 1 January 1972. The problem of the lower limit of such a cut will not be discussed here. Our interest attaches to the analogies

*Quotation marks at this and subsequent places indicate analogues to physical situations.

with physical space. Like physical space, theoretical space has a kind of "dimensions." These are the theories which differ with respect to form and content. For instance, we may regard classical mechanics, thermodynamics, theory of electricity, optics, relativity theory, cosmology, and quantum theory as "dimensions" of the "space" of theories. An event then projects upon several dimensions, because it contains statements about several theories. Thus Planck's formula of radiation projects upon thermodynamics, optics, and quantum mechanics. Indirectly, however, other theories are likewise involved. They introduce additional coordinates into the space of the c_H: we attribute to a given c_H a collection of names. Planck's formula of radiation is characterized by the names optics, electrodynamics, quantum mechanics, and thermodynamics. From the last discipline it takes the concept of entropy.

We may even introduce the idea of a spatial distance. Thus all formulas which are historically and logically connected with Planck's form a kind of "neighborhood." Together they represent a sort of cloud of "events." This cloud has a density: the more neighboring "events" that combine into a cloud, the denser will be the cloud.

Creative Leaps

Here arises the difficult problem of continuity. Do the c_H, in their spatial aspects, pass into one another? That is, is it possible to establish between them a pervasive logical connection? Do the "events" or their "clouds" emerge continuously from one another in their theoretical "space-time"? Is there a logical nexus beside the psychological one?

One might believe at first that the set of all simultaneous c_H has, at least in principle, a common logical basis, if only we could fill in the gaps between them. For instance, there is an expectation that the properties of all elementary particles can be derived from a single universal formula. If such a world formula were relatable to some fundamental cosmological equation, micro- and megaphysics and all of physical reality would be comprehensible in terms of a single principle.

No one knows whether such hopes can be fulfilled. They presuppose the possibility of perfect, definitive knowledge. At any rate, even if a world formula did exist, a thoroughgoing logical connection of all simultaneous c_H would itself be an event of the future, possibly of an infinitely distant future.

We may expect with great probability, however, that the historical emergence of the c_Hs is not continuous. To be sure, microcreativity suggests the opposite. We try, in presenting the history of science, to sketch the historical course without gaps. But this is merely an ideal task which cannot be fulfilled. There are always spaces between single discoveries which cannot be filled.

The gaps are of two kinds. On the one hand, the purely factual in the form of new experiments surges like an uncontrollable ocean, wave upon wave, into the "space" of theory formation. We never know *a priori* the result of a future experiment. We also believe in the essential identity of all results under identical conditions (we ignore quantum mechanics at this point). We thus demand, as it were, the nonoccurrence of the unexpected. But we believe equally firmly in the inexhaustible possibility of new results under presently unknown conditions. Without it, progress of experimental knowledge would be paralyzed. Hence human creativity is essentially conditioned by the desultory appearance of new factors, and is in this sense unpredictable. Its prognostic character is limited to the prediction of known events from known states. The unknown always appears unexpectedly; and this is precisely what we do expect. I shall call this *mutation of knowledge*. This expression contains a true analogy to genetics, since the known is comparable to the genes which, in the course of the scientific learning process, determine the structure of new theories (i.e., the organs of science).

Thus the "space" of the c_H is codetermined by the physical space of the c_P in an uncontrollable manner. Observations are joined to the space of the c_P in a manner both fatal and felicitous, but never completely predictable. Whether, for example, Planck's law describes the facts better than the law of Wien, in which h does not occur, is not controllable by the power of thought, but only by the facts themselves. These are related with the "space" of the c_H via procedures of measurement, and this in a way both active and challenging. There is in a sense a parallelism, a common stream of physical and theoretical creativity.

In the "space" of the c_H, this stream produces logical fractures. The formula of Rayleigh cannot be true at the same time as the formula of Planck. Through experimental verification of the latter, the former is canceled in the "space" of the c_Hs, and has no further validity. There exists something like an annihilation operator for the c_H. Every c_H is in principle threatened by it, yet the efficiency of this operator cannot be foreseen. Thus there is in addition to the normal "past" another special one for the set of canceled c_H. For this cancellation there exists no logical cause in the sense of a preceding c_H; rather, the cancellation is enforced by an external observation. This observation represents a complicated link between the space of the c_P and the c_H. Here certain events of the external world are blended with those of the psychical inner world, which we call experiences; these, too, are of course events of the theoretical "space," i.e., constructs.

Experiences (*Erlebnisse*) thus represent the c_H which lie closest to the c_P of the external space. We might say: they form the horizon of the c_H against the space of the c_P. This horizon does not arise in logical or psychological fashion from the "space" of the c_H. It is the common "product" of the physical and the theoretical space. Therefore, the correction of the Rayleigh radiation law through that of Panck represents a difficult detour, and does not follow directly from Rayleigh's law. The detour passes across additional observations within the horizon and then leads within the "space" of free constructs in Planck's derivation to the concept of entropy. The path thus suffers a double break. New observations and mobilization of the entropy concept formed fractures in the apparently continuous course of Planck's thinking.

Here we meet what is commonly called irrational, for there is no logical derivation which allows a prevision of the further course after the fracture. Jammer[(1)] describes the state of affairs concisely: "Never in the history of physics was there such an inconspicuous mathematical interpolation with such far-reaching physical and philosophical consequences" (p. 18). Let us now label the space of the c_P as external space, the "space" of the c_H as internal space. This designation has its origin in the metaphorical conception that the "events" of theory formation take place within an "interior" of consciousness before they are communicated to the outside world. This interior is subject to physical time, for there is within it a before and an after, but has no reference to physical space insofar as the contents of consciousness are totally different from the processes in the cerebral cortex, with which they are connected in a manner hitherto unknown. Thus the concept of an infinite set of cosmic events can be thought, but never realized, provided that the universe is considered to be spatially and temporally finite. For this reason, among others, it is necessary to assume in addition to an external space a special interior or inner space, although both are connected at the horizon.

This internal space likewise shows fractures which are independent of the horizon. Planck's discovery was brought about not only by new observations but also by observations plus new hypotheses, in this instance the inclusion of the concept of entropy. We remember at this point Einstein's Credo, according to which there is no *logical* path from experience to theory. For within the domain of theory formation, i.e., the part of interior space without horizon, there are likewise fractures or leaps which are not predictable. Indeed, it is often difficult to justify them later by logic and/or psychology.

It is precisely this state of affairs which leads to the experience of novelty in connection with a discovery. It is the leap which we called revolutionary, for it is not in principle foreseeable. Here we witness the incursion into the free course of thought in interior space, by way of a new observation, of an element which does not come from without, representing as it were an overlap of outer and inner spaces. We shall provisionally call this kind of element protoreason, in accordance with the postulates stated at the beginning. In analogy with the ocean of the external world, protoreason likewise forms an ocean, an internal ocean residing within the physicist.

The Fractures

In this section we examine the so-called fractures in the reasoning which led Planck to his radiation formula. Planck concentrated his attention on black body radiation, and his writings indicate that a black body meant for him something absolute, a concept of crucial importance. In a sense he thus widened his view upon the whole of physics. The logical leap from something relative to something absolute and from something partial to the whole has no counterpart in the domain of experiments. Planck started with the assumption of a large set of simple hormonic oscillators. He thus introduced the theory of dipole radiation conceived by Hertz in 1889 into thermodynamics. Furthermore, he carried the Maxwell-Boltzmann distribution from the kinetic theory of gases into the theory of cavity radiation before he made the assumption that irreversible processes converge upon a stationary state. But, in doing this, he did not employ the classical equipartition theorum, according to which in an equilibrium state the total kinetic energy is equally distributed among all degrees of freedom. For this procedure, Max Jammer cites several reasons without deciding in favor of a special one. This in itself irrational omission of a classical axiom was a matter of good fortune, for without it Planck would never have created the quantum hypothesis. He finally defined the entropy S of a harmonic oscillator somewhat arbitrarily in the form

$$S = \frac{UU}{aa\nu} \log \left(\frac{U}{eb\nu} \right) \tag{1}$$

Here a and b are constants and $U = U(\nu, T)$ represents the mean energy of the harmonic oscillator of frequency ν at temperature T. Then Planck[2] used the "happily chosen interpolation formula" (Jammer,[1] p.19)

$$\frac{\partial^2 S}{\partial U^2} = \frac{a}{U(U+b)} \tag{2}$$

Thus far, we have encountered only micro- and macrosteps of creativity. None of them was forced by external observations.

There follows, however, a creative megastep. For, departing completely from past procedures, Planck wrote, following Boltzmann, for the entropy S_N of a system of N oscillators with frequency ν the formula

$$S_N = k \log W \tag{3}$$

The symbols here have their usual meaning. Next, he investigated the distribution of energy U_N among the N resonators. Assuming U_N to be divisible only into energy quanta $h\nu = \epsilon$, he obtained as the number of energy quanta which are distributed among N oscillators the value

$$P = U_N/\epsilon \tag{4}$$

From this he obtained W as function of N and P, thence the entropy S_N, and finally the new radiation law

$$u_\nu = \frac{8\pi\,\nu^2}{c^3}\,\frac{h\nu}{\exp(h\nu/kT) - 1} \tag{5}$$

It takes the place of the radiation law of Rayleigh, which has the form

$$u_\nu = \frac{8\pi\,\nu^2}{c^3} \cdot kT \tag{6}$$

Here u_ν is the radiation density within the cavity of temperature T per unit interval of frequency at the frequency ν.

On a Curious "Metaphor"

Now we encounter a curious state of affairs. The megacreative acts leading to Planck's discovery have themselves a structure resembling the quantization of energy. In other words, there is a structural similarity between the subjective process of finding a theory and the objective phenomenon of cavity radiation.

This is wholly unexpected. The structure of cortical molecules has evidently nothing to do with the kind of thoughts they engender.

This parallelism suggests a model for megacreativity. We may regard the previous "inner space" of theoretical constructs as a kind of black body cavity. In a certain metaphorical sense, the creative unconscious, which we previously called protoreason, performs a structural comparison between thinking and a physical process. We will now lift this comparison of structures from the domain of the unconscious into the conscious. Reflection, in the sense of thinking, may then be regarded in connection with the model of cavity radiation as emission and absorption of radiation. This requires a heating of the black body, i.e., a supply of energy. Radiation itself may be regarded as a totality of electromagnetic signals. Einstein, too, used such electromagnetic signals in deriving his relativistic concept of time.

In similar manner, the initiation of thinking requires a supply of energy. Part of it surely comes from the organic energy of neuron metabolism. Whether there is in addition a special psychical energy which is capable of being measured in physical ways, or whether such an energy lies beyond physical measurability, will not be discussed at this point. The physical emission of a signal by a cavity oscillator is then to be compared with the psychical "emission" of a new idea by the theorist.

This kind of comparison is initially only of interest for the poet. He, too, places in parallel such psychical events as joy or grief with physical processes such as a sunset. But the poet is only partially in error. General linguistic consciousness likewise employs such metaphors, and these come from the unconscious. Every language is full of such metaphors. Here the human race in the course of its history has built a structural affinity between psychical and physical processes, which is possibly more than a mere metaphor. Is this structural affinity to be regarded as real?

Let us take another step. We may consider the state of the theorist before the emission of the idea as an oscillator. It is "excited" by the state of the problem, "caused to vibrate." Immediately before emission of the idea it finds itself in a highly unstable state which can perform a transition at any time into a new state, the state after the emission of the idea. This emission is not continuous but discontinuous. For this reason, we call it, equally metaphorically, an incident, an illumination, an intuition. In this way we document the discontinuity of the emission. It is preceded, of course, by a lengthy stage of preparation. There is a process of searching, of appraisal of possibilities, and a mobilization of previous ideas. All this is evident in the steps Planck took toward his discovery.

At a certain point, however, this state of preparation reaches a critical threshold. The state of emotional excitement rises to a maximum.

The "oscillator" which corresponds to Max Planck's state of mind begins to vibrate violently. And suddenly there bursts forth a new idea: the quantum of action h.

But Planck did not operate in isolation. He would never have made his discovery alone. He was dependent on the totality of all physicists interested in and concerned with the problem. He lived in a scientific "medium." This is what we might now call the creative "inner space" of the c_H. By calling to mind all relevant theoretical constructs of the world of contemporary physics, he caused within the "space" of his own c_H a large number of "oscillators" to vibrate. These are the constructs of all those physicists who either simultaneously or prior to Planck performed those experiments and established those equations and ideas which were relevant for the solution of Planck's problem. Within Planck's "inner space" there now occurred an "exchange" of ideas. This is precisely the metaphor for the exchange of ideas within the cavity. Thought processes, with their mobilization of ideas of other physicists, fill the creative "inner space" with "radiation." Again, this is meant initially as a pure metaphor. On the other hand, this process is accompanied by a purely chemical-physical process of increased energy consumption in the cerebral cortex. Metabolism is accelerated, as every researcher involved in active thought can testify. Thinking is work and consumes energy. But, metaphorically, it is an energetic process of "radiation absorption and emission." This occurs not continuously but by leaps.

All this is likewise true for Planck's contemporaries. They, too, were "excited" by his discovery. Their mental "oscillators" were set into "vibrations" and suddenly, on 14 December 1900, received the signal in discontinuous fashion. The quantum hypothesis was entirely unexpected by them. Upon hearing it, they absorbed the signal all at once, not continuously. This process can be compared with the discontinuous transition of an oscillator by absorption into a new state, the absorbed idea being analogous to the energy quantum $h\nu$.

Reason (Logos) of the Unconscious

Another feature appears in our example. Megacreativity reveals itself as magnified micro- and macrocreativity. All steps taken by Planck are leaps. In every instance, a logical and a psychological cleavage are crossed. None of them can be completely filled afterward; there is always an irrational remainder. Irrational, because logic even afterward cannot accomplish a unique derivation of Planck's radiation formula from the observed facts. This is something which was always stressed by Einstein.[3]

What is irrational in this, however, is only the leap within the logic of consciousness. Thus, for instance, the definition of entropy in accordance with equation (1) is arbitrary. But the physicist, by temporarily delving into the depth of the unconscious, apparently sacrifices his reason (logos) in order to deliver himself to the stream of the unconscious. But this is only apparently true, for it turns out later that this act is necessary in order to reach the goal. There is a special reason (logos) of the unconscious. It comes to the aid of the conscious always, or at least sometimes, when the latter has lost its way to its goal. We ordinarily call this act intuition.

Actually, the reason of the unconscious does not always enter the scene. Otherwise, every contemporary of Planck could have found the same radiation formula. There is no assurance that the unconscious will come to the scientist's aid. The physicist must wait patiently until it rises like a geyser from the depths. He has no control over the processes underlying this. Henry Margenau describes this in terms of a simile.[4] "Mathematics alone is a gallery of robots. Select one of them and connect him with the real. If you have chosen the right one, you may witness the spectacle of man-made life; blood will course through the previously empty veins of the artifact and a functioning organism has been created. No one can tell in advance which robot will cause this success to be achieved; the scientist of genius makes the proper selection."

The unconscious within the physicist is therefore a kind of scientific guardian angel. But in contrast to the supernatural beings of religion it is not a proper person. To be sure, it operates spontaneously and without control. But it accompanies the consciously acting physicist like a second self. It acts somewhat like a double with its own fate and will. It is an integrating constituent of his person yet separated from his consciousness like a second ego. In an abstract sense it exists prior to consciousness, because it prescribes his path suddenly like a pilot and leads him to his goal, which consciousness alone would not have reached. One might therefore call it the self (ego) of the physicist because it leads him where he really wishes to go but cannot go under his own power. This self or ego takes on the role of a pilot. Insofar as this pilot communicates his decisions to the logos (reason of consciousness), he may be called protologos or protoreason.

Protoreason anticipates a solution which consciousness itself is not able to find. Intuition of the quantum hypothesis came suddenly to Planck and was not suggested by any preceding observation or knowledge. Nevertheless it existed within the store of protoreason. The latter therefore anticipates the future of science. It performs the function of the prophet and the seer. Hence the mystic reserve with which the contemporary world views the inspired step of megacreativity. It is reminiscent of the reserve which met in previous times the prophet and the poet: both had access to that depth of reserves which is not accessible to the day-consciousness of the normal individual.

Protoreason cannot be manipulated. It is impossible to force it to reveal itself. In this respect it is similar to a supersensory vision, which also cannot be contrived. In all probability a future psychology of scientific creativity will here make comparisons with the phenomenon of the mystic. Even at this stage, one can recognize as common characteristic the humble waiting for illumination. Such waiting, however, is not at all passivity. On the contrary, it represents a phase of highest activity inasmuch as the waiting individual in both instances prepares himself by concentration and perhaps asceticism for the inner call. The decisive characteristic is the merging of one's self which is so well-known to the Oriental.

Indeterminism

If there are genuine leaps in the domain of the physical, then there are holes in the topological field of being where no causal connections exist. By way of the quantum jump, the radiating atom delves into an ontic depth in which we can no longer resolve the chain of causes and effects. Between the events there are dark holes which cannot be filled by a logical "fluid." No one can reconstruct *a posteriori* why an electron after its passage through a hole in the screen chooses this and no other spot upon the photoplate.

Leibniz was the philosopher who recognized with full clarity the existence of such holes. For him the mechanical motion of a mass point represented at each point of its path a process of annihilation and a new generation. He called this "transcreation," and accounted for it finally by an appeal to a series of creative acts on the part of God.(5) It is proper to cite this vision of quantum jumps because it furnishes evidence for the existence of protoreason coming from a point in time two centuries earlier than Planck's discovery.

Whoever accepts those jumps admits that the moving thing E, where it has been at the point 0 for some time, is extinguished and annihilated, then rises again after a moment and is newly created in point B. We may call this kind of motion transcreation.

Gallutius: "If one can regard this proved we would have gained something of very great importance. For we would have a proof for the Creator of all things." (Leibniz,(5) pp. 148–149)

Here Leibniz not only anticipates the physical quantum jumps but also the creativity of the physical subject. Hence physical and human creativity represent fields with topological holes. They cannot be filled any longer by conscious-rational processes, but only by those which are unconscious-protorational. Their removal, however, does not occur, in contrast to classical determinism, by an arbitrarily detailed sequence of states, but by an *ex post facto* justification of a leap by virtue of its goal: it is not causal but teleological. For Planck's discovery this is trivial, because his quantum hypothesis would never have been accepted despite its derivation if it had not afterward been justified by all of atomic physics. This may be generalized at once: physical hypotheses are not justified forward by their premises, but backward by their successes. I should like to call this the pragmatic logic of the formation of hypotheses. This justification is at least as strong as that which follows from premises: responsible for it, however, is not the logic of the conscious but that of the unconscious, what we have called protologos or protoreason. The latter does not orient itself according to premises but according to the goals of human searching.

By way of a metaphorical inverse, we may take the psychical state of affairs as model of the physical. Thus the quantum jump is not comprehensible from its causes but rather from its effects. This is nothing other than the generation of a new light quantum. Wherever creative acts are at work in the domain of the atom, the causal chain ceases to be continuous. There is an activity which lies in a plane below the causally comprehensible. Dirac was the first to recognize the existence of an electron-positron vacuum from which actual electrons and positrons can be created by supply of energy. The metaphorical inverse of this is the following: this vacuum is the physical correlate of the psychological unconscious, of protoreason. Thus the annihiliation and new creation of a moving mass point, asserted by Leibniz, ultimately refer to or converge upon the vacuum.

The passage from the domain of causal processes in classical physics to that of quantum jumps which are no longer causally comprehensible is accompanied by a loss of classical determinism, as is well known. The same is true for the passage from the sphere of what is logically comprehensible to that of the logically incomprehensible unconscious. Again, it was Leibniz[5] who recognized that in both cases classical determinism ceases: "Nothing can explain why a thing which has ceased to be in a given state begins to be in another, except a permanent substance which both destroys the former and creates the latter, because the subsequent state does not follow with necessity from the preceding" (p. 162).

Leibniz leaves no doubt about his belief that this substance is God. For him, the permanence of things represents a continuous creation.[5] He would interpret the quantum jump as annihilation of the state of a radiating atom by God and as a re-creation of the subsequent state. Similarly, for him the cessation of the psychical state of the physicist before the solution of his problem and the birth of a new idea would be the work of God.

Let us now attempt to give a depth-psychological interpretation to Leibniz's vision. There must then exist a field of the unconscious creative, c_P, which in the instance of a physical quantum jump is responsible for the cessation of the previous state and the birth of the new one, and thereby guarantees the existential identity of the radiating atom.

Similarly, there is a field of the unconscious creative, c_H, which is responsible in the case of the theorist for the cessation of the preceding state of not yet knowing and for the birth of the subsequent state of knowing, and thereby guarantees the existential identity of the physicist.

In both cases the leap is a true creation. The creative step is a leap over the abyss of nothingness into new logical territory. Nothingness is suddenly overcome. The essence of physical objects like the physical subjects is therefore creative rebirth. But this process is no longer causal but pragmatically teleological, in brief, creativistic. Causality and creativity thus stand in a relation of complementarity.

Physics presents an analogue for this also: Heisenberg's uncertainty relations. Let the mathematical code Q be a psychological "operator." Suppose this affects the "state" of the physicist in such a way that the problem at hand is given a "localization," i.e., determination of its place within certain "coordinates" in mathematical equations, experimental data, axioms, and correspondence rules. Such "localization" we shall call the "eigenvalue" of the Q-operator. This then becomes a metaphorical interpretation of the Schrödinger equation

$$Q\psi = q\psi \tag{7}$$

Here Q is the operator of a position measurement, ψ designates the quantum mechanical state of a system, and q is the measured eigenvalue of Q. Thus the theorist localizes the problem as sharply as possible within a cloud of constructs in the "space" of the c_H. Planck had to determine the entropy $S_N = k \log W$ from the number of distributions which are compatible with the energy of a system of N oscillators. In this connection, he assumed that the total energy $U_N = NU$ consists of an integral number P of energy elements ϵ, so that $U_N = P\epsilon$. The reason was that "the traditional conception of U_N as a continuous magnitude would not have admitted a combinatorial procedure for the determination of W." It was "exactly at this point that the methodological requirement for a combinatorial procedure motivated Planck's introduction of the quantum of action . . ." (Jammer,[1] pp. 19–20). Attention is to be called to the word "point": this is precisely what we shall take the eigenvalue q of the position measurement to be. Here the problem is completely circumscribed and defined. But all logical paths terminate at a limit beyond which one can not go. One encounters here a hole. The theoretician must leap over it in order to reach the connection to the further path. Planck's assumption of indivisible elements of energy followed by no means logically from the determination of W, even if one assumes the model of N radiating oscillators. For there is no way of proving that these oscillators emit or absorb only discrete energy values, $h\nu$. As in the case of sunspots, the visible lines pass into an unfathomable depth.

The logical locus of a problem, in analogy to black body radiation, can be determined only up to a minimal "plane" in the topological "space" of the C-field. The further course of thought is undetermined within this "plane." It becomes more and more indeterminate the smaller one takes the "plane." The more precisely we conceive the problem of cavity radiation the less comprehensible becomes the further path. Maximal precision of the problem and indeterminacy of the further path stand in a relation of complementarity, just like maximal position of an electron and indeterminacy of its velocity. Otherwise, Planck's contemporaries should have discovered the quantum hypothesis by analogous reflection.

But this indeterminacy of the further path can be limited afterward by joining Planck's discovery logically onto the state of the problem. Then it appears to us to be only a small step. But the uncertainty with respect to the further path is not a "horizontal" but a "vertical" uncertainty, an uncertainty which has the dimension of a depth. The clearer the problem the deeper the logical chasm which must be passed over. For regardless of the detail with which we try to reconstruct Planck's thinking, we can never find the exact logical connection between the problem and the quantum hypothesis. Here yawns the logical abyss. On this side lies the entire mountain range of traditional *a prioris*, like Hertzian dipoles of arbitrary length, the concept of energy as a fluid, and the equipartition theorem. All this had to be surmounted. The past of the theory is the mountain range that on one side surrounds the crater which must be leapt over in order that a *C*-field be reached on the other side, which is largely free of such past *a prioris*. The depth of the logical abyss is thus conditioned by the past of human thinking. Since it is the breadth and not the depth of the abyss which is important for the logical leap, protophysics is at once a depth psychology of the unknown relating to the formation of physical theories.

The "impulse" of theory formation now turns out to be the logical direction which the process of thought must take in order to reach its goal. It is driven by an *élan*, an inner "impulse." Here we encounter again the classical metaphors of ordinary speech. The theoretician seems to be guided by it during the entire course of theory formation. Planck's "impulse" was the search after an absolute, which he hoped to find in a radiating black body.*

Einstein's impulse, too, was the search for an absolute; only he believed he had found it in the principle of invariance. But these impulses themselves do not necessarily point to the correct path. Logical localization of a physical problem must always be added. In the case of Einstein, this was the interpretation of the Michelson-Morley experiment; with Planck, it was black body radiation and the law of Rayleigh-Jeans. In both instances, the direction was at first indeterminate, until Einstein found it in the definition of simultaneity, Planck in the energy quantum.

Here we encounter a kind of uncontrollable interaction between the conscious process of search and unconscious creativity. In a way, the latter is a black box in the sense of cybernetics. In contrast to the irrational decisions of human emotions, the logical path goes on at the other end of the black box, and it goes in the direction of the most reasonable path. The uncontrollable processes in the interior of the box therefore are not irrational but highly rational, only they are not conscious. For this reason we spoke of protoreason.

Turning now to the historian, he is in the same position as the physicist when he makes a position measurement and subsequently reduces

*Max Jammer(1) writes: "This problem, as he later admitted in his autobiography, had for him a special fascination, for, in view of its universal character, it represented something absolute, and he had always regarded the search for the absolute as the loftiest goal of all scientific activity" (p. 10). See Planck,(2) S. 389–390.

the wave packet: after the discovery of the quantum of action, all possible trajectories of thought are reduced to a single one, in analogy with the passage of an electron through a hole in the screen, after which it is forced to impinge upon the photoplate at a definite spot.

This shows how in the formation of theories there occurs a separation into possible and actual, latent and real, as Margenau,(6) Heisenberg,(7) and Fock(8) have proposed in their analyses of quantum mechanics.* In an analogous way, the unconscious contains only the possible values of a solution. There is no physical constraint which leads necessarily to the unique solution. Indeed, we encounter here, just as in quantum mechanics, only a probability which depends essentially on the state of the problem, the mental state of the author and the scientific climate of his world—briefly, on the theoretical "preconditions" in the "space" of the c_H.

We may therefore interpret the protologos of the unknown as the domain of the possible, the logos of the known as the domain of the real. And we note that, just as in quantum mechanics, the passage from proto-reason to reason is not uniquely predictable, not even uniquely reconstructable *ex post facto*.

Thus the theoretician operates in two worlds, a conscious one and an unconscious one. The conscious world is characterized by the actual ego as the dominating agency, the unknown world by a potential self. Self is the double of the ego. It is potential in the sense that its actions are not conscious to the ego. It demonstrates its active presence only afterward, through the reconstruction of the processes of thought, not before. At the same time, the self is more potent than the ego, for it creates new orbits of thought where the ego no longer knows where to go. One might even say that it carries the ego, like a stream. The ego is a boat upon the stream of the unknown which its pilot regards as a unique entity only because he does not see the dark depth of the stream. The self is the creative agency. It forms the source which feeds the activity of the conscious ego.

Probability, creativity, and the unconscious are related in the formation of theories in a manner similar to the relation between the square of the ψ function, the production of an actual locus of an electron by the transformation of latent observables(9) into possessed ones, and the objective uncertainty of the particle's behavior before the reduction of the wave packet.

*See pp. 175*ff*. of Margenau,(6) where the distinction between latent and possessed variables is introduced. Heisenberg(7) interprets the collapse of the wave packet as transition from the possible to the actual, as, analogously, does Fock.(8)

The Vacuum

In 1930, Dirac presented the theory of the birth of an electron by absorption of energy. A state of negative kinetic energy, conceived by Dirac as a virtual electron, is transformed from the vacuum into a state of positive energy. The process is similar to excitation of an atomic electron by a photon, causing the electron to jump from a lower to a higher energy level. The hole which remains in the vacuum manifests itself likewise as a particle, but with an opposite charge. Hence it is the positron.

We have already called attention above to the analogy of this model with theoretical activity. In this connection, the following structures have to be considered:

a. The vacuum is defined as the field of possible electrons and positrons.
b. The states of the vacuum are levels of negative kinetic energy of electrons, subject to the Pauli principle; the vacuum forms a kind of inverted hierarchy of energy levels of the actual atomic electrons.
c. The cause for the production of an electron-positron pair is the addition of radiative energy.

The physical vacuum is therefore an ontological cross between pure nothingness and observable particles. Furthermore, it is no chaos, but is well-ordered in accordance with the Pauli principle. It forms the ordered potentiality or latency of observable reality. Between the vacuum and reality there lies an intermediate zone which cannot be passed without addition of energy. Thus, for example, the actualization of a virtual electron requires the minimal energy of 1 Mev in order to bridge the difference between the highest level of the vacuum and the lowest level of the "upper world."

Theoretical creativity exhibits similar characteristics: the field of protoreason is completely occupied with hypothetically incomplete and unborn ideas.* They form a store of information for the theorist; out of it he draws his intuitions. Therefore, the word "idea" has a double significance: before the birth or the incidence of an intuition it represents a possible information but an information nevertheless existing in proto-

*In using the world "idea," we employ modern linguistic usage which makes it a scientific or technical innovation, which we have also termed "intuition." It is important to draw attention to the change in the meaning of idea from the static pictures of Plato to the creative intuition of modern science.

reason; after birth it is conscious information which not only can be realized but also can be communicated.

We now assert: between the physical vacuum of all possible unborn particles and the field of the unborn c_H there exists not only a qualitative but also a quantitative similarity. That is to say, the field of the unborn c_Hs contains all structures of physical reality, just as the physical vacuum contains all possible particles. Protoreason is therefore omniscient on the modal plane of possibility; i.e., it is in possession of all information concerning the structure of the physical universe. Only on this basis does it become clear why knowledge concerning the universe is possible. For we do not glean it from experience alone, as empiricism erroneously assumes, but from experience plus new ideas. The latter, however, we create out of ourselves.

An idea like Planck's quantum of action was not logically entailed by experiment, nor could it be derived from previous theories. Planck conceived it out of his own self. The self of the theorist thus reveals itself as a store of information concerning the structures of the universe.*

References

1. Max Jammer. *The Conceptual Development of Quantum Mechanics.* New York: McGraw-Hill (1964).
2. M. Planck. *Physikalische Abhandlungen und Vorträge.* Braunschweig: Vieweg (1958), Vol. 3, p. 125.
3. S. Müller-Markus. Einstein's philosophisches Sebstverständnis. In R. Klibanski (ed.). *Contemporary Philosophy: A Survey.* Firenze: La Nuova Italia (1968), pp. 260–276.
4. Henry Margenau. In P. Schilpp. (ed.). *Albert Einstein: Philosopher-Scientist.* Library of Living Philosophers (1949).
5. G. W. Leibniz. Pacidius an Philaletes, Grundphilosophie der Bewegung (Oct. 1676). In W. von Engelhardt (ed.). *G. S. Leibniz, Schöpferische Vernunft, Schriftenaus den Jahren 1668–1686.* Marburg: Simons (1951).
6. H. Margenau. *The Nature of Physical Reality.* New York: McGraw-Hill (1950).
7. W. Heisenberg. In W. Pauli (ed.). *Niels Bohr and the Development of Physics.* London: Pergamon Press (1955).
8. V. A. Fock. Ob interpretacii kvantovoj mechaniki. In Filosofskie problemy sovremennogo estestvoznanija. *Akad. Nauk SSSR* (1959).
9. H. Margenau. *Open Vistas.* New Haven: Yale University Press (1961). J. Park and H. Margenau. *Int. J. Theor. Phys. 1* (1968), 211.
10. H. Margenau. Science and faith. In *Integrative Principles of Modern Thought.* New York: Gordon and Breach (1972).

*A sequel to this article, presenting a further elaboration of the role played by man's creative unconscious in science (see also Margenau[10]), is planned for publication in *Main Currents in Modern Thought.*

John Bacon

A QUASI-SCIENTIFIC METHOD FOR ETHICS

All too commonly nowadays, the scientific and the humanistic spirits are felt to be embodied in "two cultures" faced off against one another. The split is widened by the advocacy of the one spirit to the exclusion of the other or *vice versa* by technocrats and counterculturists, respectively. Nothing could be farther from the position of Henry Margenau, who in his long career as philosopher and scientist has combined the values of a deeply humanistic commitment with an unabashed admiration for the capacity of natural science.

So far from claiming that science can solve everything, Margenau holds that

> the large amorphous matrix of human experience contains only a small region which is occupied by the crystalline structure we call science. . . . it seems as if the crystalline, scientific portion of knowledge, in terms of our metaphor, while growing forever, will always be small in comparison with the potentially infinite domain of man's total knowledge. [pp. 55*f.*]*

Yet he goes on to say

> Science is precious, not because of any claim it makes to universality, but because its successful function in human experience is rare, uncommon, in some respects miraculous. [p. 56]

*All quotations followed by page numbers in square brackets are from Margenau.[(1)] Page numbers in parentheses refer to other sources, as indicated.

John Bacon • Department of History and Philosophy, York College (CUNY).

Thus Margenau puts science in its place without ultimately circumscribing that place: like a crystal it may grow to encompass an ever-widening area of human concern.

For the humanist as for Margenau, ethics is a central, if not the central, human concern. It was perhaps his mediating conception of the place of science in human affairs that led Margenau to the insight that

> There is a complete parallelism between the formal structure of science and that of ethics. [p. 138]

Or, slightly tongue-in-cheek,

> I plead guilty at once to the charge of using science as a model for ethics, albeit as a formal, methodological model. [p. 142]

In this insight, Margenau is of course a true scion of the pragmatism in which he came of age as a philosopher. Indeed, he credits similar views to Ducasse and to Northrop [pp. 180*f*], and it is arguable that a consciousness of such a parallel informed the work of Lewis.(2) But as Henle had already convincingly pointed out, a parallel conception of epistemological and ethical method need not be conceded as a special prerogative to pragmatist philosophies.(3) Whether or not such a parallel can be demonstrated—and these authors go far toward making it plausible—I believe it provides the key to a metaethical theory more capable than any other of handling certain problems.

Depending on one's conception of scientific methodology, the key will yield somewhat differing conceptions of ethics. Although I share with Margenau a broadly hypotheticodeductive view of natural science, I part with him on certain details, for reasons to be made clear presently. Also, I will fill in important aspects of the ethical parallel differently. But the view I am defending supports Margenau's basic contention that a method like that of science can be fruitfully applied to ethical matters.

Scientific Method

Fundamental to Margenau's epistemology is a distinction between the *P*- (protocol, primary, perceptual) components and the *C*- (constructual, conceptual) components of cognitive experience. *Cognitive* experience includes not only observation but also abstraction, reasoning, inference, and induction [p. 11] and excludes "purely affective moods, appreciative or depreciative attitudes, decisions and actions" [p. 10]. Its *P*-components are those which force themselves upon us regardless of any choice or synthesis on our part; they are "immediate" [p. 48], "coercive, indubitable, or simply given" [p. 45]. The *C*-components, on the other hand, "are of our own making" [p. 14]. The distinction is not dichotomous, however, but gradual. The "*P*-plane," as Margenau sometimes speaks, is the lower extreme of a range or scale, the rest of which constitutes the "*C*-field."

It is not immediately clear whether *P*- and *C*-components are individuals, properties, or propositions; presumably they may include all three. Thus among *P*-components Margenau mentions sense data [p. 13], subjective force [pp. 25, 31], "external observations, and the results of experimentation" [p. 45], and "*P*-facts" [e.g., p. 33]. As *C*-components he mentions among others decay constants, atomic numbers, "concepts like elementary particle, a universal charge *e*, nuclear forces, . . ." [p. 35], "axiomatic propositions" [p. 38], and verifacts. "Concepts" or "constructs" apparently include properties as well as propositions or facts. Presumably the latter are meant when Margenau speaks of confirmation or verification of concepts.

The *P-C* distinction is not linguistic [pp. 23*f*.]. We might nevertheless expect that it would permit a derivative distinction of something like observational language from theoretical language. This is not in the spirit of Margenau, however; for him, general empirical laws already pertain to the *C*-field.* Indeed, Margenau even seems to suggest that *all* language is constructual, including reports of *P*-experiences:

> With respect to the facts themselves there is no truth, there is presence or absence, existence or non-existence. To be sure, a *report* of facts may be true or false, and this is precisely because the report transcends the facts, associates them with words or symbols which are constructs linked to the *P*-counterparts by rules of correspondence. [p. 126]

If, as the context seems to imply, the "facts" referred to here are *P*-facts, this would mean that the *P*-domain of experience is strictly speaking ineffable. No sentence could be a protocol sentence. Yet, according to Margenau,

> most ineffable experiences, especially when they recur, clamor for symbolic formulation, usually for expression in language. And this drive normally succeeds . . . . [p. 24]

---

*Actually, Margenau could be said to distinguish two kinds or, more precisely, two status of general laws corresponding to the two kinds of induction he recognizes [pp. 41*ff*]. In both status the laws are conceptual, but as mere generalizations they are "near *P*, . . . avoiding any significant reference to the *C*-field" [p. 41], while for laws as explanatory "the whole connective tissue in the *C*-field extending to the axioms must be filled in" [p. 42].

Thus *P*-experiences can be effed, but only in at least inchoate *C*-language. Perhaps, then, we may conceive of "observational language" as that part of language which comes "closest" to treating exclusively of *P*-experiences.

The *P*- and the *C*-fields are connected for Margenau by rules of (epistemic) correspondence and principles of verification The correspondence rules enable the scientist to confirm or verify his *C*-propositions (to a certain degree) by deducing *P*-consequences from them and the correspondence rules and checking the result against his *P*-experiences. *C*-propositions with a sufficiently high degree of confirmation Margenau calls "verifacts." In deciding whether the *P*-experiences agree, the scientist also has recourse to *principles of verification.* These "are deliberate, arbitrary, and conventional" [p. 51] stipulations of a range of tolerance or probable observational error; e.g., "An observed clock-reading of $t \pm 10$ ms is to count as a confirming instance of a clock-reading of *t*."* The linguistic difference between rules of correspondence and principles of verification, then, would be that the former contain both *C*- and near-*P*-terms, while the latter are couched in exclusively observational or near-*P*-language (or metalanguage?).

As already mentioned, laws are for Margenau already up into the *C*-field [p. 37]. The rules of correspondence relating such low-level constructs to *P*-experiences include so-called operational definitions. General laws may then in turn be connected by further correspondence rules† to more abstract propositions, and so on up. There may be several such levels, but at the farthest *C*-reach are found terminal constructs, the axioms or postulates constituting a theory [p. 36]. As science progresses, these postulates may themselves be subsumed under a still more abstract or general theory.

Thus Margenau, like most contemporary philosophers of science, sees science as a "hypotheticodeductive" structure, and in this I agree

*As a more general principle of verification, Margenau cites "science intends to explain, or foresee, factual or datal happenings" [p. 172]. However, that would seem to be a methodological statement about science rather than a scientific statement.

†My interpretation; I nowhere find Margenau referring to these higher-level connections as epistemic correspondences. He says that concepts in the abstract region of the *C*-field are "endowed with logical relations and so chosen that the concepts of lower order are entailed by them" [p. 35]. Diagram 2 (p. 353) in Margenau(4) would suggest that the connections in question are not epistemic correspondences but "logico-mathematical relations, postulated for, or implied by, the concepts" (p. 349). If so, then to the extent that they interrelate levels, such relations would presumably have the structure of bridge laws and hence a logical form similar to that of correspondence rules.

with him. But, following Sellers[5] (pp. 106–126), I would dissent from Margenau's multilevel picture of science and knowledge. To be more specific, let me refer to the relation of a *C*-level to the next lower or less abstract level in Margenau's methodology as "subsumption." Subsumption thus includes roughly the same relations that have at one time or another been referred to as "explanation." It seems to me that subsumption thus conceived conflates five distinct relations: (a) meaning, (b) the relation of conceptualized observables or physical objects to pure *P* or sense-data, (c) the relation of general empirical laws to particular observations, (d) the relation of theory to empirical laws, and (e) the relation of reducing to reduced theory.

Relations (c)–(e) can be construed either as relations between sets of concepts and propositions or as relations between sets of predicates and sentences. The meaning-relation (a), by contrast, admits of neither construction but rather relates linguistic expressions to concepts and propositions. The relation is laid down not by rules of correspondence but by semantic rules.

Relation (b) is not parallel to (c)–(e) either, since, as we have already noted, the elements in its converse domain are not strictly effable. A would-be correspondence rule establishing an equivalence between an observational concept and a pure *P*-quality would have to use the same predicate on both sides of the equivalence, there being no purely *P* predicates. If such a rule cannot be stated, we may doubt whether it can be thought, either. In any case, I have doubts about the usefulness of the *P*-plane for scientific methodology, or for epistemology either, for that matter. For my present purposes (and, it seems to me, for Margenau's as well), it is sufficient to start at the lowest *C*-level, which I will call the observational framework. Indeed, there is some indication that this is what Margenau has in mind when he refers to *P*. Thus he speaks of "*P*-facts or, more precisely, concepts near *P* . . ." [p. 41]; and he equates *P*-experiences approximately with Northrop's "concepts by intuition" [p. 14]. Elsewhere he includes external objects in *P*-experience, even though he sees them strictly as constructs out of immediate sensations (Margenau,[4] p. 348).

The relation of general empirical laws to observations (c) differs from (d) and (e) in requiring no correspondence rules. For, with the possible exception of modal words, general laws are formulated in the same vocabulary as observation statements.

Finally, I admit that there may be a resemblance between the relation of reducing to reduced theory (e) and the relation of theory to general laws (d) (or to the observational framework generally). The "bridge laws" needed for the former seem to have the same structure as at least some of the correspondence rules used for the latter. The difference is that correspondence rules contain both theoretical and observational terms, while bridge laws contain terms from two different theories. To be sure, a complete parallelism would be restored if we conceded that the observational framework is *itself* just another theory. But that would raise again the question of what the next lower level was that the observational framework subsumed or explained, and we should be back on the *P*-plane with sense impressions or what not (*cf.* Sellars,[5] pp. 119, 124). More seriously, such a concession would obscure the fact that a reducing theory, like the theory reduced, already has correspondence rules of its own connecting it to observables.* Otherwise it wouldn't have been a theory in the first place. The bridge laws linking it to the other theory are then added. Once the reduction has succeeded, of course, some tidying up may be in order, but that is another story.

Upon taking account of these distinctions, we are left with a two- (or perhaps three-)† level picture of science in place of Margenau's potentially multilevel conception. Yet when all is said and done, there remains a rough analogy between Margenau's *C-P* distinction and the theoretical-observational distinction. Two such levels are all that are needed for his ethical analogue anyhow.

Extension to Ethics

In his ethical methodology, Margenau makes no use of the possibility of intervening *C*-levels between the axiomatic *C*-extreme and the *P*-plane. Corresponding to the postulates of a scientific theory are the *commands* of an ethical code. The ethical *P*-experiences "are observed or historically established human actions" [p. 165] taken for the most part in a collective or social context.

Right here we notice a contrast to the scientific situation. In science, the *P*- (or near-*P*-) statements are *deducible* from the postulates of a well-confirmed (two-level) theory together with correspondence rules and statements of initial conditions. In Margenau's ethics, on the other hand,

*Margenau himself mentions this fact in connection with the reduction of thermodynamics to the kinetic theory of matter [p. 38].

†Depending on whether empirical laws are reckoned as a separate level. Against this view, compare Sellars,[5] (p. 121).

statements of observable human actions are *not* deducible from the commands plus initial conditions. For example, from the general command

No person shall kill

and the initial condition

Jack is a person

we cannot deduce the observation statement

Jack doesn't kill,

but at best the specific precept

Jack, don't kill!

The indicative does not in general follow from the imperative: commands are not always obeyed. Thus for Margenau ethics, unlike science, is not a hypothetico*deductive* structure, but at best a hypotheticocasuistic one.

It may seem that we have forgotten correspondence rules. That is another difference between Margenau's pictures of science and of ethics. He has no analogue of correspondence rules connecting the commands to the *P*-actions. It is fairly clear why not: aside from the imperative neustic (or deontic operators), the commands are already formulated in the same terms as the *P*-statements. In this respect, commands are more like general laws than theoretical postulates.

If actions do not follow from commands, and commands cannot be induced from actions, how can observable actions ever confirm an ethical code? It is here that the *principles of verification*, which in science play a rather marginal role, come into their own. Confirmation in ethics Margenau calls validation; the corresponding principles are accordingly *principles of validation*. They are arbitrary selections of certain possible "natural qualities" of collective human behavior as "primary values" [pp. 166, 171, 175]. An ethical code is validated if its espousal by a social group results in the realization of the qualities laid down by the principles of validation. As examples of eligible qualities, Margenau lists "cultural longevity, . . . , freedom, health, wealth, self-fulfillment, altruistic love, education, refinement, self-denial, austerity, . . . , saintliness" [p. 166], "power over others, personal happiness, collective happiness, . . . , the peace that passeth understanding," "self-survival" [p. 173], "benevolence of fellow men" [p. 175], "group survival" [p. 178], and "the ineffable nirvana" [p. 139]. Margenau summarizes the process of validation as follows:

> The commands engender specific precepts by casuistic explication; the precepts are lived by a group of men, producing determinate actions. In these actions, certain qualities, often called primary values, appear. Some of these are chosen as validating principles. If the initial specific commands produce the primary values chosen, the entire living complex is a proper, a valid, ethical system . . . . [pp. 139, 175]

At this point the parallel to science has become tenuous indeed. The gap between a specific precept or the corresponding action (if carried out) and, say, collective happiness is of an altogether different order from the gap between a predicted clock-reading of 9:07 and an observed reading of 9:07:00.006 [*cf.* p. 49] mediated by a principle of verification. What's more, the qualities expressed in principles of validation, unlike those involved in principles of verification, require *correspondence rules* to relate them to the observational plane [p. 166]. Thus correspondence rules occupy an entirely different position in ethical method than in scientific method. This architectonic inelegance is not a conclusive objection, however, for it was never established that a complete parallel had to obtain. I merely asserted that it furnished a useful key for construing ethical method.

It seems clear that *something* important is confirmed by essentially the procedure Margenau delineates as validation, but I question that it is an ethical system. What is confirmed is precisely that, if followed, the "commands produce the primary values chosen." In other words, the commands are subject to the hypothesis that the primary values are desirable: they are *hypothetical imperatives*. As such they are equivalent to indicative causal laws of sociology and psychology, devoid of essential ethical import. That does not detract from their importance to us in making ethical judgments, in particular judgments of entire social systems. The *ethical* question, meanwhile, reverts to the desirability of the primary values. Here Margenau's only advice is "Choose!" It is hard to exonerate him from the charge of Blanshard's[(6)] against Hare that Margenau quotes [p. 174]: "What we have here is a free use of reason on the lower slopes with a surrender to unreasoned decision at the summit."

Validation in Margenau's sense is thus not a sufficient condition for the categorical validity of an ethical code. Nor is it a necessary condition. It is essential for Margenau's procedure to get off the ground that the commands be followed, at least a good part of the time. Yet we might defend the validity of an ethical code even if no one followed it.

Margenau is nevertheless optimistic:

> let us merely note the overwhelming unanimity that exists among people in all parts of the world with respect to the validating principles of ethics. I think

> humanity could agree without prompting that happiness, benevolence of fellow men, and peace are among them . . . . [p. 175]

The consequences of Nazism, he says, "met no conceivable standard of validation" [p. 178]. The qualities eligible as primary values thus seem to be limited to those a human group could in fact desire; they are "vaguely connected with 'human nature' " [p. 199]. Unfortunately, "human nature" is more protean than Margenau here envisions. Societies that have valued war above peace are depressingly familiar from history. The Dobu Islanders find benevolence ludicrous; paranoia and treachery are their primary values.[7] If there is nevertheless substantial unanimity among men, especially those of similar cultural background, on certain basic values, it is the task of ethics to *explain* those values, just as it is the task of science to explain observed facts.

An Alternative

A more satisfactory account of ethics is to be found, I believe, by keeping closer to the scientific analogy. From here on I shall assume a scientific methodology like that developed on pages 159*f* above. The corresponding ethical methodology I will defend is similar in its essentials to that of Ducasse, although I will not follow him in his applications of the methodology*:

> the primitive, basic facts of ethics consist of the spontaneous appraisals which a person or group of persons *P* makes of various concrete acts.[8]

> just as the function of theory in the natural sciences is to explain some given set of facts and to predict additional ones not yet observed, so the function of an ethical theory is to explain the spontaneous appraisals which it was devised to explain, and to predict additional ones that had not been considered in the tailoring of the theory.[8]

By "appraisal" here is meant not the act of appraising but its content. Appraisals are thus the ethical counterpart of observations. I would extend Ducasse's conception to include the possibility that things of other sorts besides concrete acts might be spontaneously appraised. Thus appraisal statements might contain axiological as well as deontic (or imperative) terms.

It will be objected that appraisals are simply not "given" to us in the way observations are. In fact, I doubt that either are given in any immediate sense, although it is not my purpose here to enter upon episte-

**Cf.* also Ducasse.[9]

mological questions of how perceptions and appraisals arise. It is clear in any case that both are manifest in our day-to-day experience. Both may be confident or doubtful, both may be private or shared. In ethics as in science, only those that are confident and in principle sharable provide a reliable basis for confirmation and action. Nothing more need be "given."

Reliable spontaneous appraisals are generalized in *moral laws*. Corresponding to, say, "Other things being equal, salt dissolves in water" we have "Other things being equal, thou shalt not kill" or "Other things being equal, compassion is better than indifference." The "Other things being equal" clause is as essential to informal statements of general moral laws as it is to general natural laws. This is the kernel of truth in the point so insisted upon by the situationists in their attack on what they call "legalism." Moral laws do certainly admit of exceptions in application to particular situations. But in this they are logically *no different* from physical laws. Explicitly to expound all the necessary antecedent conditions lazily abbreviated by the "Other things being equal" clause takes some doing even for "Salt dissolves in water." Or try enumerating all the necessary antecedent conditions for a lamp's lighting when you throw the switch. It is not clear that the list of qualifications is any longer than for "Thou shalt not kill." If the situationists take umbrage at such nit-picking, they ought to object just as strenuously to the recognition of hard and fast general laws in science. Their methodological assumption that no general rule, however precisified, can be free from exceptions is just the opposite of the methodological assumption that has served science so well.

The accepted or tested corpus of general moral laws is in turn subsumed under and thus explained by a (normative) *ethical theory*. Ideally, such a theory, together with correspondence rules and initial data, ought to permit the deduction of all particular appraisal statements. On this methodology, unlike Margenau's, ethics is thus genuinely hypothetico-deductive. If the deduced appraisals pertain to a future occasion, they are the ethical counterpart of predictions, *viz*. "prescriptions." Like predictions, prescriptions can be of the utmost practical relevance. Like predictions, they often go unheeded. But whereas predictions tell us with a certain degree of probability *what will be*, prescriptions may tell us (again with but a certain degree of probability) *what to do* or *how good* something will be. The reliability of the prescriptions, like that of the predictions, is of course contingent on the confirmation of the theory, something that is perhaps never fully achieved. Another possible explanation of the situationists' antilegalism may be that they have hypostatized this *epistemic* limitation as an intrinsic limitation on ethical matters.

I taxed Margenau with failing to provide a place for correspondence rules in ethics similar to the one they occupy in science. At first glance, the view just proposed would seem to fare no better. For if we look at ethical theories we find a comparative continuity of theoretical and appraisal terminology. A need for correspondence rules is not immediately apparent. Although *the good* of Plato is of a different order from particular goods, the same term was nonetheless used for both. Where is the ethical counterpart of "neutrino" or "quark"? This circumstance, I submit, is symptomatic of the rudimentary state of development of ethical theory. It is instructive to compare the situation in less developed observational sciences, e.g., sociology. While the "working class" is not simply a class of working people, much less a class that works, the same term is still used to characterize both. The relation of "ideal types" to particular exemplars is much nearer to that of the good to goods than it is to the relation of fundamental particles to billiard balls. It is not surprising that with the rise of particle theory ethical theorists thought of such things as "hedons," which sounds more like "mesons" than it does like pleasure. But the attempt at quantification was premature. It is rather as if the superior vitality of animals over plants had been "explained" as an excess proportion of "bions" to mass before anyone got around to observing cells.

By the same token, we find no sharp split between a theoretical and a manifest image of ethical man. There is no demand for a "stereoscopic" vision as yet simply because the commonsense monocle is the well-nigh universally accepted way of looking at values. Intimations of a gap can nevertheless be discerned. Kant was consistent enough to deduce from his theory that, contrary to common sense, one should not lie to divert a murderer.[10] Even a little conscientious examination of sexual morality suggests the acute need for a departure from common sense. The protracted convulsions of this country over civil rights testify to the bankruptcy of the manifest image of liberty, whether in its reactionary or its liberal guise.*

One contemporary type of "ethical theory" that I would not count as such is noncognitivism, e.g., emotivism. There is no point in constructing a theory if your appraisal terms are meaningless. Perhaps, however, emotivism is best construed as a proposed informal metaethical semantics or pragmatics (asemantics?) for moral language. As such it is unillumi-

*Cf. the alternative image of the South African Lovedu, in Lee,[11] (pp. 48–52).

nating and needlessly newfangled. Why, for example, say that certain kinds of sentences "express" rather than "assert"? To elucidate assertion is job enough for pragmatics. Once that is done, it is not at all clear that the result might not extend to "expressive" or "motivative" discourse as well.

Stripped of its methatheoretic trappings, however, emotivism may still be appreciated as calling attention in a confused way to the primacy of spontaneous appraisals, as we might interpret the "feelings" or "emotions" it finds expressed. This was certainly a salutary antidote to the rationalistic tone of intuitionism.

Viewed in light of the proposed methodology, a number of ethical theories proper, as well as situationism and emotivism, fall into place. Intuitionism, e.g., was right in positing a faculty of moral intuition. But the faculty should be conceived not along the lines of reason or understanding (in Kant's terms) but of empirical intuition—not as *Einleuchten* but as *Anschauung*. In mistaking appraising for some sort of reason, the intuitionists lost sight of the need to go beyond what is intuited and form hypotheses. Intuition does not reveal to us, say, the practical imperative (unless, as might be argued, it is analytic) or the goodness of pleasure, but only the deontic and value qualities in particular acts, person-slices, things, or events.

The various forms of naturalism may be seen as theoretical attempts to relate appraisal to observation, or ethical to physical theory. That is a natural thing to try to do in view of the necessity for factual premises in ethical inferences. Indeed, our view of ethical theory leaves open the possibility that it might one day be subsumed under or unified with physical theory. The objection that a "naturalistic fallacy" is thereby committed appears to be a methodological dogma of the worst sort (the sort that would block the road to inquiry). While it is undeniable that ethical terms are not *now defined* naturalistically for most speakers, the possibility remains that observations and valuations might be best systematized by the recognition of *extensional* (not necessarily truth-functional) *coimplications* between ethical and factual terms. And we must accustom ourselves to the possibility that yesterday's novel extensional coimplication may upon revolution become tomorrow's conceptual identification. As no subsumption is as yet visible on the horizon, I have introduced separate methodological terms to characterize ethics for the time being. Before we can contemplate a "unified field theory" embracing ethics, I think we must work at developing the field of ethics itself.

Ethical theories are often taken to task for not accounting for moral disagreement. What seems to have been overlooked is the great extent of *agreement* among men on practical questions, particularly but not only when they share a common cultural background. From the prevailing viewpoint of metaethics, this can only be accounted for in one of two ways. Either the great mass of men, most ethicists included, implicitly share a common unacknowledged normative ethical theory (socially instilled perhaps). Or else they go about their practical affairs "by the seat of their pants," voluntaristically or situationally or whatever, i.e., at bottom irrationally. Thus we should be driven either to anti-intellectualism (thinkers fool themselves) or misanthropy (the herd doesn't think).

On the view here advanced, practical agreement is understandable. Nontheorists don't perhaps theorize very much, but they do appraise. Theorists may hold widely differing theories but the touchstone again is practical appraisal, and they're all in that together.

Limits of the Parallel

One plausible way of contrasting the factual with the ethical is Sellars' distinction of "language-entry" and "language-departure" moves in language games.[(12)] Upon observing, say, a red object, I make the language-entry move of asserting "This is red." Having deliberated *in foro interno* or elsewhere about the considerations I deem relevant, I then perhaps arrive at the resolution "May this be black," whereupon I make the language-departure transition to painting the object black. Thus an observational sentence would *follow* an observation, whereas an appraisal sentence would precede an action. This evident asymmetry would seem to cast doubt on the parallelism claimed here.

But let us look in more detail at what goes on when we are in the intralinguistic part of the game. We move from assertion to assertion on the basis of deductive, inductive, and empirical laws, as well as perhaps wishful thinking, rationalizations, and other relevant or extraneous considerations. Typically, both factual and moral assertions will appear in this chain of deliberation. Both may come by inference, by spontaneous positing, or by language-entry moves. Thus far there is no asymmetry.

On the other hand, a language-departure move *does* presuppose a concluding optative sentence. The apparent exceptions are actually cases of dropping out of the game and plunging into spontaneous action. Whether subconsciously motivated behavior is thus impulsive or whether it is rather a language-exit move out of a subconscious language game *in eo* (in the *id*) is an interesting question that need not be settled here. With regard to language-departure moves, then, there is an asymmetry between factual and optative sentences. The latter, however, form a wider class than moral statements. For we all recognize the possibility that we might resolve to do something we considered wrong.

Another *prima facie* contrast between observation and appraisal is that observational terms might possibly be hoped to be extensional, whereas appraisal terms seem typically to be intensional (to form referentially opaque contexts). If intensional terms are needed at all in natural science, it would seem to be at the nomological or theoretical rather than the observational level. If we turn to the social sciences, however, the contrast evaporates, for it is a fact that the observation statements of sociology and psychology include intensional verbs. Any attempt to explain them away, even if it could get by without dispositional (and hence intensional) terms, would already be on the theoretical level.

Finally, it might be objected that morals are culturally relative, while science is at least potentially and ideally transcultural. That would suggest that appraisal is a product of cultural conditioning while science deals with what is. The contrast is largely apparent, however. People of widely divergent cultural backgrounds not only appraise differently and value different things, they also *see* differently and different things. We may believe that our way of looking and listening is better or more useful than, say, that of an Oglala Sioux.* Witness our television sets and spaceships, which were not invented by Sioux. Such a belief is certainly ethnocentric. It begs the question of what is good for an individual or a people.

As we know, the Sioux don't hear differently because of any difference in their auditory faculty. By the same token, we don't appraise differently because of any difference in our "moral faculty." The differences are occasioned rather by the contrasting rudimentary physical and ethical theories constituting our respective "common" sense. Natural scientists from contrasting cultures can learn to perform similar observations because they have to some extent superimposed a common (and more highly confirmed) theory on their disparate commonsensical conceptions of nature. It is of interest to note that they also show signs of being capable of adopting a common ethical outlook.

A hypotheticodeductive approach to ethics would for the first time put it on a self-consciously rational footing—"rational" in the sense not

Cf. Luther Standing Bear,[13] quoted in Lee[11] (p. 55).

of *a priori* reason but of empirically probabilistically confirmable rational belief. The reasonable direction for mankind to take is neither the insular insistence on provincial moralities nor the obliteration of diverse forms of moral sensitivity in favor of a globally received code. What is rather needed is cooperation to construct and confirm a normative ethical theory that would explain and prescribe in accordance with the tenable practical appraisals of all cultures. Thus the movement for ethical unity would embrace both theoretical controversy and practical pluralism.

References

1. H. Margenau. *Ethics and Science*. Princeton, N.J.: Van Nostrand (1964).
2. C. I. Lewis. *An Analysis of Knowledge and Valuation*. LaSalle, Ill.: Open Court (1946).
3. Paul Henle. Method in ethics. In W. Sellars and J. Hospers (eds.). *Readings in Ethical Theory*. New York: Appleton-Century-Crofts (1970), 2nd ed., pp. 188–189. [Not in 1st ed.].
4. H. Margenau. Is the mathematical explanation of physical data unique? In E. Nagel, P. Suppes, and A. Tarski (eds.). *Logic, Methodology, Philosophy of Science: Proceedings of the 1960 International Congress*. Stanford, Calif.: Stanford University Press (1962), pp. 348–355.
5. Wilfrid Sellars. The language of theories. In *Science, Perception and Reality*. New York: Humanities Press (1963), pp. 106–126.
6. Brand Blanshard. *Reason and Goodness*. London: Allen and Unwin (1961), p. 254.
7. R. F. Fortune. *Sorcerers of Dobu*. New York: E. P. Dutton (1932).
8. C. J. Ducasse. Scientific method in ethics. *Philosophy and Phenomenological Research 14* (1953), 72.
9. C. J. Ducasse. *Philosophy as a Science*. New York: Piest (1941).
10. I. Kant. Über ein vermeintes Recht, aus Menschenliebe zu lügen. *Berliner Blätter* (1797), pp. 301–314.
11. Dorothy Lee. *Freedom and Culture*. Englewood Cliffs, N.J.: Prentice-Hall (1959).
12. Wilfrid Sellars. Some reflections on language games. In *Science, Perception and Reality*. New York: Humanities Press (1963), pp. 327*ff*.
13. L. Standing Bear. *My People the Sioux*. Boston: Houghton Mifflin (1928).

11

Lee Thayer

THE FUNCTIONS OF INCOMPETENCE*

I have long been a fan of Henry Margenau, not only because of what he has said, but also because of what he is. In ways that cannot easily be recounted, Henry Margenau has done much for all of us who want to understand science without disunderstanding man. It was in the late 1940s that Margenau first discussed his "theory of latent variables"—later to become "epistemological feedback." He was, I believe, pointing toward what had to be reckoned with if there ever was to be a science of man. Schrödinger's own great contributions were developed in part from these early exchanges with Henry Margenau.

In what follows, I had no expectation at all of resolving, neatly or otherwise, the fundamental difficulties which Margenau and Schrödinger and others have pointed to, and which still plague the social and behavioral sciences. What I did propose to do was to point to some social functions of but one of Margenau's possible "latent variables"—in this case, incompetence. At the time I prepared this, I didn't know if Henry Margenau would either approve of or agree with what I propose. That was but another of those "latent variables" which comprise human life.

The conclusions I reach regarding a possible science of man are not likely to be those which Margenau would himself set forth. But it is gratifying to think that they may be in the spirit of that which Henry Margenau stands for.

*Originally given as a Henry Margenau Lecture at Hartwick College, New York, November 11, 1971.

Lee Thayer • Simon Fraser University.

> Alas! What are we doing all through life, both as a necessity and as a duty, but unlearning the world's poetry, and attaining to its prose! (Cardinal Newman, *Idea of a University Defined*, 1873)

Man is, seemingly by nature, a maker of paradoxes. To know a thing, he must name it. To name it, he must distinguish it from other things. In distinguishing it from other things, he makes it into something it is not.

Imagine, if you will, standing high on a bluff overlooking a beach. Focus, if you will, on that meandering line along which sand and water meet. Are you looking at the shoreline or at the water's edge?

We know, of course, that there could not be the one without the other. Yet when we look, we see the one, or the other. Which one we see depends on who we are and what difference it makes to us. If thoroughgoing landlubbers, we are likely to see the shoreline. To a snorkeling buff, the point at which land and water meet will likely be the water's edge.

The difficulty is not so much that we make these paradoxes when we name the things we want to look at and talk about. It is that we want to believe that these verbal creations of ours serve always and only to enlighten us. We want to forget that the ways in which we make things over by our words inevitably obscure at least as much as they reveal.

The difficulty is that once we have a satisfactory way of talking about a thing, we cease looking at it. Alfred North Whitehead once suggested that "it requires a very unusual mind to undertake the analysis of the obvious." Said Publilius Syrus in 50 B.C.: "Familiarity breeds contempt." I think that if he were alive today, he would change that to read "Familiarity breeds *content*."

The difficulty is that there may come a time when our familiar and customary ways of seeing a thing may no longer be adequate. Every unhappy marriage is testimony to what happens when one's persistence in seeing things as they *were* runs afoul of things as they *are*. Every modern war has been based on just such misperceptions.

It is this paradox and its implications which give origin and cause to my proposition that mass incompetence is more functional than individual competence in the United States today.

I

My argument is simple enough—perhaps deceptively so. In its briefest form, it is this:

It seems obvious that the evolution of any complex society requires the development of ever more specialized competencies. It seems perhaps less obvious, but no less true, that the machinery of any richly complex society requires that one man's special competency be complemented by the *in*competencies of one or hundreds or thousands of other men. What is not at all obvious is that there comes inevitably that point at which incompetence surpasses competence as the real basis of any further social evolution.

Thus there comes that point at which we need to stop looking at the shoreline in order to see the water's edge—in order to see how the functions of incompetence came to exceed in importance the functions of competence in the workings of our present society. It is time we stopped treating the jockey's stiff finger for the horse's broken leg. It is time to look again, not so much at what we have done as at what we have failed to see.

What I hope to do in what follows is, first, to make evident that this is so; second, to offer some explanation as to why we are faced with the left-handed end of this paradox; and, third, to speculate on what it may mean for the future of man and of men.

II

My point of departure was that

the continued evolution of any complex society requires the emergence and accommodation of more and more specialized competencies.

What could be more obvious? Any human society which does not continuously produce new specializations will neither be evolving nor increasing its complexity. It will stagnate.

Any society in which the alternatives for its children are no greater than the alternatives that existed for their parents is closed, unchanging. In primitive societies, there were typically but four or five clearly distinguishable roles: food-gatherer/warrior, family-bearer, dominant leader, and healer. Most of the rest were irrelevant, indistinguishable parasites. In the last century, there were hundreds of occupations and professions in the United States. Today, there are thousands. That's the way complexity evolves.

But the plot begins to thicken at the very outset. What we largely fail to see is that most of this apparent growth is merely internal elaboration.

A New World explorer and an astronaut are novel additions to the list of specialized competencies. They add a competency that didn't exist before. But most of the increase in complexity of modern societies comes not from new and novel competencies but from dividing up an existing competency into parts. The great Industrial Revolution—which should perhaps have been called the Industrial Involution—championed once and for all the notion that the best way to get things done fast, efficiently, and reliably was to divide one man's skills into logical or expedient parts.* The shoreline view is that this notion worked so admirably well because one man could be more predictable and interchangeable at a piece of a task than at the whole. The water's edge view is different: it is that what a man once did for himself now *had* to be done by two or six or scores of men.

The issue, therefore, from this water's edge view is not that our human competencies were actually increased but that we achieved a new power (in this case industrial) by radically impoverishing one man's competence, thereby giving usefulness to scores of other men. Veblen saw this clearly. But what most men saw was the efficiency, not the "trained incapacity" on which that efficiency depended.

III

This brings me to my second step. It was that

> *the machinery of any richly complex society requires that one man's competency be complemented by the incompetencies of one or hundreds or thousands of other men.*

If that seems not so obvious, consider this: If two men wanted to divide the work involved in hunting down a bear or building a barn, it is likely that each would be familiar and competent enough at the other's task to be able to substitute for the other. But what happens when there are three or ten or a hundred? While it might be possible for one man to be competent enough to build an automobile *in his own way*, if the

*It seems unnecessary to assign any special role to "technology" in this regard. When tools and techniques are used to augment a man's competence, they are enriching. When they are employed merely as substitutes for one or more of his competencies, then they have the effect of impoverishing that man's competencies. If the man-machine system is not qualitatively novel, then the man is simply being incapacitated in order to give utility to the machine.

product of such an enterprise is to be predictable (efficient, standardized, etc.) then each man's particular competencies depend upon every other man's "trained incompetencies" and *vice versa.* How many of the necessary specialized competencies can any one man provide as the enterprise gets more and more complex? Up to a point, a man can perform several different tasks. But as each task becomes more sophisticated and refined, and as the number of complementary tasks grows and grows, his utility changes. The man of rangy competence is obsolete. The economy can no longer support a man who can do several things well. It can only support men who do one thing predictably, interchangeably. To be predictable, interchangeable, to be standardized, men must, as humans, be incapacitated.

The traditional view of this whole process—and let's take this as the shoreline view—is that it permits us to accomplish what would otherwise be impossible. One could hardly argue the fantastic power of the idea and the even more fantastic accomplishments it has brought us. Neither heart surgery nor space flight nor the working parts of our kitchens nor our daily economy would be possible without specialization. Our affluence no less than the scope of our consciousness depends on it.

But there is the water's edge view as well. From this perspective, we can see some conditions and consequences which are not altogether so marvelous.

First, we can see that once a particular pattern of specialization is set, then future generations must be fitted into it. While we must try to endow future generations with certain kinds of skills and competencies that were *not* required by their forefathers, we must also endeavor to make them selectively incompetent and selectively ignorant. This is as simple a matter as it is certain. People do what they do first because they *can*, and then because they *must*—i.e., because there is some personal or social necessity for doing so. You will never observe a man doing what he cannot do. The structure and machinery of any society depend not only on enabling men to do what is humanly or socially necessary but also on disabling them from being able to do what is, for that economy or that society, neither necessary nor useful. Overcompetence and excess competence can become greater burdens to a society or social institution than is undercompetence or scarcity of competence. Specialization alters society, and society alters man.

Second, from this water's edge perspective we can hardly fail to see that breaking large competencies down into smaller and smaller pieces means, eventually, not only that it will be economically unfeasible for one man ever again to accomplish the whole task. It means as well that it will be less and less likely that any one man will ever again be competent to accomplish the whole task. It means as well that one man's capacities, never developed, never tested, will never develop.* "When you care enough to send [say?] the very best . . . ," you send a mass-produced greeting card. Here's a billion-dollar industry based on the fact that when we have to choose which to send to say the "very best"—ourselves or a canned slogan—we no longer have to hesitate. For 25 cents we can buy a better sentiment than we can have. We know quality when we see it.

Recall the famous East Coast blackout of 1965. Incredible as it seemed at the time, there was apparently no one person anywhere who knew exactly how the whole system was put together. There may be a dozen or more highly sophisticated competencies represented in a complex operating room procedure, and the power of this sort of specialization has made possible hitherto undreamt-of surgical achievements, such as heart transplants. But now the patient's life hangs in balance not only on how competent these other people are to perform their delicate specialties but also on how reliably those delicate specialties work together at crucial moments. Not only must the patient be incompetent to repair himself, but the surgeon must be less competent than the anesthetist at *his* job, the anesthetist less competent than the surgical nurse at *her* job, and so on. A surgeon who knows how to operate only under those sophisticated conditions wouldn't be much help to one who had a ruptured appendix on a hunting expedition. This is one reason why we are learning, in our ever more specialized role as patients, to be sick only within reach of equipped hospitals, and only during the day, when the senior staff members are aboard.

In our complex and affluent society, it is clear that one man's competency must be complemented by the incompetencies of one or hundreds or thousands of other men. Any specialization requires social utility. For one man to be uniquely competent to do something, other men in the same social system must be relatively incompetent to do the same thing. The one defines the other.

The functions of competence seem obvious enough. What is not so obvious are the functions of incompetence.

*Even domesticated (civilized?) mice have a markedly limited capacity for learning compared to their wild counterparts. Wild mice explore an unfamiliar environment in order to "get on top of the situation." Domesticated mice are apathetic. See, e.g., Lee Kavanu[1] and *cf.* Klausner.[2]

IV

This brings me to the heart of my thesis, which is that

there is a point at which incompetence surpasses competence as the real basis of any further evolution of a complex, affluent society such as ours.

There comes that point at which the functions of incompetence exceed in importance the functions of competence in the further growth and development of any society as complex and as affluent as our own. We have passed that point, and it is time we began to see what it means to us today, and for the future.

To see what has happened we need to neutralize the terms we want to use: in this case, *competence* and *incompetence*. By *competence*, I want to mean only that it would be possible to measure the capacity or the ability that people have, relative to each other, to perform certain tasks, to know certain things, and so forth. Like competence, incompetence is a relative measure. By *incompetence*, I want to mean two things: first, that people are relatively less competent than other people at certain tasks; and, second, that there is often a discrepancy between the way people perform any specialized task and the way in which that task should be, or could be, performed. Both of these are forms of incompetence. It is in these two basic ways that incompetence becomes socially functional—and, hence, indispensable.

People who are less competent than they might be or ought to be to perform certain social roles or tasks give rise to the need for all sorts of specialized new jobs and services to compensate for the incompetence with which they perform. Here, it is best to go to example.

Take parenthood. It is typically only in "advanced" societies such as our own that the consequences of parental incompetencies are subtly redefined as faults of "society," to be corrected or atoned for by that vague abstraction. To be affluent, we must abide mass incompetence. In abiding mass incompetence, we raise intention to a virtue. Like Charlie Brown says, "How can we fail when we're so sincere?" The parent who is well-intentioned is forgiven all the humanity he can destroy. We disagree with Disraeli, who believed that "Man is not the creature of circumstances. Circumstances are the creatures of man." Federal assistance to children has risen from \$3.5 billion in 1960, to more than \$18 billion in 1971, an increase of more than 500%. Two particulars: Of the 6 million cases of mental retardation in the United States, about three-fourths are considered to be attributable to inadequate parental care and home environment. Since 1953, juvenile delinquency has been increasing faster than juvenile population. That which parents are considered incompetent or incapable of achieving has given rise to thousands of new occupational specialties—in juvenile courts, in education, in the health services, and elsewhere. If these thousands are not to be thrown out of work, future parents must be at least as incompetent as present parents with respect to these inadequacies and deficiencies in their children.

Consider reproduction. As a species, we seem to prefer starvation to competence. In this literate, sophisticated, freedom-loving, liberated, pill-conscious nation of ours, nearly 60% of all married women report more pregnancies than wanted. Unwanted children, however, provide increasing employment for counselors, psychiatrists, detention home personnel, and special police and court services. To the extent that a child is personally or socially incompetent, there is an increasing need for various kinds of specialized competencies, either to take care of him or to take care of the consequences.

Then there is the recently blossomed how-to-do-it sex and love industry. Some of the biggest best sellers in recent years have been how-to-do-it sex or love books, and more and more men's and women's magazines are based on how-to-do-it (right or good or better) philosophies. It would be a serious blow to our economy if we were all to become, overnight, competent lovers! Our pilgrimage to Esalen is not for the purpose of making ourselves competent as human beings, but for the purpose of learning how to wallow (and take "joy" in) our incompetencies. Our cultural hangup about "love" is not outgoing, but in-looking. "Men fly into each other's arms," says Demian (in Hermann Hesse's *Demian*), "because they are afraid of each other."

Consider the matter of health itself. Now that more information and more resources and more money are available to more people with respect to their basic health and hygiene, how are we faring? Between 1950 and 1967, what the American people spent on health rose from about $14 billion to more than $50 billion, a threefold increase in less than 17 years. Are we better off for having spent all of that money for health and medical care? We are not, says Lewis Herber, who argues that "we are exchanging health for mere survival," that, if it were not for our artificial environments and our extraordinary medical remedies, "today's adult might well have a much shorter life span than his grandparents had. This . . . suggests that modern man would find it very difficult to survive outside a medical and pharmaceutical hothouse."[3] The shoreline view of this phenomenon is, of course, that those medical and pharmaceutical specialties which produce new medicines and new living conditions are saving us. But from what? The water's edge view, on the other hand, reveals to us that people have become less and less competent to provide for their own robust good health, and that this increasing incompetence now indirectly employs hundreds of thousands of people and contributes ever more mightily to our total national economy.

And what of education? The cost just of maintaining school facilities increased more than 700% between 1939 and 1965. When the teachers and the students of the old one-room schoolhouses became incompetent to clean and maintain those complex and elaborate new buildings with controlled environments, there had to be specialists, with all sorts of special new machinery and equipment. No longer does the janitor fix the furniture; now there must be a technician called by the specialist who runs the projector, who in turn is notified by a special consultant who has been especially trained to help specially trained teachers to use the latest equipment, who in turn was hired and evaluated by an education personnel specialist. And so on. Our children are considered incompetent either to learn by themselves or to design the curriculums or operate the equipment by which they might do so. Can the modern child be other than irresponsible in our modern educational systems, where there is no responsibility for his own education which he can rightfully even claim as his own? As Jonathan Baumbach said in opening his review of *The Love Machine* (in *The Nation*, September 1, 1969), "It us no secret that education in America has a vested interest in stupidity."

James Conant once said that only about 15% of all children are competent to learn well by books and study in school. It is obviously the incompetence of the other 85% that provides for the employment and concern of thousands of education specialists, critics, commissions, researchers, and crisis-watchers. It is these relative and enforced *in*competencies in the educational system—including the student's—which have contributed most centrally to the fantastic growth of this industry in our economy. To keep it growing, we will need to decrease the competence needed to perform any of the existing roles in the system—including the student's. If this industry is to continue to grow, what will be necessary is less competence and more incompetence on the part of those who perform in it, with respect to the whole.

Or consider that new and rapidly growing industry, "consumerism." There are no figures available on the total costs of all activities now devoted to protecting the consumer, but my guess is that the bill already exceeds the maximum cost to the consumer of having no "protection" at all. For example, it is very rare that an automobile crash is caused by a mechanical defect. What accounts for most automobile accidents is driver incompetence, whatever its source. Yet driver incompetence provides employment for ambulance drivers, patrolmen, automobile designers, a large part of the insurance industry, a salvage industry, a highway safety industry, the several skilled personnel of emergency wards, and so on. There are more than five fatalities every 100 million passenger miles in motor vehicles in the United States, but less than one fatality every *300* million passenger miles in domestic air travel. Is that solely because the equipment and the roads are that much better, or does the fact that the pilot can't drive unless he is competent to do so have something to do with it? "Naderism" in all of its many forms is now a multimillion dollar business. This modern phenomenon has a classic form. First, it is assumed that the consumer is incompetent to fend for himself. All sorts of new jobs and agencies and commissions are created to protect him. Soon he has no choice but to *be* incompetent. He can be as irresponsible as he likes; someone else is looking out for him. After a while, he ceases to be able to look out for himself. After a while, he can't claim or reclaim his competence to look out for himself; if he did, the whole economy would be upset, and thousands of his self-professed protectors would be out of work, disemployed not by automation but by competence.

Where once it was man's competence to manufacture and distribute and use tobacco products that was a significant factor in our economy, it is now the incompetence of large numbers of people to quit smoking which contributes an increasing share. The economic value of those industries involved in curing and researching the consequences of smoking, and of attempts at prevention, may one day exceed the economic value of the tobacco industry itself. The "helping" and "caring" and "protecting" and "saving" industries will be the giant industries of future America. But how can we be protected from those who would "save" us? Montaigne said it: "God save us from the man who wants to save us."

Crime affords an interesting example. By definition, criminals could be considered social misfits or social incompetents of one sort or another. Not only does crime cost us directly—some $21 billion in 1965—it costs us more and more indirectly. In the 2 years between 1969 and 1971, federal outlays alone for crime reduction—and this means all sorts of new jobs for people—increased 200% from around a half-billion dollars to nearly $1.5 billion. Yet the crime rate itself continues to rise, up almost 200% since 1960. Yet, if this part of our economy is to continue to grow, we need more and more social misfits.

In 1955, there were about a million and a half "mentally ill" being directly cared for in institutions of all sorts. In 1968, that figure was up to almost 3.5 million, and the rate is on the increase. People who can't cope with themselves or each other create the need for new occupations, new skills, new facilities, and new and growing expenditures to care for them. Mental and emotional incompetence are increasing contributors to our economy, and to our changing political and social structures. Carl Rogers has said that "sensitivity training . . . is one of the most rapidly growing social phenomena in the United States."[(4)] I do not intend either to deny what Rogers deems the "utterly beyond belief" demand for it (as there was an "utterly beyond belief" demand for hula hoops) or to belittle the great good that encounter groups may do. But I ask you to note that it takes a lot of people who are or who feel incompetent (or less than "fully functioning") as human beings to create the need for an industry that will minister to their "needs." What happens to that huge and growing industry when everyone has been encountered and sensitized? My guess is that it will become a permanent fixture of our society and our economy, and that the next generation will assume that their children "need" it as a matter of course. People of the next generation will have to be at least as incompetent as we are to cope with each other if these giant industries are to survive and grow.

Today, we are "ecology" conscious. In fact, we are just about as likely to be overcome by talk on this crisis as we are by the crisis itself. The long-run problem has little to do with industry and major pollutants. It has to do with the fact that humans are, as a species, just about the untidiest, dirtiest creatures on earth. How did it happen that the tin can industry got blamed for the littering that people do because they are seemingly incompetent to do otherwise? A Yiddish proverb says that a girl who can't dance says the band can't play. Because people don't know or don't care how to "pick up after themselves" (we used to say), there must now be special occupations ranging all the way from pollution control abatement specialists and waste disposal engineers (who used to be called garbage men) to the more mundane, but increasingly ubiquitous government hirelings who pick up after picnickers. Households and commercial establishments account for almost four times as much solid waste per day as do industries. The incompetence and ignorance and indifference of the average guy with respect to trash and waste and dirt are going to choke us, not a belching smokestack. And, in the long run, that incompetence and indifference are going to be the major factors in the continued growth of this part of our economy.

Closely related are all of the burgeoning social problem-solving and crisis-resolving industries. On the assumption that the really big problems and crises can be seen only by experts, whole new industries have been created by all sorts of new think-tank specializations which invent new crises for those of us who are incompetent to see them. These social repair and prognostication industries rely on the continued incompetence of the man in the street to perceive either those problems on which they depend or the esoteric and scientistic ways in which they operate. The hidden formula: to be saved, a man must be made incompetent to save himself!

According to Gilbert Y. Steiner,[5] there were about 14 million people in the United States in 1970 who were incompetent to support themselves and were willing to prove it.* Social incompetence or social inutility takes many forms. But the bill adds up. In just the 2 years between 1968 and 1970, the total bill for all social welfare expenditures in the United States increased about 34%, from around $11 billion to more than $14 billion. The shoreline view reveals to us the need, and justifies an even greater rate of increase. But the water's edge view reveals something else: that thousands of specialized jobs have been created to deal with all kinds of social welfare problems, from health to housing to public aid, and that the continued growth of this part of our total economy will depend upon increasing—not decreasing—the incompetence of the citizenry in all of those dimensions for which there is a welfare program, or for which a program might be invented and funded.

I hope I have offered sufficient evidence for my proposition: that there is some point at which the functions of incompetence exceed the functions of competence in a complex, affluent society, and that, for all practical purposes, we have in this society passed that point.

It could easily be argued, of course, that it is the new competencies and specializations which provide the growth factor in our economy and the change factor in our social and political structures. But it can just as

*In his *Essay on Pauperism* (1835), Alexis de Tocqueville wrote: "There are two incentives to work: the need to live and the desire to improve the conditions of life. Experience has proven that the majority of men can be sufficiently motivated to work only by the first of these incentives. The second is only effective with a small minority . . . ."

easily be shown that no additional specializations are needed where no further incompetencies are nurtured.

The two go hand in hand, just as the shoreline and the water's edge complement each other. Yet they do not contribute equally. Truly new competencies or skills contribute independently to the growth and evolution of any society. But when gains are made by splitting existing competencies into parts, then there comes that point at which the sheer *mass* of incompetencies necessary to sustain the economy swings the scale over to the functions of incompetence as the most important factor in the further growth or evolution of a society.

In a simple, unaffluent society, there could be a simple one-to-one ratio, with a complementary competence supporting every incompetence. Not so in our modern, complex, affluent society. By default or by design, we've created a system in which there is a multiplier. More and more people can be adequately maintained with smaller and smaller competencies (with respect to the whole), and it is this broad and deep and growing accumulation of incompetencies across the whole population which is now the vital mechanism in our further growth and evolution as a society.

We are, thus, faced with a radically different perspective from which to think about the human condition in the foreseeable future: that perspective in which incompetence will become increasingly functional and competence decreasingly so. This water's edge perspective on what has come to pass—and the paradox it presents—is only the starting point for thinking about what is yet to come.

V

There is one (at least logical) consequence which overshadows all the rest. It completes the paradigm by which Western man's mania for growth and progress continues to fill his hands and empty his soul. e. e. cummings wrote,

> pity this busy monster, manunkind, not.
> Progress is a comfortable disease . . .

Is a disease not to be treated just because it is "comfortable"?

Robinson Jeffers put it in another way:

> We must adapt our economy to the new abundance . . .
> Of what? Toys, motors, music-boxes,
> Papers, fine clothes, leisure, diversion.

I honestly believe (but really an alien here; trust me not)
Blind war compared to this kind of life
Has nobility, famine has dignity.

To the extent that a man is ignorant of how to do something, or the need for doing it, he will be incompetent to do it. The social value of an incompetency is not to be found in the incompetency itself but in its social utility. How much utility a particular incompetency has depends on how functional it is to a given society, on how much a given society requires that incompetency.

In the United States today, mass incompetence is more functional to the society than individual competence. What this means is that incompetence has more social utility because it is more functional. It is more necessary, more indispensable, than competence. In an "underdeveloped" society, it is often thought that the more competent a man is, the more relevant he is. In those "advanced" societies in which individual competencies are increasingly less functional to the economy than mass incompetence, the more competent a man is, the less relevant he is to the functioning of those societies.

Specialized ignorances make possible mass incompetence, and mass incompetence has the ultimate consequence of making individual human beings irrelevant. This is the path we're on.

Any competency which has no social utility of any kind is an irrelevant competency. As less and less is required of more and more of us in this society, what increases faster is not our leisure, but our irrelevance. It is not that human societies will become irrelevant, but that individual humans will cease to be relevant to them. We can no longer assume that what is functional to any modern society is humanizing to man. Or *vice versa*.

The more incompetent a man is, the less relevance he has to himself as a human being. Yet the more mass incompetence there is in societies such as our own, the more functional it is. This is so not because men are either evil or stupid, but because all of our lives we have looked either at the shoreline or the water's edge, not at what both taken together comprise.

Our social and economic and political survival literally depend on extending and expanding those functional incompetencies that we have managed thus far to develop and maintain. Yet each increment of growth and progress now costs us some further loss of our human relevance. It was Mark Twain who said, "All you need in this life is ignorance and confidence, and then success is sure." Times have changed. A man no longer needs confidence. We've turned the world around. Once men had little to live for. Now they have nothing to die for. In our society, a man has to learn how to be incompetent even to die by or for himself. "Conviction," wrote Brian Moore, "even a wrong conviction, makes the rest of us uneasy."(6) Today, we prefer the ease of social utility to the dignity of human relevance.

VI

Kenneth Boulding once said that waste is a key factor in economic development. But our basic waste is not of our natural resources; it is of human nature.

It is inevitably some concept of human nature that we must confront. Because we have had popular science as our model of what progress and development and growth should be like, we have avoided that confrontation for too long.

The popularized version of science would have us believe that the behavior of men is to be understood only in relation to other men, that the measure of a man is the society in which he lives, in the same way that the spin of an electron is to be understood in relation to the field in which it is observed. This posture was reasonable enough until it became apparent that human societies are artifacts of man's own making, and that they, therefore, reflect quite as readily what is worst in man as what is best.

Men can no longer be measured by what they *are* within a given society. Rather, our societies must come to be measured in terms of what men *might be*, a criterion not reachable by the traditional methods of science. Calling social studies the "social sciences" or the study of politics "political science" is word magic. The measure of the one is the measure of the other: their humanizing contribution to individual men and to mankind. The ultimate aim of the sciences of social man, wrote Max Weber, is the enhancement of one's "sense of responsibility." We move in precisely the opposite direction. To avoid the very onus of that responsibility, we devise ever more means of political, economic, and social egalitarianism, standardization, interchangeability. Kant called "inequality among men" a "rich source of much that is evil, but also of everything that is good."(7) Being incompetent to cope with either, we've managed to throw the baby out with the bath water. In *Critique of Pure Reason*, Kant remarked as well on the illusion of the dove, given to suppose that it could fly even better in a vacuum:

> The light dove, clearing the air in its free flight, and feeling its resistance, might imagine that its flight would be still easier in empty space.

In recognizing the limits of popular science for constructing a concept of human nature, we must also avoid the humanistic fallacy—assuming that if we could just get our social institutions fixed right, people would be all that they might be. This is patently unsound. In this age of mass affluence and mass leisure, when more men than ever before in history have both the time and the wherewithal to pursue the best that is in them, what do they do? In the United States, there are only about 30,000 persons involved in symphony orchestras (including the professionals). There are only some 40,000–50,000 persons involved in amateur theater. Few adults in our society read even one book a year. Less than half even crack the cover. Can a man truly achieve his own highest humanity on a snowmobile or watching a prize fight on television? If the alphabet and the printing press and the transistor didn't enhance the humanness of the common man, how is television going to do it?

Society may lay us waste. But neither so callously nor so efficiently as we do so ourselves. Most working Americans today spend less than one-fifth of their lives at work. But with all of the alternatives available to us, with all of our modern wonders of electronics and power and medicine and leisure, we are further than ever from learning how to be useful to ourselves.* Half of all the drugs manufactured in the United States are tranquilizers. The ultimate affront to human nature is not that we be irrelevant to society but that we be irrelevant to ourselves.

We don't know what we want to be when we grow up, so we never do. We languish in the semi-dependent, semi-virile, semi-conscious, semi-irrelevant self-destructiveness of perpetual adolescence.

One of the functions of incompetence is that it provides us with a rationale for remaining thus stillborn as humans. Really out of fear, although we speak of science and of "human rights" and such, we dawdle. Unable to bear the uncertain thought of being fully human, we settle for being partly useful. We organize our dawdling. Surely, we hope, progress will soon deliver us from our great burdens, and we will be able to live without trouble, and die without pain.

Such are the functions of incompetence.

*Susan Ertz wrote, "Millions long for immortality who do not know what to do with themselves on a rainy Sunday afternoon." There are other kinds of tranquilizers, of course. G. L. S. Shackle wrote, "Money is dope, a tranquilizer against the effects of not knowing what to do."[8]

References

1. Lee Kavanu. Behavior: Confinement, adaptation, and compulsory regimes in laboratory study. *Science 143* (1964), p. 490.
2. Samuel Z. Klausner (ed.). *Why Man Takes Chances.* New York: Anchor Books (1968).
3. Lewis Herber. *Our Synthetic Environment.* New York: Knopf (1962), pp. 196–200.
4. Carl Rogers. Interpersonal relationships U.S.A. *J. Appl. Behav. Sci. 4*, No. 3 (1968).
5. Gilbert Y. Steiner. *The State of Welfare.* Washington, D.C.: Brookings Institution), 1971.
6. Brian Moore. *An Answer from Limbo.* Boston: Little, Brown (1962), p. 6.
7. Immanuel Kant. *Populäre Schriften.* P. Menzer (ed.). Berlin: Reimer (1911), p. 325.
8. G. L. S. Shackle. The economist's model of man. *Occup. Psychol. 32*, No. 3 (1958).

III

VISTAS IN EDUCATION

12

Harold G. Cassidy

HENRY MARGENAU'S PHILOSOPHY AS FRAME OF REFERENCE IN TEACHING SCIENCE

Knowledge, experience, and action encompass most of the academic side of education.* By *knowledge* I mean rational mental activities (activities in the brain) and the products of such activities as they may be written in articles and books; by *experience* I mean direct sensuous stimuli as well as the perceptions that arise when these are amplified by memory and connections to other brain processes; and by *action* I mean the application to practice of knowledge and experience. The three are not sharply categorizable, and in some cases may not be distinguishable at all. The power and usefulness of Henry Margenau's philosophy, as I understand it from his *The Nature of Physical Reality*,[1] lie in its compellingly sensible approach which can usefully be applied to these three educational concerns.[2] Its importance lies in providing us with a compre-

*Margenau uses the word *experience* to include thought, conjecture, and feeling as well as sensation. I have listed *knowledge* and *experience* (which here has a more restricted meaning) separately for purposes that will become apparent, I hope. I do not mean to tear the immediately given from the rational, but only to make it somewhat more convenient to use these two for my purposes.

Harold G. Cassidy • Professor-at-Large, Hanover College.

hensive frame of reference of the kind that has been called for by T. R. Blackburn in his essay on sensuous-intellectual complementarity.[3] By *comprehensive frame of reference* I mean one that can give meaning to modern science as well as to the contemporary countercultures.[4] Use of this frame of reference in teaching physical science is *philosophy by example*, and it must be acknowledged that in such use there is less than enough room for the explicit qualifications and necessary subtleties that accompany his discussion in Margenau's work. There is added the certainty that when an ideal is reduced to practice the results never meet the absolute standards of the ideal. I hope, however, that I have done no great violence to his thoughts.

This essay is based on many experiments with Yale undergraduates over a period of years*; with Wesleyan undergraduates; with Hanover undergraduates; with selected graduate students; with some high school students; and in occasional trials during American Chemical Society, Sigma Xi, and Association of American Universities lecture tours, and with The National Humanities Faculty.†

The approach I have used employs Margenau's[1] Fig. 5.1 taken as a metaphor (Fig. 1). I interpret this diagram with as great as possible fidelity to Margenau, and I recommend his book to the class. A great advantage of the diagram is its concreteness, which tends to desensitize students who are put off by abstractions. Actually, of course, they use abstractions all the time in their daily lives and in other classes ("the law," "Democrats," "Republicans," and so on) but these are familiar words. Unfamiliar abstracts, especially in a science course, are almost threatening, and invoke counterculture reactions. How helpful it is, then, to acknowledge the importance of direct sensuous experience and to show its place at the "plane of perceptions" (*P*), to display the double "pathways" to constructs, and to discuss the role of intuition and of openness to nature in the largest sense that leads to new knowledge, the network of constructs in the "field of constructs" (*C*).

I realize, of course, that in this approach I am using certain concepts rather loosely: specifically in correlating the sense data of immediate experience with things "out there," the material objects, the purely sensory

*Carnegie Corporation supported a 5-year experiment through a grant to Dean William C. DeVane of Yale College.

†Earlier reports of these experiments are not easily available. Two short announcements are in *Superior Student 2*, No. 7 (November 1959), 9, and *Sc. Educ. News* (December 1959), 3. Progress reports are in *Yale Sci. Mag. 37*, No. 2 (November 1962), 6–12, and in *Yale Alum. Mag.* (February 1965), 12–15. More recent and accessible reports are in *J. Chem. Educ. 46* (1969), 64, and *48* (1971), 212, and in *Physics Teacher 9* (1971), 267.

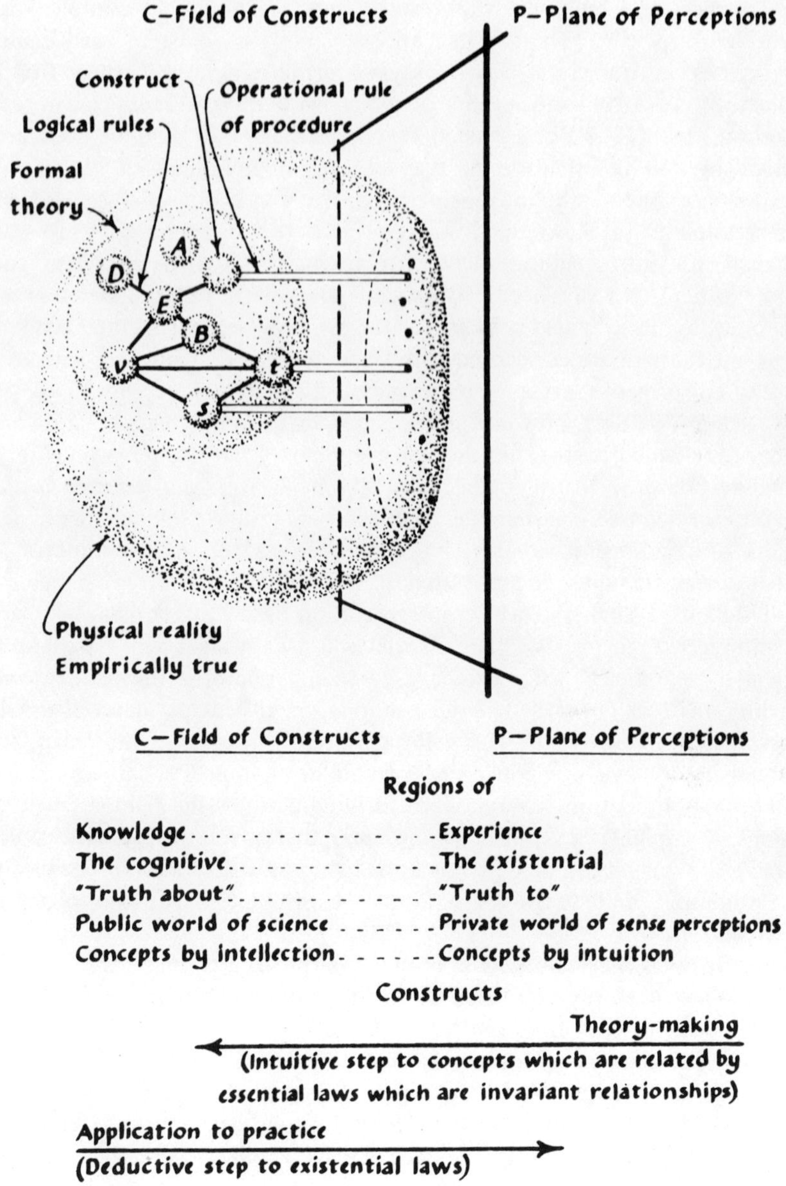

Fig. 1. A metaphor of the structure of formal theory and physical (scientific) reality, using as the basic figure that of Henry Margenau[1] (p. 85), used with permission and considerably expanded and modified. From Cassidy [15] (Fig. 1–4).

experience of which, as Margenau[1] says, is "ineffably complex and multiple" (p. 51). [The sensible approach here is to begin with mental processes in ourselves. These, says neurophysiologist Arturo Rosenblueth,[5] "are the only events of which we can be directly aware and certain" (p. 114).] But I hasten to complement this possible source of error by calling attention to the subjective component of all sensory experience: the symbolic transformation (to use a powerfully suggestive expression from Susanne K. Langer[6]) of these experiences into constructs, and into relations between constructs and to nature. This goes on "in here," in our heads. Our heads are clearly not the same as our habitat, but clearly also neither isolated nor independent. It then becomes evident that we are concerned with open systems that interact with each other to form the larger system symbolized in Fig. 1.

Consider, now, how useful this approach can be in teaching a laboratory for premedical organic chemistry students. I put this diagram (Fig. 1) on the board and turn to the class: "Somewhere along the way in this laboratory course you may find yourself turning to the person at the next desk and inquiring with suitable expressions what, in the name of all that's sensible, does the nitration of phenol have to do with being a physician?" I then give a brief interpretation of the Fig. 1 metaphor, and continue: "Your professional activities will usually start with a patient—a person 'out there,' who will have come to you for some reason, with symptoms real or fancied. Your examination of this patient will invoke certain techniques which you will have been at some pains to learn, and you will also have been concerned to avoid being fooled. What you observe in your examination is symbolized as lying close to the plane of perceptions. We might think, also, of the manipulations and observations you make at your laboratory desk as corresponding with such a procedure of examination and diagnosis, and as being 'plotted' at the plane of perceptions.

"It takes a certain degree of creative thinking to make the requisite symbolic transformation from those things out there to the ideas and perceptions in your head. By 'perceptions' I mean that the raw sensuous data become in your head connected to recollections, and other kinds of ideas and theories, and then may be termed more richly 'perceptions.' These are wrought into symbolic form in your head; indeed, the sense organs that you rely on for contact with your habitat already begin this transformation by encoding light, and sound, and odor, taste, and touch into electrical and chemical phenomena in nerves. In class you are learning how to make this transformation, and how to manipulate these symbols logically according to the theory that you are learning (which is mostly in the *C*-field). In a corresponding manner you will diagnose and prescribe for your patient. In both cases you may carry out rote procedures and associations when the evidence leads along well-worn paths. But we shall help you to surmount this temptation by providing unknowns; and irrelevant data that must be discarded. Then finally, having moved into the field of constructs, and navigated that network of relations until a conclusion, or diagnosis and prescription, has been reached, you apply this with a return to the plane of perceptions to see if you were right.

"Admittedly we can only introduce you vestigially to this process. This is an introductory course. But this is what we would like you to learn: a scientific approach that welds together careful observation, cognitive manipulation, and validation of your conclusions; that unites experience, knowledge, and action. We also hope that you will find, as you do this work, and connect yourself through your experiences and thoughts to this grand tapestry that we call 'science,' that you will experience pleasure, and a sense of meaning, and a realization that you are becoming a better and fuller person all the time."

The effects of this brief talk, accompanied of course at the blackboard, are usually noticeable, and sometimes striking. One student came up afterward, glowing. "Thanks, thanks! That's the first time anyone *ever* told me what I was doing in the lab!"

This is only one aspect of the use of this metaphor. It enables a clarifying and delicate balance to be reached within the dualism implied by "out there" and "in here," and it confirms the impossibility of isolating one or the other. Further, it is possible to discuss aspects of sensation and perception that must intrigue even the antiscience student. Recent essays by David Bohm[(7)] and by Arturo Rosenblueth[(5)] are helpful here. Bohm uses the findings of Piaget and others to show how the young child develops a concept of his habitat in relativistic, frame-of-reference, and invariant-relation terms. Rosenblueth describes sensations and mental processes from a physiological standpoint. The raw stimulus, arising from the material universe and impinging on our receptors, is coded as a message that travels along afferent nerve fibers to specific anatomical central connections in a unidirectional manner. He says, "Our only possibility for decoding the afferent messages is via the conscious processes correlated to the central neuronal events. We cannot infer from the coded messages anything about the 'nature' of the material universe. Any statement or speculation about this 'nature,' whether made by a physicist or a physiologist or a philosopher, is doomed to be utterly meaningless.

The purpose of science is to construct a material universe with a structure compatible with our conscious decodings" (p. 52). It seems to me that this position is not in full accord with that of Margenau in that it is too strong: one surely can infer *relationships*. But if students could be led to really understand its implications, their cultural literacy would have been advanced by a large quantum jump.

It is possible to use this metaphor to show how analogous are creative and scholarly activities in the sciences and the humanities. In the latter, too, one can construct a metaphor that corresponds to Fig. 1. However, because of the highly individual emphasis and the sought-after personal messages in many of the data and inferences from them, we cannot speak of constructs or of mathematical relations in the same sense as in science.[8]

Scientists are beginning to realize that there is a deep in origin and now increasingly strong current of antiscience sentiment surfacing in the world.[9] It appears to gain strength from many sources, to judge from the numerous theories that have appeared. A central source of difficulty is confusion between science and technology, especially with respect to the bomb and to pollution. I remember being challenged at a lecture by someone who shouted that "science is evil because it made the atomic bomb." My reply, that the decision to make the bomb was made by a historian and a politician, not by scientists, served to arouse his attention, so I could distinguish between science and technology, and between theories and the uses made of them by people. But the misunderstanding runs deep.

My experience in working with nonscience (as well as preprofessional science) students leads me to formulate the problem of teaching them in the following way. Broadly speaking, we are concerned with nonrational experiences and rational knowledge. The former arises for us at the plane of perceptions, and melds with, and gives rise in some cases to, the latter. Science begins with experience; with the response of our sense organs to stimuli. These sense organs, transducers at the ends of extensions of our brains, may already censor incoming stimuli. Not only are there thresholds above and below which they fail to respond, or change their response[10]—to that of pain, for example, or discomfort—but also in some cases they actively intervene: as an example, we learn what and how to see, and what to ignore. Broadly speaking, still, there are multitudes of experiences that are amenable to the methods of science; equally there are vast ranges of experiences not amenable (or at least not yet amenable) to the scientific approach. Both are necessary to the life of the whole person.

In addition, we must be concerned with—or at least aware of—the existence of irrationality. This is an ever-present threat to reason and the nonrational components of existence. To categorize irrationality as sickness does not make it go away. Our challenge and opportunity is to offer the therapy of better solutions to the problems of existence. These are what this approach is about.

As Margenau's metaphor displays to us, science is concerned with the invention of "constructs" and with the discovery or invention of invariant relations between them (the single lines in Fig. 1 represent these relations, for example, $F = ma$). It is of the greatest consequence to help the student to grasp this, for consider what can be brought home to him through this metaphor. Science does not question that there is an existential side to life. Insofar as those existential experiences that concern the scientist when he is doing science are concerned, they lie at or close to the *P*-plane. That there are existential issues other than those presently amenable to science is not to be doubted. The scientist himself does not behave scientifically at all times. He has other societal roles to fulfill: parent, perhaps, or taxpayer, and so forth. Thus the scientist has no war with the sensible aspects of existentialism. He begins with existing things and returns to the *P*-plane for validation. His war must be against the irrational and nihilistic types of existentialism, and their evil ideological consequences.

On this firm-enough base, then, by the operational pathways of rules, recipes, and so on, the scientist arrives at his constructs (many persons prefer to speak of "concepts"). Throughout the science course these pathways can repeatedly be pointed out as they are traversed. They lead to constructs which, to be fruitful, as Margenau has shown, must be connected to others. The connections are clearly invariant relations. In a sense these are absolute: we made them so. From the bringing together of two or more constructs in a train of reasoning a prediction may issue which can be tested against nature (at the *P*-plane). If a "hit" is made that is sufficiently close, the train of reasoning is validated, as also the network of theory (Margenau,[(1)] pp. 102, 103).

We have delineated, then, a realm of existential concerns, and a realm of essential concerns. The former is the seat of the rich, varied, little known and less understood, potentially infinite realm of nature outside of our minds; the latter, in science, is the orderly field of absolutes and invariant relations between them; in the humanities, too, there are defined absolutes—concepts of style, beauty, truth, grace, good, and evil—and relations between them, with canons of validation, but these clearly differ from the analogous mental tools in science. Together they all are needed in the education of a cultured person.

What is additionally important is that we have in this formulation one of the "better products" with which to compete for the minds and hearts of our students.* We know, that is, that our brains are pattern-forming organs (unless ill or abnormal in some way). A newborn infant appears not to have the well-developed network of cerebral connections that are correlated to "mind" and mental processes. But after birth it rapidly develops organized responses to its habitat: making sense out of its experiences; giving them meaning. We and our students can understand that this fantastic network of linked cognitive relationships connected to nature through experience, presented metaphorically in Fig. 1, *is a bearer of meaning*. Each, himself, can verify this in the delight that comes as a well-worked-out chain of reasoning brings a fitting conclusion; or as some experience of his fits into this body of knowledge and experience and so validates his own experience; or as he observes the intelligent behaviors of others.

Science is, indeed, our most comprehensive and powerful bearer of rational meaning. Figure 1 is a metalhorical gestalt, so to speak, for this insight. It follows, then, that whoever—practicing scientist or student—contributes to this grand nexus is himself creator of something meaningful. He gives meaning to his own activities and in this respect he becomes a more whole person in the frame of reference of his science. Activities of this meaningful kind are of the greatest social and cultural, as well as personal, consequence. They challenge the bases of the nihilistic philosophies and associated ideologies of the absurd and of meaninglessness that seduce, through the duplicity of words, inexperienced, ahistorical, and naive people.† One recalls in this connection the strictures of the great French philosopher Maurice Merleau-Ponty[13] against what he characterizes as Hegelian "romantic irony which is equivocal, tricky, and self-conceited. It relies on the power which we can use, if we wish, to give any kind of meaning to anything whatsoever; it levels things down; it plays with them and permits anything . . . this kind of madness . . ." (p. 39).

*Another group of "better products" are the theory and consequences of unified science as described in several recent publications (e.g., Cassidy,[2] Chapter 7; Haskell[11]).

†I cannot recommend too strongly here Bettelheim.[12]

In culmination, I cannot resist quoting Merleau-Ponty[13] once more. "Because we live in one of those situations that Hegel called diplomatic," he writes, "in which every initiative risks being changed in meaning, we sometimes believe that we are serving the cause of philosophy by isolating it from the problems of the day . . ." (p. 61). The work that I have described here is, as I said, *philosophy by example*. It confronts with the constructive philosophy of Margenau the problems that are disturbing students today. This philosophy is utilized as a frame of reference within which (cognitive) knowledge, (subjective) experience, and action in applying these to practice can unify what we are doing throughout the college or university, and thus give meaning to education and to the lives of ourselves and our students.*

I cannot close without acknowledging that in taking this metaphor to the physical science classroom certain compromises occur which detract from the purity of the original. This may already have been apparent to the reader. The ideal may be held in mind as a framework for test† and essay on education[2]; it is by no means easy to assure that it is communicated to the student in its proper form. Indeed, it is certain that in any class the students will end with a variety of interpretations. It is the old problem of the academy and the marketplace. One expects only a good enough hit at the *P*-plane when the idealized constructs and absolute (small "a") relationships are used to lead to some application. And one can always retire from the market place to the academy to reestablish integrity. Thus I find myself repeatedly, after applying these ideas in the classroom, returning to Margenau's work for refreshment.

Acknowledgment

I am pleased to acknowledge the assistance of Battelle Institute in preparing the manuscript.

References

1. Henry Margenau. *The Nature of Physical Reality: A Philosophy of Modern Physics.* New York: McGraw-Hill (1950).

---

*It is rare that a textbook in science states explicitly the intellectual, philosophic framework upon which it is constructed. That of Eastman and Rollefson[14] does so.

†I have written a textbook[15] which is built on the themes discussed here.

2. Harold G. Cassidy. *Knowledge, Experience and Action: An Essay on Education.* New York: Teachers College Press, Columbia University (1969).
3. Thomas R. Blackburn. *Science 172* (1971), 1003.
4. T. Roszak. *The Making of a Counter Culture.* Garden City, N.Y.: Doubleday (1969).
5. Arturo Rosenblueth. *Mind and Brain: A Philosophy of Science.* Cambridge, Mass.: MIT Press (1970).
6. Susanne K. Langer. *Philosophy in a New Key: A Study of the Symbolism of Reason, Rite, and Art.* New York: New American Library (1951).
7. David Bohm. Physics and perception. In *The Special Theory of Relativity.* New York: W. A. Benjamin (1965), pp. 185–230.
8. Harold G. Cassidy. *The Sciences and the Arts: A New Alliance.* New York: Harper (1962).
9. A Tiselius and S. Nilsson (eds.). *The Place of Value in a World of Facts:* Proceedings of the Fourteenth Nobel Symposium, Stockholm, 1969. New York: Wiley-Interscience (1970). A. Koestler. Rebellion in a vacuum, p. 221; The urge to self-destruction, p. 297.
10. Georg Von Bekesy. *Sensory Inhibition:* Princeton, N.J.: Princeton University Press (1967).
11. Edward F. Haskell. Assembly of the sciences into a single discipline. *Sci. Teacher 37* (1970), supplement. E. F. Haskell (ed.). *Full Circle.* New York: Gordon and Breach (1972).
12. Bruno Bettelheim. Obsolete youth: Towards a psychograph of adolescent rebellion. *Encounter 32* (September 1969), 29–42.
13. Maurice Merleau-Ponty. *In Praise of Philosophy.* Tr. John Wild and James M. Edie. Evanston, Ill.: Northwestern University Press (1963).
14. E. D. Eastman and G. K. Rollefson. *Physical Chemistry.* New York: McGraw-Hill (1947).
15. Harold G. Cassidy. *Science Restated: Physics and Chemistry for the Non-Scientist.* San Francisco: Freeman, Cooper & Co. (1970).

Emily B. Sellon

INTEGRATIVE EDUCATION: TEACHING AS A HUMANE ENTERPRISE

During the past quarter century, American education has undergone a far-reaching and often painful process of self-criticism—a process that has resulted in profound changes, both in the definition of educational goals and in attitudes toward teaching. There is no need to expatiate upon the causes for this reevaluation, for we are only too familiar with the extensive dislocations in (and disenchantment with) our educational system, whose lack of relevance to the problems and life experiences of young people finally culminated in widespread protest and acts of violence in the 1960s. The outcome has been a program of administrative and curricular reform, whose intent was to make education more responsive to both the individual and the social needs of students. Less obvious but more important, however, have been the changes in academic attitudes toward what might be called the human or humane implications of education in the context of twentieth-century life and culture.

The philosophy of education in the United States, an outgrowth of British and (to a lesser extent) European systems, was of course originally oriented toward the classics, long considered to be the indispensable foundation of a cultivated mind. But the growing dominance of the sciences in the Western world, as society became industrialized, radically revised

Emily B. Sellon • Editor, *Main Currents in Modern Thought.*

this point of view. Education was directed toward a more limited and pragmatic goal: the training of skilled, scientifically competent specialists, whose importance in a technologically advanced culture was held to be self-evident. As in the definition of C. P. Snow, the world was to be increasingly divided into two classes: the dominant scientific community and all the rest of humanity—those "others" whose lack of scientific knowledge was thought to render them incapable of understanding, much less of solving, the problems confronting a technological society.

This was an extreme view to which many scientists themselves did not subscribe. Nevertheless, the increasing importance of scientific discovery, the growing sophistication of scientific techniques, and the proliferation of specialties within the sciences all had a profound effect on the academic world. Not only did higher education come to be dominated by the sciences; it also became powerfully oriented toward the development of "experts" whose education lay within narrow fields of specialization. This unquestionably equipped them well for specific jobs, but hardly fitted them at all to participate in the "other" world, which is, in the final analysis, the world of human being—of human life in its widest sense. As a result, science was often accepted uncritically as an ultimate, almost absolute, value, while the humanities, being merely for "enhancement" or enjoyment, were relegated to the kind of dilettante role suited to off-hours pursuits.

During this period, the liberal arts university rapidly declined from the status of *universitas*—a genuine "universe" of knowledge—to that of a collection of increasingly isolated departmental disciplines whose criterion of excellence was their ability to imitate the objectivity and analytic methods of the sciences. Unfortunately, the wider purposes of creative inquiry in the sciences—the opening of new vistas to the human mind as it searches the face of nature for deeper understanding and a more embracive vision of the whole—were often ignored in the current intoxication with technique, method, and measurement. For the prevailing philosophy was pragmatic, positivist, and behaviorist, inexorably rooted in the primacy of facts and things. The perspective was still further limited, however: recognition of the fundamental values in all things, a view which Western education inherited from the Greeks, gave way to an insistence on usefulness—a sphere in which technology was obviously paramount. The sciences were pictured as completely objective and value free, which removed them even farther from the subjective values of the humanities. This state of affairs sharpened the divisions between Snow's two worlds, and increased the danger of scientific irresponsibility (or irresponsibility in the name of science, to be more exact), for the commitment of scientists to objectivity and the "purity" of the scientific enterprise was thought to free them from concern with the social consequences of their work.

Such an artificial situation could scarcely endure, however. Some of the obvious faults and weaknesses in the monolithic facade of American education soon began to show themselves, often detected by members of the scientific community itself. There were many who had grave doubts about the morality (to say nothing of the truth) of scientific detachment and the myth of scientific objectivity. There was growing awareness that the very nature of twentieth-century science, which had outmoded the mechanistic, deterministic concepts of the nineteenth century, was also rendering obsolete the philosophical consequences of classical science—in particular, the crude physicalism which stemmed from the view that phenomena of consciousness and life were reducible to their physicochemical constituents, and thus amenable to manipulation. The new science which arose from relativity, quantum and field theory, indeterminacy, and the like, demanded a much more comprehensive and holistic view of nature and of man.

Henry Margenau was one of the earliest and most articulate of the scientists and teachers of science who argued that the philosophical and cultural consequences of contemporary science required an entirely new approach in education—an approach which would take into consideration the whole range of implications for human understanding that science offers. He well understood that it is not only dangerous but also unrealistic to treat science as an arcane subject completely beyond the reach of the ordinary man, since it is just this ordinary man who is ultimately the important factor in all social action. He believed that the real character of the scientific enterprise, that is, its fundamental principles, goals, and methods, can be made comprehensible to any intelligent person, and, further, that the sciences offer a common language which can be immensely useful in bridging cultural gaps, since science has no national or ethnic boundaries. Taking issue with the view that insisted on the philosophical neutrality of science, then prevalent in the West, as well as with the Communist doctrine that science must serve or promote a specific ideology, he urged that the cultural implications of science in terms of values be carefully sought and rigorously examined. And he held that a careful scrutiny of the method of science would verify its fundamental unity with all valid learning, since it reveals itself as an effort to rationalize and generalize our experience in terms of universals—a basic goal of philosophy.

It soon became evident that this new perspective, which was steadily gaining in support, would have to transform the teaching of science not only for scientists but also for the whole of liberal education. Over-specialization was increasingly under attack as not only inadequate but also actually inimical to the goals of a general education. The immediate result was a movement to introduce some form of interdisciplinary studies into the college curriculum. Well-intentioned as this effort has been, however, it has failed to solve the fundamental problem which arises from the separation of the disciplines into tightly segregated divisions, lacking conceptual unity as well as shared goals and values. Interdisciplinary studies at best furnish a program which is grafted onto the academic corpus—an additive, which effectuates no fundamental changes in the basic departmental structure or emphasis.

Interdisciplinary education has therefore failed to meet Margenau's requirements for a radical reevaluation and reorganization of university teaching. He has argued that the root of the problem lies deep within the epistemological foundations of education, and therefore is not amenable to partial or superficial solutions. A genuinely integrative approach is needed: an approach that can unify the disparate fields of knowledge in terms of the universal, reliable principles they hold in common, principles which are so comprehensive, so intrinsic, and so indispensable that they constitute both the integrative nexus and the bridge of communication between divergent disciplines, ideologies, and cultural modes. The proper delineation and application of such principles, Margenau holds, could not only integrate the world of academic studies but also serve as a means of cultural unification, since the "existence and universal use of scientific language acts as a bond between different cultures, as a means of communication between East and West."(1)

The possibilities which such an integrative approach to education could offer, with respect to healing the deep schisms in our society, encouraged Margenau to join with other educators and professionals who shared his concern about the divisive tendencies engendered by overspecialization, and were seeking ways to ameliorate them: most especially F. S. C. Northrop, the philosopher of law and culture, Clyde Kluckhohn, the anthropologist, Pitirim Sorokin, the sociologist, Edmund Sinnott, the biologist, Kirtley Mather, the geologist, and F. L. Kunz, who had in 1940 founded the journal *Main Currents in Modern Thought* to provide a forum for the reappraisal and reintegration of knowledge in terms of "the whole of things—nature, man and society." This group, together with other scholars and members of the business-industrial community, collaborated in 1947 in founding the Center for Integrative Education, which Margenau serves today as president.

The initial purpose of the Center was to promote the development of integrative educational programs by exposing the conceptual roots of the arts and sciences—those fundamental truths or postulates which ground the presuppositions on which all knowledge rests, and which are therefore shared by all disciplines, no matter how diversified their content and methods. This conceptual base was described by Margenau in 1953:

> It seems axiomatic that unified knowledge and understanding have roots that go below facts and go below language. . . . The facts of our knowledge, the visible excrescences of our culture, are in a sense less important than the roots from which they spring. Lack of continuity and organization between facts, gaps between disciplines of learning, failures of understanding between people and peoples, can be viewed as signs indicating places where conceptual roots, though present in the deeper soil, have for some reason not pushed up to the surface. Discontinuities are in many instances confined to the surface, and a probing for basic concepts may eliminate them. For it is the nature of concepts to be extensible, while facts can only fill dead spaces in knowledge. Concepts guide the researcher when he is looking for facts, they provide the texture in which facts are embedded. . . . [Therefore, this] is a plea for integration in the realm of fundamental concepts.(2)

This plea was primarily addressed to teachers and institutions, and directed at the reform of educational policies and programs. But the larger social and human implications of integrative education—the promotion of "understanding between people and peoples"—gave the Center its basic impetus, and enlisted a wide variety of individuals in its cause. For the concept of integration can obviously extend to include the unification of many disparate fields of human interest and activity. In this sense, integrative education may be thought of as providing a method which not only can help to bridge gaps between disciplines but also can help specific disciplines, such as the life sciences, to see their own work in more holistic terms.

> Integration is a living process, something that occurs in a living system, in a person, in a group of persons . . . . Integration arises out of an effort, out of a reverence for the universe, aroused and maintained. The work of integration, synthesis or design has a special nature which needs to be understood. It differs from the work of reduction and analysis. Excitement, success, haste, priority of discovery, newness, victory, secrecy, goals . . . are not its counterparts. For integration is concerned with bringing together, tying together, resolving contradictions, healing, making whole, connecting that which was considered unconnected. . . . And this is one mark of such an approach, that its fruits can be shared by and be meaningful to everyone.(3)

This statement identifies the dimensions of the integrative task, and reveals the difficulties of the synthetic method in contrast with the analytic. It is quick and comparatively easy to sever, to separate, even to dissect, but the gradual process of healing a body or mind, or a fragmented discipline, or the fabric of society, is subtler and more complex, for it involves the restoration of meaning as well as the union of parts. It is, indeed, the work of giving wholeness and significance to human existence.

Initially, the task of unification was seen to lie in uncovering the meanings inherent within the disciplines, and in relating these fundamentals to the other fields of knowledge. Thus the first efforts of Margenau, Northrop, Kunz, and their colleagues in the Center for Integrative Education had two aims. The first of these was to expose to view and encourage the study of concepts and principles which are basic to many different fields of knowledge; the second was to make these concepts useful within the educative process. These crucial ideas, which reach across the whole spectrum of learning, were seen as the engines of progress within the specific disciplines and, at the same time, the means of integration among them; the critical importance of such concepts to all fields of knowledge lies in the fact that they are fundamental to our notions of what constitutes reality.

Obviously, in attempting to integrate, to restore unity to the sciences and the humanities, it was necessary to confront some basic questions. How, for example, is one to reconcile science's idea of reality, which is abstracted from the sensed world, with the experienced and shared reality which is the primary concern of the arts? Do not these seem to be separate orders: one theoretical, mathematical, rational, generalized; the other aesthetic, immediate, experiential, and personal? This question leads directly into what has been a central epistemological concern of the members of the Center for Integrative Education. If integration is to be achieved, it is necessary to distinguish whether these two orders represent genuine, valid modes of knowing, and, if so, how they are related. In other words, how are sense impressions, which deliver to us the moving world of color, beauty, and feeling, related to ideas—those abstractions from nature which can be formulated into language, communicated, and universally shared?

The epistemology which was formulated, and later extended and applied both generally and specifically for the sciences and for cultural studies, was essentially correlational. It embodied a method whereby the two fundamental terms of experience, the rational or theoretic realm on the one hand, and the perceptual or aesthetic world of sense impressions on the other, can be seen to relate to one another. To exhibit this relationship, Margenau introduced his method, "rules of correspondence,"[4] while Northrop designated it by the term "epistemic correlations," but the epistemology was essentially the same. Both men have stressed that the two modes of knowing constitute essential values, irreducible to one another and equal in importance, even though they are not always emphasized equally in particular cultural settings.

It has become widely recognized that the heavy stress placed in the West on the rational mode, to the detriment of the aesthetic—to the "propositional" over the "appositional," to thinking over feeling—has resulted in the desiccation and attenuation of knowledge. It is this deprivation that primarily alienated American youth in the 1950s and 1960s, and caused their violent reaction in the opposite direction—which tended to embrace everything in our culture that might be called nonrational. The West's exclusive concern with the theoretic component of knowledge has also, as Northrop pointed out, divided us from those non-Western cultures whose essentially aesthetic mode of knowing has been so different from our own, and which we have accordingly tended to disparage as well as misunderstand.* It seems evident that if we are to heal the schisms within our own culture, and also create bridges of understanding between Western and non-Western societies, we must seek a true balance between these two modes of knowledge and perception.

Recent studies on the functions of the two hemispheres of the brain corroborate the Margenau-Northrop epistemology and reveal its physiological base.[(6)] Although the terminology may vary, a variety of sources suggest that there are indeed two fundamental modes of the human mind, very similar to those which Northrop characterized as theoretic and aesthetic and which Margenau identified with the construct field and the perceptual plane wherein the "protocol facts" of our experience reside. It is now recognized that the dominant or left side of the brain includes the speech center as well as the locus of logical mathematical thought, while the inferior or right side is given over to pattern recognition, music, imagery, network associations, and the like. Most people would now agree that the unity of the person is best realized by a higher level of functioning which gives expression to both these modes; it has been advocated that all educational programs, even the technical, should include appropriate teaching situations which encourage equivalent opportunities for developing both functions.[(7)]

The Center for Integrative Education has long held that the whole of

*See Northrop[(5)] for cultural exemplifications of dominance of the theoretic and aesthetic components.

human knowledge should be explored in ways which seek to reveal both the theoretic structures within the various disciplines (which ensure their development as well as their universality) and their aesthetic dimensions (which endow them with personal immediacy and human significance). It thus becomes apparent that the work of integration has indeed a "special nature." It aims to unify the various fields of knowledge and thus reintegrate the world of learning, but its deeper purpose is to unify the knower with the objects of his knowledge and thus foster the integration of the human person.

Margenau himself, as a teacher of science and of the philosophy of science, has been uniquely situated to apply this epistemology to science teaching within the larger perspective of a holistic education. He has thus been able to confront directly many of the criticisms which were inevitably raised against so novel a method, and indeed against the whole concept of integrative education. He addressed himself specifically to the argument that attention to the human dimensions of science teaching would dilute the pure and rigorous character of science, thus reducing the level of professional competence required of students. Furthermore, the question was raised as to whether students who were encouraged to perceive the limits of the scientific approach—that is, the inapplicability of science to certain areas of human life and experience, and the consequent need for enrichment from the humanities—might be tempted to reject science entirely. As a result of his own teaching experience, he was able to reply:

> There is mounting evidence that any serious discussions which open a student's eyes to the virtues and the limitations of the scientific approach and disclose at the same time the dogmatism inherent in some of its claims, which distinguish carefully between cognitive experience that is subject to scientific systematization and non-cognitive, e.g., aesthetic encounters with reality that are not—such discussions conducted in prescientific everyday language *intensify* the student's desire to learn science, *enhance* its later understanding and, so the record seems to show, make him a more successful scientist even in a pragmatic sense.[8]

For these reasons, Margenau continues to advocate the introduction of an integrative course—designed to orient the student to the historical and cultural dimensions of science, as well as to its fundamental presuppositions, methods, requirements, possibilities, and limitations—early in the student's training, in fact, in the freshman year. Such a course, drawing upon both the history and the philosophy of science, would focus on the essentially human dimensions of the struggle for scientific understanding. It could therefore furnish a broad context within which specialized studies undertaken later could be seen as contributions to, particulars of, the scientific enterprise as a whole, and this enterprise in turn related to the most enduring concerns of mankind. For this larger, more comprehensive view of science reveals it as a special way of organizing experience which is valid and valuable for every human being. This places it properly as embodying a fundamental mode of knowing—a perspective which is essential to the development of imaginative, responsible, and creative scientists, rather than mere technicians.

The central purpose of the Center for Integrative Education, therefore, originally enunciated as the reunification of the sciences and the humanities, has been sharpened, refined, and expanded to include a widening range of human concerns and their educational implications. These seem to group themselves naturally under the three great horizons which were identified early on by Margenau and Kunz[9] as nature, man, and society, placed deliberately in that order, since it is to our new understanding of the nature of the physical world and its basic nonmateriality and ambiguity that we owe our emergent perception of man's unique character and role in the universe.

Scientists today are compelled to consider very deeply such questions as the extent to which their observations depend on the nature of their own perception, as well as on the external world, and indeed to question the meaning and relation of "inwardness" and "outwardness."[10] Thus the hard lines between objectivity and subjectivity have dissolved, just as the ultimate materiality of the world has dissolved into patterns of energy which are seen as increasingly complex and finely graded. For this reason, there is now room in science to consider problems of life and consciousness on their own terms. We owe this freedom, in the final analysis, to the twentieth-century revolution in science, for it was the new physics of Einstein, Planck, and Heisenberg that brought about the changes in Western conceptions of reality, thereby transforming the arts as well as the sciences, as Waddington[11] has shown.* The social consequences, which promise to be yet more far-reaching, are still largely in the making, although their signs are everywhere apparent.

The integrative perspective therefore offers the possibility of querying

*"After the [Second World] war there was an efflorescence of new varieties of painting, many of which show affinities to various aspects of science which had been of little influence earlier. In particular, a whole complex of ideas deriving from quantum physics, both about the nature of the material world and about the character of our perception of it, began at this time to colour the thoughts of non-scientists in general and painters in particular" (Waddington,[11] p. 3).

the fundamental nature of things as well as the complex systems of the living world in terms of the underlying order, harmony, and unity of the universe. It acknowledges that every thing and every creature embodies this order in its own unique fashion, thus establishing individuality as an indispensable value. From the point of view of education, this means that we can now study the world—that is, nature, man, and society—as one finely articulated, interdependent, and interrelated whole.

References

1. H. Margenau. Intercultural communication of ethical judgments. Appendix to Wenner Lecture (November 17, 1966).
2. H. Margenau. Integrative education in the sciences. *Am. Assoc. Colleges Teacher Educ.* (1953), 134.
3. Joseph Engelberg. *To Life*. Unpublished Ms. Department of Physiology and Biophysics, University of Kentucky (1974), 8.
4. H. Margenau. *Phil. Sci. 2* (1935), 48; *2* (1935), 164. H. Margenau. *The Nature of Physical Reality*. New York: McGraw-Hill (1950).
5. F. S. C. Northrop. *The Meeting of East and West*. New York: Macmillan (1946).
6. Joseph E. Bogen. The other side of the brain (three papers). *Bull. Los Angeles Neurol. Soc. 34*, No. 2 (April 1969), 73–105; *34*, No. 3 (July 1969), 135–161; *34*, No. 4 (October 1969), 191–219.
7. Jean LeMée. Educating the professional. *Main Currents Modern Thought 31*, No. 3 (January–February 1975), 88–94.
8. H. Margenau. Thoughts on an integrative science curriculum for American colleges. *Main Currents Modern Thought 27*, No. 5 (May–June 1971), 164.
9. F. L. Kunz. The reality of the non-material, consequences to life and man of field physics. *Main Currents Modern Thought 20*, No. 2 (November–December 1963). F. L. Kunz. The metric of the living orders. In H. Margenau (ed.). *Integrative Principles of Modern Thought*. New York: Gordon and Breach (1972).
10. Adolf Portmann. *Animal Forms and Patterns*. New York: Schocken (1967).
11. C. H. Waddington. *Behind Appearance*. Cambridge, Mass.: MIT Press (1969).

PUBLICATIONS BY HENRY MARGENAU

The Zeeman effect in the cerium spectrum between 3000 and 5000A units. *Phys. Rev. 30* (1927), 458–465.

The problem of physical explanation. *The Monist 39* (1929), 321–349.

Dependence of ultra-violet reflection of silver on plastic deformation. *Phys. Rev. 33* (1929), 1035–1045.

Die Abweichungen vom Ohmschen Gesetz bei hohen Stromdichten im Lichte der Sommerfeldschen Elektronentheorie. *Z. Phys. 56* (1929), 230–231.

Zur Abweichung vom Ohmschen Gesetz bei hohen Feldstärken. *Z. Phys.* 60 (1930), 234–236.

Ueber die Veränderlichkeit der Anzahl freier Metallelektronen mit der Temperatur. *Phys. Z. 31* (1930), 540–546.

Zur Theorie der Molkekularkräfte bei Dipolgasen. *Z. Phys. 64* (1930), 584–597.

Second virial coefficient for gases: A critical Comparison between theoretical and experimental results. *Phys. Rev. 36* (1930), 1782–1790.

Causality and modern physics. *The Monist 41* (1931), 1–36.

Note on the calculation of Van der Waals forces. *Phys. Rev. 57* (1931), 1425–1430.

Surface energy of liquids. *Phys. Rev. 38* (1931), 365–371.

Role of quadrupole forces in Van der Waals attractions. *Phys. Rev. 38* (1931), 747–756.

The equation of state of real gases. *Phys. Rev. 38* (1931), 1785–1796.

The uncertainty principle and free will. *Science 74* (1931), 596.

Quantum dynamical correction of the equation of state of real gases. *Proc. Natl. Acad. Sci. 18* (1932), 56–62, 230.

Probability and causality in quantum physics. *The Monist 42* (1932), 161–168.

Pressure shift and broadening of spectral lines. *Phys. Rev. 40* (1932), 387–408.

Shift of the transmission band of silver by cold working. *Phys. Rev. 40* (1932), 800–801.

Pressure broadening of spectral lines, II. *Phys. Rev. 43* (1933), 129–134.

Pressure effects of foreign gases on the sodium D-lines. *Phys. Rev. 44* (1933), 92–98 (with W. W. Watson).

Pressure effects of nitrogen on potassium absorption lines. *Phys. Rev. 44* (1933), 748–752 (with W. W. Watson).

Zur Theorie der Verbreiterung von Spektrallinien. *Z. Phys. 86* (1933), 523–529.

Asymmetries of pressure broadened lines. *Phys. Rev. 44* (1933), 931–934.

Application of many-valued systems of logic to physics. *J. Phil. Sci. 1* (1934), 118–121.

Meaning and scientific status of causality. *J. Phil. Sci. 1* (1934), 133–148.

The complex neutron. *Phys. Rev.* (1934), 107–110.

Progression of nuclear resonance levels with atomic number. *Phys. Rev. 46* (1934), 228 (with Ernest Pollard).

Nuclear energy levels and the model of a potential hole. *Phys. Rev. 46* (1934), 613–615.

Flexibility of scientific truth. *Phil. Sci. 1* No. 4 (1934).

Methodology of modern physics, I. *Phil. Sci. 2* (1935), 48–72.

Methodology of modern physics, II. *Phil. Sci 2* (1935), 164–187.

Evidence regarding the field of the deuteron. *Nature 135* (1935), 393 (with Ernest Pollard).

Resonance interaction between deuterons and alpha-particles. *Phys. Rev. 47* (1935), 571 (with E. Pollard).

Collisions of alpha-particles in deuterium. *Phys. Rev. 47* (1935), 833 (with E. Pollard).

Natural width of K-α lines. *Phys. Rev. 47* (1935), 89.

Collisions of alpha-particles in hydrogen. *Phys. Rev. 48* (1935), 402.

Theory of pressure effects of foreign gases on spectral lines. *Phys. Rev. 48* (1935) 755.

Pressure effects on spectral lines. *Rev. Mod. Phys. 8* (1936), 22 (with W. W. Watson).

Quantum mechanical description. *Phys. Rev. 49* (1936), 240.
Note on pressure effects in band spectra. *Phys. Rev. 49* (1936), 596.
Foundations of Physics. New York: (1936) (with R. B. Lindsay).
Discussion on Nagel's paper. *J. Am. Stat. Assoc. 31* (1936), 27.
Relativity and nuclear forces. *Phys. Rev. 50* (1936), 342.
The teaching of intermediate physics. Invited paper read at the Brunswick Meeting of the New England Section of the American Physical Society. *Phys. Rev. 50* (1936), 872.
Critical points in modern physical theory. *Phil. Sci. 4* (1937), 337.
Long-range interactions between dipole molecules. *Phys. Rev. 51* (1937), 748 (with D. T. Warren).
Pressure broadening in bands of dipole molecules. *Phys. Rev. 51* (1937), 48 (with W. W. Watson).
Pressure shifts of krypton lines. *Phys. Rev. 52* (1937), 384 (with W. W. Watson).
Normal states of nuclear three and four body systems. *Phys. Rev. 52* (1937), 790 (with D. T. Warren).
Note on the validity of methods in nuclear calculations. *Phys. Rev. 52* (1937), 1027 (with D. T. Warren).
Excited states of the alpha particle. *Phys. Rev. 53* (1938), 198.
Pressure effects on spectral lines. Proceedings of 5th Conference on Spectroscopy, M.I.T. (1938), p. 123 (with W. W. Watson).
The teaching of intermediate physics. *Am. Phys. Teacher 6* (1938), 295.
Variational theory of the alpha-particle. *Phys. Rev. 54* (1938), 422 (with W. A. Tyrrell, Jr.).
Binding energy of Li^6. *Phys. Rev. 54* (1938), 705 (with K. G. Carroll).
Quadrupole contributions to London's dispersion forces. *J. Chem. Phys. 9* (1938), 896.
Van der Waals forces. *Rev. Mod. Phys. 11* (1939), 1.
Binding energies of light nuclei. *Phys. Rev. 55* (1939), 790 (with W. A. Tyrrell, Jr., and K. G. Carroll).
Binding energy of He^6 and nuclear forces. *Phys. Rev. 55* (1939), 1173.
Probability in modern physics. Invited paper read before 5th International Congress for Unity of Sciences, Harvard (1939).
Probability, many-values logics and physics. *Phil Sci. 6* (1939), 65.
Van der Waals potential in helium. *Phys. Rev. 56* (1939), 1000.
Relativistic magnetic moment of a charged particle. *Phys. Rev. 57* (1940), 383.
Magnetic moments of odd nuclei. *Phys. Rev. 58* (1940), 103 (with E. Wigner).
Interaction of alpha-particles. *Phys. Rev. 59* (1941), 37.

Statistics of excited energy states of nuclei. *Phys. Rev. 59* (1941), 627.
Development of the sciences. New Haven: Yale University Press (1941), pp. 91–120.
Foundations of the unity of science. *Phil. Rev. 1* (1941), 431.
Forces between neutral molecules and a metallic surface. *Phys. Rev. 60* (1941), 128 (with W. G. Pollard).
Metaphysical elements in physics. *Rev. Mod. Phys. 13* (1941), 176.
On the forces between positive ions and neutral molecules. *Phil. Sci. Tech. Sect. 4* (1941), 603.
The role of definitions in physical science, with remarks on the frequency definition of probability. *Am. J. Phys. 10* (1942), 224.
The development of the sciences. In *Physics*. New Haven: Yale University Press (1942), Chapter III.
The Mathematics of Physics and Chemistry. Princeton, N.J.: Van Nostrand (1943) (with G. Murphy).
Theory and scientific development. *Sci. Monthly 57* (1943), 63.
The Forces between hydrogen molecules.
Phys. Rev. 63 (1943), 385.
Phys. Rev. 63 (1943), 131.
The forces between water molecules and the second virial coefficient for water. *Phys. Rev. 66* (1944), 307 (with V. W. Myers).
The forces between a hydrogen molecule and a hydrogen atom. *Phys. Rev. 66* (1944), 303.
Phenomenology and physics. Invited paper, American Physical Society, Pittsburgh, 1943. *Phil. Phenomenol. Res. 5* (1944), 286.
The exclusion principle and its philosophical importance. *J. Phil. Sci. 11* (1944), 187.
Atomic and molecular theory since Bohr.
Article I: Historical survey. *Am. J. Phys. 12* (1944), 119 (with A. Wightman).
Article II: Logical and mathematical survey. *Am. J. Phys. 12* (1944), 247 (with A. Wightman).
Article III: Résumé of specific results. *Am. J. Phys. 13* (1945), 73 (with R. B. Setlow).
Film formation of water flowing through thin cracks. *Am. J. Sci. 243* (1945), 192 (with R. E. Meyerott).
On the frequency theory of probability. *Phil. Phenomenol. Res. 6* (1945), 11.
Conduction and dispersion of ionized gases at high frequencies. *Phys. Rev. 69* (1946), 508.
Physical processes in the recovery of TR tubes. *Phys. Rev. 70* (1946), 349

(with F. McMillan, Jr., I. H. Dearnley, C. S. Pearsall, and C. G. Montgomery).

Fysikkens Oprinnelse og Utvikling. Hefte I–II. Saertrykk av fra Fysikkens Verden (1947), 49–97.

Particle and field concepts in biology. *Sci. Monthly 44*, No. 3 (March 1947).

Western culture, scientific method and the problem of ethics. *Am. J. Phys. 15* (1947), 218.

Theory of high frequency gas discharges. I. Methods for calculating electron distribution functions. *Phys. Rev. 73* (1948), 297.

Theory of high frequency gas discharges. II. Harmonic components of the distribution function. *Phys. Rev. 73* (1948), 309 (with L. Hartmann).

Theory of high frequency gas discharges. IV. Note on the similarity principle. *Phys. Rev. 73* (1948), 326.

Pressure broadening in the inversion spectrum of ammonia. *Phys. Rev. 76* (1949), 131.

Collision theories of pressure broadening of spectral lines. *Phys. Rev. 76* (1949), 1211.

Inversion frequency of ammonia and molecular interaction. *Phys. Rev. 76* (1949), 1423.

Reality in quantum mechanics. *Phil. Sci. 16* (1949), 287.

Einstein's conception of reality. In Einstein volume, Library of Living Philosophers (1949). Paul Schilpp, Ed.

Ethical science. *Sci. Monthly 69* (1949), 290.

Theory of magnetic resonance in nitric oxide. *Phys. Rev. 78* (1950), 587 (with A. F. Henry).

Electric conductivity and mean free paths. *Phys. Rev. 79* (1950), 970.

The Nature of Physical Reality. New York: McGraw-Hill (1950).

The Nature of Concepts (two chapters). Norman, Okla.: University of Oklahoma Press (1950) (ed. with F. S. C. Northrop).

The meaning of "elementary particle." *Am. Scientist 39* (1951), 422.

Statistical theory of pressure broadening. *Phys. Rev. 82* (1951), 156.

Conceptual foundations of the quantum theory. *Science 113* (1951), 95.

Report on recent developments in philosophy of quantum mechanics. (with J. Compton).

Reply to Professor Beck on Kantianism. *Phil. Phenomenol. Res.* 11, No. 4 (1951).

Book reviews and minor articles replying to comments elicted by *The Nature of Physical Reality*.

Physics and ontology. *Phil. Sci. 19* (1952), 342.

Physical versus historical reality. *Phil. Sci. 19* (1952), 193.

Intermolecular forces in helium. Note 21. Conference on Quantum Mechanical Methods in Valence Theory, September 8–10, 1951 (published 1952).

On a possible method for estimating the repulsive potential between closed shells. Note 22. Conference on Quantum Mechanical Methods in Valence Theory, September 8–10, 1951 (published 1952).

Note on the calculation of exchange forces. Note 23. Conference on Quantum Mechanical Method in Valence Theory, September 8–10, 1951 (published 1952).

Ion clustering. *Phys. Rev. 85* (1952), 670 (with S. Bloom).

Scientific bases of ethics: Lecture. *Main Currents Modern Thought* (1952).

Quantum theory of spectral line broadening. *Phys. Rev. 90* (1953), 791 (with S. Bloom).

On the interaction of closed shells. *J. Chem. Phys. 21* (1953), 394 (with P. Rosen).

The deductive method in the physical sciences. *Electr. Eng.* (April 1953).

The forces between hydrogen molecules.
Phys. Rev. 90 (1953), 1021.
J. Chem. Phys. 21 (1953), 958 (with A. A. Evett).

Integrative education in the sciences. *Am. Assoc. Colleges Teacher Educ.* (1953), 132.

The new faith of science. Carleton College (1953), pp. 1–20.

The deductive method in the physical sciences. *Main Currents* (March 1953), 131.

Physics. New York: McGraw-Hill (1953), 2nd ed. (with W. W. Watson and C. C. Montgomery).

Can time flow backwards? *Phil. Sci. 21* (1954), 79.

Advantages and disadvantages of various interpretations of the quantum theory.
J. Wash. Acad. *Sci. 44* (1954), 265.
Physics Today 7 (1954), 6.

On interpretations and misinterpretations of operationalism. *Sci. Monthly 79* (1954), 209.

Causality in quantum electrodynamics. *Diogenes 6* (1954).

Quantum theory of line broadening by an ionic plasma. *Astrophys. J. 121* (1955), 194 (with R. Meyerott).

Electron impact broadening of spectral lines. *Phys. Rev. 98* (1955), 495 (with B. Knivel and S. Bloom).

Knowledge, faith and physics. *Main Currents in Modern Thought 11* (1955), 108.

The competence and limitations of scientific method. *J. Oper. Res. Soc. Am. 3*, No. 2 (1955), 135.

Effects of electron collisions on the width of spectral lines. *Phys. Rev. 98* (1955), 1822 (with B. Kivel).

Line broadening by electrons: Validity of simple theories. *Phys. Rev. 99* (1955), 1851 (with R. Meyerott).

Facts and values. *Brown University Papers 31* (1955), 1–21 (commencement address).

Present status and needs of the philosophy of science. *Proc. Am. Phil. Soc. 99* (1955), 334.

The Mathematics of Physics and Chemistry. Princeton, N.J.: Van Nostrand (1956), 2nd ed. (with G. Murphy).

Estimate of pressure effects on No-band lines. *USAF Rand Res. Memo. 1669* (1956), I-S.

The validity of the statistical theory of pressure broadening. *Rand Res. Memo. RM-1670-AEC* (1956).

Stark effects in line broadening. *Rand Res. Memo. RM-1779-AEC* (1956).

A unified method of science: Can it be applied to religion? *Christian Register* (1956).

Line broadening of an impurity spectrum in silicon. *Phys. Rev. 103* (1956), 879 (with D. Sampson).

Why teach philosophy of science? *Age of Science*, Vol. 2 (1956).

Moderna Primer Congreso de la Sociedad Interamericana de Filosofia. Santiago, Chile (1956).

The new view of man in his physical environment. *Centennial Rev. Arts Sci. 1*, No. 1 (1957).

The meaning and the faith of science. *Main Currents in Modern Thought 13*, No. 3 (1957).

Stark effects of line broadening. *Phys. Rev. 106* (1957), 244 (with M. Lewis).

Philosophy of science in the twentieth century. *Jr. World Hist.* (Summer 1957).

Relativity: An epistemological appraisal. *Phil. Sci. 24* (1957), 297.

Van der Waals forces. In G. Clark (ed.). *Encyclopedia of Chemistry*. New York: Reinhold (1957).

The modern predicament. *Main Currents in Modern Thought 13* (1957).

El nuevo concepto de hombre en su ambiente fisico. Episteme, Anuario de Filosofia, Universidad Central de Venezuela, Caracas (1957), 312.

Cyclotron resonance: Method for determining collision cross sections for low energy electrons. *Phys. Rev. 108* (1957), 1368 (with D. Kelly and S. C. Brown).

Philosophical problems concerning the meaning of measurement in physics. *Phil. Sci. 25* (1958). 23.

Conductivity of plasmas to microwaves. *Phys. Rev. 109* (1958), 6.

Statistical broadening of spectral lines emitted by ions in a plasma. *Phys. Rev. 109* (1958), 842.

Conductivity of plasmas to microwaves. *Phys. Rev. 112* (1958), 1437.

Structure of spectral lines from plasmas. *Rev. Mod. Phys. 31* (1959), 569 (with M. Lewis).

Perspectives of science. *KEY Reporter, Phi Beta Kappa* (July 1959).

Philosophy of physical science in the twentieth century. *J. World Hist. 4*, No. 3 (1958).

Fields in physics and biology. *Main Currents in Modern Thought 15* (1959).

Theory of pressure effects on alkali doublet lines. *J. Chem. Phys. 30* (1959), 1556 (with L. Klein).

Frequency shifts in hyperfine splitting of alkalis caused by foreign gases. *Phys. Rev. 115* (1959), 87 (with P. Fontana and L. Klein).

Microwave conductivity of slightly ionized air. *J. Appl. Phys. 30* (1959), 1385 (with D. Stillinger).

Naturphilosophie: Die Philosophie im XX. Jahrhundert. Stuttgart: Ernst Klett Verlag (1959).

Scientific basis of value theory. In A. Maslow (ed.). *New Knowledge in Human Values*. New York: Harpers (1959).

The structure of spectral lines from plasmas. Proceedings of IV International Conference on Ionization Phenomena in Gases, Uppsala (August 17–21, 1959).

Formulas for estimating widths of spectral lines emitted from plasmas and their limits of validity. Proceedings of IV International Conference on Ionization Phenomena in Gases, Uppsala (August 17–21, 1959).

Elargissement des raies spectrales produites par des ions dans un plasma. *Coll. Int. CNRS 77* (Bellevue, July. 1–6, 1957, published 1959) (with M. Lewis).

Essais d'interpretation theorique des bandes satellites induites par la pression. *Coll. Int. CNRS 77* (Bellevue, July 1–6, 1957, published 1959) (with L. Klein).

High frequency breakdown of air. *J. Appl. Phys. 31* (1960), 1617 (with D. Kelly).

Forbidden helium line in a plasma spectrum. *J. Quant. Spectros. Radiat. Transfer 1* (1960), 46 (with L. Sadjian and H. Wimmel).

Meaning and scientific status of causality. In A. Danto (ed.). *Philosophy of Science*. New York: Meridian Books (1960).

Science, philosophy and religion. *Dalhousie Rev. 39* (1960), 447.

Does physical "knowledge" require *a priori* or undemonstrable presuppositions? In L. W. Friedrich, S.J. (ed.). *The Nature of Physical Knowledge*. Milwaukee: Marquette University Press (1960).

Causality. *Main Currents in Modern Thought 16* (1960).

Frequency shifts in hyperfine splitting of alkalis: A correction. *Phys. Rev. 122* (1961), 1204 (with R. Herman).

Bacon and modern physics: A confrontation. *Proc. Am. Phil. Soc. 105* (1961), 5.

Correlations between measurements in quantum theory. *Prog. Theor. Phys. 26* (1961), 722 (with R. Hill).

Preliminaries. In D. R. Bates (ed.). *Quantum Theory*. Vol. I. New York: Academic Press (1961).

Fundamental principles of quantum mechanics. In D. R. Bates (ed.). *Quantum Theory*. Vol. I. New York: Academic Press (1961).

Causality in quantum electrodynamics. In E. Ryan (ed.). *A Science Reader*. New York: Holt, Rinehart and Winston (1961).

Open Vistas. New Haven: Yale University Press (1961).

Is the mathematical explanation of physical data unique? Proceedings of 1960 International Congress (published 1962).

Discussion: Comments on Professor Putman's comments. *Phil. Sci. 29* (1962), 292 (with E. Wigner).

The new style of science.
Main Currents 19 (1963).
Yale Alum. Mag. (1963).

Theory of pressure shifts of HCl lines caused by noble gases.
J. Chem. Phys. 38 (1963), 1 (with H. Jacobson).
J. Quant. Spectros. Radiat. Transfer 3 (1963), 35 (with H. Jacobson).

Measurements and quantum states.
Part I. *Phil. Sci. 30* (1963), 1.
Part II. *Phil. Sci. 30* (1963), 138.

Intercultural communication of ethical judgments. XIII Congreso Internacional de Filosofia, Universidad Nacional Autonoma de Mexico (1963), 65.

Absorption coefficient and microwave conductivity of plasmas. *Astrophys. J. 137* (1963), 851 (with P. Mallozzi).

Measurement in quantum mechanics. *Ann. Phys. 23* (1963), 469.

Theory of pressure shifts of HCl lines caused by noble gases. *J. Quant. Spectros. Radiat. Transfer 3* (1963), 445.

Philosophy of physical science in the twentieth century. In G. Melraux (ed.). *Evolution of Science.* New York: Mentor Books (1963) (with J. Smith).

Disucssion: Reply to Professor Putnam. *Phil. Sci. 31* (1964), 7 (with E. Wigner).

Mathematics of Physics and Chemistry. Vol. II. Princeton, N.J.: Van Nostrand (1964) (with G. Murphy).

Ethics and Science. Princeton, N.J.: Van Nostrand (1964).

Book review of "Physics in the Soviet Union" by A. S. Kompaneyets. *Am. J. Sci. 263* (1965).

On interpretations and misinterpretations of operationalism. In F. T. Severin (ed.). *Humanistic Viewpoints in Psychology.* New York: McGraw-Hill (1965).

The Scientist. Life Science Series. New York: Time, Inc. (1965) (with D. Bergamini).

Role of scholarship in the age of science. Speech on inauguration of new president of University of Wyoming (published by University of Wyoming, Laramie, 1965).

Man in nature. *Hartwick Rev.* (*Oneonta, N.Y.*) *1*, No. 1 (1965).

Modern science in philosophical perspective. Speech given at Symposium at University of Western Ontario (published in *Science in Industry and the Universities*, 1965).

What is a theory. In S. Krupp (ed.). *The Structure of Economic Science.* Englewood Cliffs, N.J.: Prentice-Hall (1966).

The philosophical legacy of contemporary quantum theory. In R. Colodny (ed.). *Mind and Cosmos.* Vol. III. Pittsburgh: University of Pittsburgh Press (1966).

Theory of plasma radiation. Part I: Mathematical development. Proceedings of the 7th International Conference on Phenomena in Ionized Gases, Beograd, Yugoslavia (1966) (with P. Mallozzi).

Theory of plasma radiation. Part II: Physical applications. Proceedings of the 7th International Conference on Phenomena in Ionized Gases, Beograd, Yugoslavia (1966) (with P. Mallozzi).

Statistical theory of radiative processes in plasmas. *Ann. Phys. 38* (1966), 177 (with P. Mallozzi).

Simultaneous action of resonance and van der Waals forces. *Technical Report No. AFWL-TR-66-35*, Vol. II (1966).

Exclusion principle and measurement theory. In P. O. Löwdin (ed.). *Quantum Theory of Atoms, Molecules and the Solid State*. New York: Academic Press (1966).

ESP in the framework of modern science. *J. Am. Soc. Psychical Res. 6* (1966), 214.

Science and philosophical perspectives. Proceedings of the 1966 National Junior Science and Humanities Symposium, Ft. Monmouth, N.J. (1966).

Nonadditivity of intermolecular forces. In P. O. Löwdin (ed.). *Quantum Theory of Chemistry*. Vol. 3. New York: Academic Press (1967) (with James Stamper).

Pursuit of significance. *Main Currents 23*, No. 3 (1967).

Objectivity in quantum mechanics. Chapter 10 In Mario Bunge (ed.). *Foundations of Physics*. Berlin: Springer-Verlag (1967) (with James Park).

Quantum mechanics, free will and determinism. *J. Phil. 64* (1967), 714.

Integrative education in the sciences. *Main Currents in Modern Thought 24* (1967).

The search for meaning in a senseless world. *Hartwick Rev. 4* (1968), 56.

Scientific indeterminism and human freedom: Wimmer lecture. Latrobe, Pa.: Archabby Press (1968).

Causality in contemporary philosophy. In R. Klibanski (ed.). *La Nova Italia Editrice*. Firenze (1968) (with B. van Fraassen).

Simultaneous measurability in quantum theory. *Int. J. Theor. Phys. 1* (1968), 211 (with J. Park).

Theory of Intermolecular Forces. Oxford: Pergamon (1969), 2nd ed. 1971) (with N. Kestner).

Integrative Principles of Modern Thought. New York: Gordon and Breach (1971) (ed.).

Emergence of integrative concepts in contemporary science. *J. Phil. Sci. 39* (1972), 252 (with E. Laszlo).

Physics and semantics of quantum measurement. *Found. Phys. 3* (1973), 19 (with J. L. Park).

On Popper's Philosophy of Science. Popper Volume, *Library of Living Philosophers* (P. Schilpp, ed.), 1974.

On Blanshard's Ethics. Blanshard Volume, *Library of Living Philosophers* (P. Schilpp, ed.), 1975.

INDEX

AFTER THERAPY WHAT?

AFTER THERAPY WHAT?

Lay Therapeutic Resources in Religious Perspective

The Second John G. Finch Symposium on Psychology and Religion

Finch Lectures by
THOMAS C. ODEN, Ph.D.

with Responses by
NEIL C. WARREN, Ph.D.
KENNETH B. MULHOLLAND, D.Min.
CALVIN R. SCHOONHOVEN, D.Theol.
CHARLES H. KRAFT, Ph.D.
WARREN WALKER, S.T.B.

Edited by
NEIL C. WARREN, Ph.D.

CHARLES C THOMAS • PUBLISHER
Springfield, Illinois, U.S.A.

Published and Distributed Throughout the World by
CHARLES C THOMAS • PUBLISHER
Bannerstone House
301-327 East Lawrence Avenue, Springfield, Illinois, U.S.A.

ISBN 0-398-03105-3

Library of Congress Catalog Card Number: 73-21983

With THOMAS BOOKS careful attention is given to all details of manufacturing and design. It is the Publisher's desire to present books that are satisfactory as to their physical qualities and artistic possibilities and appropriate for their particular use. THOMAS BOOKS will be true to those laws of quality that assure a good name and good will.

Printed in the United States of America
PP-22

Library of Congress Cataloging in Publication Data

John G. Finch Symposium on Psychology and Religion, 2d, Pasadena, Calif., 1973.
After therapy what?

1. Psychiatry and religion—Congresses. 2. Psychotherapy—Congresses. 3. Pastoral counseling—Congresses. I. Oden, Thomas C. II. Title. [DNIM: 1. Psychology, Pastoral—Congresses. 2. Psychotherapy—Congresses. 3. Religion and psychology—Congresses. W3J012 1973a/WM61 J632 1973a]

RC480.5.J63 1973 261.8'32'2 73-21983
ISBN 0-398-03105-3

CONTRIBUTORS

Thomas C. Oden, Ph.D.: Professor of Theology and Ethics, Theological School, Drew University, Madison, New Jersey.

Neil Clark Warren, Ph.D.: Associate Professor of Psychology, Graduate School of Psychology, Fuller Theological Seminary, Pasadena, California.

Kenneth B. Mulholland, D.Min.: Formerly a graduate student in the School of Theology, Fuller Theological Seminary, Pasadena, California. Presently a Professor, Seminario Biblico Latinamericano, San Jose, Costa Rica.

Calvin R. Schoonhoven, D.Theol.: Assistant to the President in Academic Affairs, Director of Library and Associate Professor of Biblical Interpretation, School of Theology, Fuller Theological Seminary, Pasadena, California.

Charles H. Kraft, Ph.D.: Associate Professor of Missionary Anthropology and African Affairs, School of World Mission, Fuller Theological Seminary, Pasadena, California.

Warren Walker, S.T.B., Student in the Graduate School of Psychology, Fuller Theological Seminary, Pasadena, California.

OTHER BOOKS BY THOMAS C. ODEN

THE COMMUNITY OF CELEBRATION

RADICAL OBEDIENCE: THE ETHICS OF RUDOLF BULTMANN

KERYGMA AND COUNSELING: TOWARD A COVENANT ONTOLOGY FOR SECULAR PSYCHOTHERAPY

CONTEMPORARY THEOLOGY AND PSYCHOTHERAPY

THE STRUCTURE OF AWARENESS

THE PROMISE OF BARTH: THE ETHICS OF FREEDOM

BEYOND REVOLUTION: A RESPONSE TO THE UNDERGROUND CHURCH

THE INTENSIVE GROUP EXPERIENCE: THE NEW PIETISM

GAME FREE: THE MEANING OF INTIMACY

FOREWORD

It is gratifying to know that the John G. Finch Symposium has become a reality. This volume is the fruit of the second series of lectures. A faculty never knows whether its ideas will stand the test of time, but it appears that this symposium inaugurated by the Graduate School of Psychology of Fuller Theological Seminary will, indeed, make a continuing contribution to the better understanding of the relationships among theology, psychology, and anthropology.

The name of John G. Finch is one that has special meaning for us but in a wider sense stands for the combination of the "great men" and "Zeitgeist" theories of history. His long-standing interest in the training of psychotherapists and his efforts toward the establishment of a school wherein psychology and theology could be studied together are recalled by the naming of this symposium in his honor. We are gratified that this effort is bearing fruition and that his work is being dignified in this way.

We are appreciative of the scholarship of Thomas C. Oden and of the efforts of Neil C. Warren which have led to making this volume a reality. At this point in time, we perceive that this will be but one of many significant contributions which come from these yearly symposia.

H. NEWTON MALONY

PREFACE

Psychotherapy and the Christian faith can be related on many levels. Thomas C. Oden's involvement with a theoretical analysis of the two fields led to the publishing of *Kerygma and Counseling* in 1966 and *Contemporary Theology and Psychotherapy* in 1967.

Now he returns to the task of interrelating these disciplines, and he "hones in" at a different level. In focusing on the *practice* of psychotherapy he argues that the church can and should become more vitally involved. The thrust of his position is that psychotherapy has become excessively professionalized and increasingly ineffective in meeting the widespread need, and that the therapeutic agency must be gradually transferred to laymen.

But beyond his proposals for the laicization of psychotherapy, Professor Oden argues that the Christian faith provides answers to the basic questions of life and death which continue to gnaw at many who have completed therapy and experienced all that it can do.

The most controversial aspect of the lectures is their attack on the current delivery systems for psychotherapy and their proposal to make the church a primary source of emotional guidance and a primary context for emotional growth and change.

But some of the most exciting discoveries which Oden makes involve an application of the work of theologian Wolfhart Pannenberg to the questions to which psychotherapy is ordinarily addressed. The proposals which emerge from this effort create new potential for meaningfully relating psychotherapy and Christian faith.

The book contains five responses to Oden's lectures. The discussion is joined by a theologian, an anthropologist, and a psychologist. In addition to these three the text is greatly enriched by the inclusion of two other responses—one from a doctoral student in the School of Theology and the other from a student in Fuller's Graduate School of Psychology. Professor Oden, in a final chapter, has an opportunity to answer the "responders."

The John G. Finch Symposium on Psychology and Religion is

made possible through the efforts of many persons. This year's lecture series was planned and managed by a student committee which spent scores of hours. David Bock, Bruce Crapuchettes, John Davis, Edward Rettig and Robert Richard served on this committee. The entire psychology faculty, but in particular H. Newton Malony, the originator of this series and Lee Edward Travis, Dean of the Graduate School of Psychology, contributed in a very supportive way. Of critical importance was the constant encouragement and assistance of the Administrative Assistant of the School of Psychology, Marnie Frederickson.

Finally, I am especially grateful to Chris Welch for her many contributions to the task of readying the manuscript, and to Winifred Bock for her assistance with the typing.

The task of relating the disciplines of theology and psychology requires our best effort. The Finch Symposium is dedicated to this process. It is hoped that this volume, when read in its entirety, will stimulate new discussion and provoke additional insights which will further our understanding and increase our momentum.

NEIL CLARK WARREN

CONTENTS

AFTER THERAPY WHAT?

CHAPTER 1

THE DIFFUSION OF THERAPEUTIC AGENCY

THOMAS C. ODEN

EVERY TIME I venture into the dialogue between psychotherapy and theology, I experience a sense of risk. As I now venture again, after a five-year moratorium on this subject following two previous books,[1] I experience a sense of risk even more keenly, since some of my current proposals are likely to be distasteful to persons with whom I otherwise find broad areas of agreement and mutual interest. Particularly I anticipate heavy flack from certain professional psychotherapists who are relatively pleased with what is currently happening in therapeutic theory and practice, from clinical pastoral supervisors who are trying to intensify their professionalism, and from certain accommodationists who would wish more easily (and uncritically) to absorb psychotherapy into religion.

In this essay I am going to explore the theme of the transmutation of psychotherapy through its laicization. Like a magnet we are drawn toward the question of the alleged incipient death of psychotherapy. Already Erich Fromm has raised this question from the post-Freudian perspective in *The Crisis of Psychoanalysis* when he speaks of psychoanalysis "stagnating," "losing its character," being "on the defensive" and "losing its therapeutic monopoly." "Its death can be predicted," he says, "*unless* it changes its direction."[2] Fromm makes no attempt, however, to probe the more decisive sociological aspects of the laicization movement which appears to be the most serious crisis which all psychotherapies, and not psychoanalysis only, are now facing.

Is it possible that we are now witnessing the senility and demise of a great tradition of healing from Freud to Fromm, Rank to Rogers,

Jung to Jourard, Binswanger to Laing? I believe the hegemony of the typical *one-to-one* pattern of professional psychotherapy following the medical model is waning. The language and myths of the great classical schools of Freud, Jung, Rank, Adler, and even client-centered therapy today have a very archaic sound. I suspect that much (although not all) of this revered tradition is now in its senile stage. It is now in a defensive posture, but in my opinion, the larger therapeutic tradition is far from dead. It is in fact in process of being transmuted.

Thesis: the excessive professionalization of psychotherapy following the medical model has allowed therapeutic agents to get themselves trapped in a style of professional transactions which are counterproductive to the task of being a liberating, growthproducing therapeutic agent. Introverted professionalism not only fails to meet widely acknowledged therapeutic objectives, but actually works against them. Professional time frames, barriers, rituals and games may prevent the self-disclosure and empathy necessary for a genuine therapeutic agent. The psychiatrist, a medical specialist, is ill-advisedly taken as the prototype of the therapeutic agent. No vocation is more disastrously and excessively professionalized in our society than medicine. It is the worst available example, and yet the one which most deeply shapes professional psychotherapeutic practice. If the ministry or priesthood had not been burdened with quaint and archaic role expectations, it might have served as a more viable model than medicine.

THERAPY AS AUTHENTICITY

In previous discussions I have assumed the value and efficaciousness of psychotherapy, and argued that it can be best understood by means of theological analogies. I have even defended fee basis counseling.[3] Now I find myself experiencing serious reservations about the practice of psychotherapy professionally. One force which is changing my mind is the laicization of therapy which is transmuting the entire therapeutic scene in a way which religious communities need carefully to assess.

I still have no doubt that there is such a thing as a therapeutic relationship, and that it is still well-described as a relationship in which self-disclosure, congruence, empathy, unconditional positive regard and other processes elicit openness, accountability, behavior modification and positive growth. What has shifted in my thinking is the strengthening intuition that the therapeutic agent need not be a professional. He may just as easily, and sometimes even better, be a friend, brother, husband, partner, sister, father, son, teacher, neighbor.

In a previous essay I have proposed the thesis that the "therapeutic relationship" is nothing other than a relationship of authenticity with another.[4] It is that earlier intuition that I now wish to develop in more depth and inquire into its consequences for the practice of psychotherapy.

Psychological growth may be achieved in any authentic human relationship and not psychotherapy only. Psychotherapy just happens to be a place where clients have a right to expect that it will happen in decisive ways. The growth-producing aspects of the therapeutic agent are not finally his diagnostic capacities, his professional education, his technical expertise, or his understanding of developmental psychology, and certainly not the placebo effect, but rather hinge more finally upon his personal authenticity, openness, capacity for self-disclosure, and his ability to mediate actual redemptive possibilities to the neighbor.

Many persons are discovering that this sort of authentic relationship can be and is being mediated through ordinary people, not professional therapists, in leaderless groups, or in permissively led groups where the function of the leader is to enable and release the therapeutic power of the ordinary persons in the group.[5] No longer can the professional make an unchallenged claim that he can mediate a therapy or elicit constructive change in a capacity entirely out of the range of the nonprofessional.[6]

When I have spoken of the therapeutic transaction as analogous to the therapeutic action of God toward history, I have never intended to refer merely to professional therapy, but more particularly to any therapeutic agent, whether he appears in professional or lay form.

I do not mean to be saying that we no longer need professional helpers in our society to take care of persons who are deeply conflicted and neurotically unable to function. I think it would be disastrous if the professional psychotherapists would cease functioning altogether, although I would like to see their professionalism redefined and humanized, and the delivery systems for it reformed. My intent is not to discredit the professionals, but to call them to a more modest assessment of their function, and an awareness that they are not the only ones who mediate therapeutic relationships, as they themselves have come to define therapy.[7]

My misgivings are not directed toward any particular school of psychotherapy, but instead more toward a general trend among therapeutic professionals of differing orientations. For amid even the theoretical orientations with which I personally would most agree (namely, the humanistic psychologists, Gestalt therapists and transactional analysts),[8] it is not difficult to find depressing examples of therapists who operate sociologically still in terms of the doctor-patient model of a human relationship and with the mental illness imagery persisting. Likewise among Freudian and behaviorist orientations from which one might expect a more authoritarian relationship, one can also find many therapists who operate on more democratic, humanized models. So my caveat is not directed against any particular theoretical orientation, but against the dysfunctional aspects of professionalization in any and all schools.

THE MOMENTUM OF CRITICISM

It is not my intention to review (though such a review might be useful) all of the substantial and persuasive criticisms of therapy. Part of the health of the therapeutic tradition is its ability to generate self-criticism, even though it does not always digest that criticism fully. Among the most provocative critics, I would note:

THE CRITIQUE OF THE MEDICAL ANALOGY FOR PSYCHOTHERAPY. Psychiatrist Thomas Szasz has delivered a startling and decisive critique of concepts of mental health and mental illness upon which so much therapeutic activity hinges.

> It is customary to define psychiatry as a medical specialty concerned with the study, diagnosis, and treatment of mental illnesses. This is a worthless and misleading definition. Mental illness is a myth. Psychiatrists are not concerned with mental illnesses and their treatments. In actual practice they deal with personal, social, and ethical problems in living. . . The notion of a person "having mental illness" is scientifically crippling. It provides professional assent to a popular rationalization, namely, that

problems in human living. . . are significantly similar to *diseases of the body.* It also undermines the principle of personal responsibility. . . by assigning to an external source (i.e. the "illness") the blame for anti-social behavior.[9]

BEHAVIORIST DOUBTS. Therapy has not proved as successful as it was once generally thought. Constructive change in traditional psychotherapies, according to careful studies by Eysenck and others,[10] does not occur any faster than the normal rate of cure with the simple passing of time.

Eminent behaviorists such as B. F. Skinner, O. H. Mowrer, J. Wolpe, P. London, and others [11] have made a significant charge against all therapies which begin with an analysis of feelings and insight, hoping for behavioral change on that basis. We have come to the end of a therapeutic tradition which can remain completely uncritical about such a charge, or which would continue to do analysis and therapy oblivious of the behaviorist criticism. Insight therapy may not be collapsing, but it may no longer be taken as a normative approach. Behaviorists such as Mowrer challenge the naïve but widely held assumption that simply the release of the superego from coercive constraints will be health producing.[12] Although no respected therapist would openly admit to that assumption, many clients come away from therapy somehow learning that dubious assumption from the therapeutic process as if it were an elementary behavioral maxim.

THE POSITIVE DIMENSION OF MADNESS. In a sharp attack on mainstream psychiatry, psychiatrists Roland Laing and David Cooper ironically affirm and even prize the experience of so-called mental illness as a profound human quest, a deep moment of human consciousness. If they are right, we have come to an end of the paternalistic tradition of medical therapy which would assume that the therapist is well and the patient is sick and a clear line can be drawn between the two, with no sense of common grappling with the human predicament by both.

"Madness", says Laing, "need not be all breakdown . . . It is potential liberation and renewal. . . . Our sanity is not 'true' sanity. The madness of our patients is an artifact of the destruction wreaked on them by us and by them on themselves."[13]

Antipsychiatric psychiatrist David Cooper writes: "If one is to speak of violence in psychiatry, the violence that stares out screaming, proclaiming itself as such so loudly that it is rarely heard, is the subtle, tortuous violence that other people, the 'sane ones,' perpetrate against the labelled madmen. In so far as psychiatry represents the interests or pretended interests of the sane ones, we may discover that, in fact, violence in psychiatry is preeminently the violence of psychiatry."[14] Although I do not join all aspects of this abrasive critique, it is at least important to recognize that psychiatry is under attack as never before, from both within and without.

THE SOCIAL CONSEQUENCES OF PSYCHOTHERAPY. The alleged disastrous effects of psychoanalysis upon Western culture are beginning to be pointed out, and need to be better assessed by religious communities. If Richard LaPiere's argument is correct, then much therapy is on a self-destructing path, since it deliberately undermines the very social fabric on which it parasitically bases itself.

Sociologist LaPiere believes that:

> The Freudian ethic does sanction everyone's living irresponsibly upon society, and the Freudian theory of man assures us that no one can really be expected to be socially responsible. But neither Freud nor any of his disciples or lay followers has ever explained how it is possible for society to provide security—psychological and otherwise—for each of its several members if no one of them can endure the strain of contributing to the support of society.[15]

Although overstated, the deeper issue cannot be lightly dismissed.

The fragile pedestal upon which some have tried to place psychotherapy is showing signs of severe stress. Therapeutic hopes are being disappointed. Wherever the tradition has been idolized, it is appearing to have clay feet. An increasingly wary and sophisticated public is unconvinced of the Freudian mythic apparatus which would explain football watching as sexual sublimation, goal posts as phallic symbols and touchdowns as orgasms.

Therapy can never again be the same. It has come to the end of its hegemony, its unquestioned place as an arbiter of cultural attitudes and values in certain quarters where it has stood with minimal challenge for a quarter of a century.

It is not my purpose to amplify or even restate such criticisms, though they are many and varied, and sometimes persuasive, but to shift now to a description of a phenomenon which more powerfully addresses a new genre of questions to professional psychotherapy, viz. the laicization process occurring in encounter groups and the so-called encounter culture.

THE RESOURCES OF ORDINARY PEOPLE

Psychotherapy is now in the process of being laicized, and the religious communities have an important role to play in making more constructive this volatile process of laicization.

THE CHALLENGE TO ELITISM. It is difficult for professionals, whether in medicine, religion or psychology, to make sense out of the fact that ordinary people in fact have become significant resources for personal growth for other ordinary people whose communication systems are not functioning well. Professional elitism, whether in the clergy or in therapeutic circles, tends to assume that ordinary people cannot be therapeutic agents. Such assumptions are being challenged today by the encounter culture with its heavy reliance upon ordinary people as agents of constructive behavioral change.

Groups usually need facilitators, but the real healing competencies reside in the ordinary people in the group, not in the professional therapeutic agent. The task of the facilitator is to mobilize the therapeutic competencies of the ordinary people.

THE POPULIST TRADITION. The encounter culture's discovery of the trustworthiness of ordinary people is analogous politically to the populist movement's conviction that ordinary people are capable of shaping their own destinies, and that professional elites too often work primarily for themselves and therefore subtly against the interests of defenseless clients. There is a deep do-it-yourself populist suspicion in the American tradition. In the nineteenth century its distrust was focused upon the wily lawyer, the shrewd banker, the corrupt politician. Wherever a strong professional elite has arisen in American culture, the populist critique has gone to work. There is a corollary trust in the competence of ordinary people to help themselves. This tradition is still alive among American young people who today are raising their own organic food instead of buying manufactured products, taking their own vitamins and doing their own preventive medicine instead of appealing to medical professionals, shaping their own religious lives instead of looking to the professional religious elites.[16] They regard their own competencies as equal or better than the partially self-interested professional elites. Although this tradition has hazards and limitations, none can deny that it is deeply engrained in the frontier American spirit and that it in fact has borne some impressive fruits.

THE PURCHASE OF FRIENDSHIP. The particular advantage which lay therapeutic agency has over professional is that it cannot be bought. When psychotherapy becomes "the purchase of friendship," as Schofield described it, its ingenuousness is suspect.[17]

One group encounter participant wrote, "This is different than personal therapy. None of you *had* to care for me. None of you had to seek me out and let me know of things you thought would help me . . . yet you did, and as a result it has far more meaning than anything I have so far experienced."[18] This points out a decisive difference between professional psychotherapy and group encounter. One does not buy the resources, the strength, the support and trust of the group. One can only receive it in a relationship of mutual trust and risk.

THE DEFENSIVE RUMOR. There is a widely held assumption that participants should be professionally screened or given psychological tests to determine normalcy before entering encounter groups. Gibb rightly argues that there is no reliable basis for making any restrictions whatever as to group membership.[19] Rogers pointedly says, "No one should be admitted unless he is a person."[20]

There does seem to be a rumor phenomenon which aims its gossip at the disturbing, emotively bizarre and potentially dangerous aspects of intensive group experiencing.[21] Gibb however reports a careful study of 1200 YMCA directors who had been involved in groups. "Rumors were widespread in the organization of 'severe psychological disruption due to training.' It was found in fact that only four of twelve hundred had felt their experience was negative. By the time the investigators had reached these four, three had decided the experience had actually been helpful."[22] Rogers explains the rumor phenomenon in this way: "Many individuals are threatened by the possibility of change . . . they are dimly aware that the major outcome of a group experience is *change*. Thus when they hear that someone wept in the group or spent a sleepless night . . . they leap to the conclusion that groups are bad and psychologically destructive.

They thus save themselves from being exposed to the possibility of change."[23]

THE DIFFUSION OF THERAPEUTIC AGENCY

The resistance which the encounter culture has received from the psychotherapeutic professionals reminds me of the resistance which Protestant evangelical pietism received from the religious professionals of the eighteenth century. Today, as in the heyday of Hassidic Judaism and Protestant pietism, we are witnessing the dispersal and diffusion of therapeutic agency to the laity.[24]

There is a confusion of tongues today in professional circles regarding criteria of competency for doing therapy and group processes. Amid this confusion, I celebrate the freedom such a situation gives to religious ministries to reclaim functions which they have tended to lose the last two centuries. Amid the maze of jurisdictional disputes about who should be doing what kind of therapy, I say to my theological students that ordinary people should be viewed as behavior change agents and that the pastoral task is to mobilize the growth-eliciting capacities of the laity.

Analogies with ordination. The tension between professional and lay resources for therapeutic change is similar to and reminiscent of the religious traditions' tension between the ordained ministry and ministries of the laity. That legitimate tension, and the dialectic which informs it, deserves to be applied to the question of who can mediate therapeutic relationships.

The function of the ordained ministry is to mobilize the resources of the laity. The ordained ministry exists only in order to implement and facilitate the growth of the body of Christ, so that ordinary people, the *laos,* will be able to manifest the healing power of God in the world.[25]

What is being discovered, to the embarrassment of some therapeutic professionals, is that there are in fact immensely creative resources for constructive behavioral change resident and available in ordinary people unlettered and uninstructed in developmental psychology and therapeutic skills—in people who know nothing of diagnostic categories and who have not the least thought of themselves as therapeutic agents. The religious communities, amid this discovery, must not be found exclusively on the side of the defensive professionals, but must rather be present at the point of this fresh rediscovery of the power of ordinary human authenticity to elicit extraordinary human growth.

This does not imply that there is no longer any need for professional facilitators of groups. In order for the lay resources to be mobilized, often a skillful facilitator is needed. But once mobilized, the therapeutic agency lies chiefly within the group and not the leader. This is exactly parallel to the concept of the letter to the Ephesians which views the task of ministry as that of equipping the laity for their serving ministry in the world.[26]

THE HIDDEN AFFINITIES. It is not surprising that psychotherapy as a one-to-one relationship has been immensely attractive to Protestant, Catholic, and Jewish clergy. It was assimilated relatively easily into the older image of pastoral care and the cure of souls. It is surprising, however, that the human potential movement is not often recognized essentially as a lay movement, which by its very nature has many more subtle affinities with a theology of the laity and notions of the priesthood of all believers than we could ever expect from psychotherapy.[27]

Clearly the pastoral notion of spiritual director and enabler predates the human potential movement's notion of facilitator.[28] The similarity between them is intriguing, however, because it is another instance where language from the Christian tradition of community building has been borrowed and applied in the human potential movement without any awareness of its historic rootage. I do not think Christian and Jewish traditionalists need necessarily be possessive about their own language. The highest form of flattery still is imitation. But this need not prevent our hypothesizing that much of the language, assumptions and procedures of the human potential movement today are unselfconsciously rooted in religious traditions from which that movement ordinarily considers itself to be estranged, particularly Protestant and Jewish pietism. I do not intend that as a pejorative judgment, but as a description of a historical flow which deserves more serious attention both from religious historians and human potential advocates.

THERAPY OR COMMUNICATION?

Is the intensive group experience a cheap populist substitute for professional therapy, and therefore a poor man's Freud, or on the contrary, is it superior, in some ways, to professional psychotherapy?

LEADERLESS GROUPS. It is a common misunderstanding of leaderless groups and encounter groups generally to assume that not very much research has been done in this area.[29] Over one hundred separate studies and dissertations have been analyzed by Jack R. Gibb who concludes that the "evidence is strong that intensive group training experiences have therapeutic effects." What he means by therapeutic effects are such behavioral changes as greater sensitivity, awareness of one's own feelings, openness, spontaneity, better management of feelings, better ownership of feelings, congruence, greater self-actualization, commitment, self-determination, self-acceptance, self-esteem, less authoritarianism, greater acceptance of others, more interdependence. These studies indicate that "groups without leaders are effective as training media."[30] Rogers reflects the same viewpoint: "The evidence was clear that the group process in leaderless and leaderled groups was very similar. In my own judgment a leaderless group is definitely preferable to a leaderled group in which the leader displays the negative characteristics. . . ." Rogers says, "It has been found, however, that in leaderless groups, where individuals simply meet intensively without any one person named as facilitator or leader, the process that occurs is similar."[31]

AVOIDING JURISDICTIONAL DISPUTES. Ordinary people, it is being discovered, have potent resources for feedback, insight, criticism, forgiveness, care and affirmation. Thus it becomes important for professional therapists to compare what they are doing with what is happening in the intensive group experience.[32]

Part of me wants to say forthrightly (yet with considerable risk to credibility among psychotherapeutic friends and colleagues) that therapy despite certain achievements has been vastly overrated. Therapists often extend themselves far beyond their competence into hidden or implicit moral, political and religious judgments, precisely where their self-critical capacities and historical awareness are, by their training, extremely limited.

Another part of me wants more rationally and compliantly to try to accommodate current therapies to the religious community and theology. It warns me against getting into an endless hassle about the boundary lines between religion and psychology, minister and therapist, church and clinic, which sometimes remind me of those jurisdictional disputes which certain labor unions have with other labor unions whose services sometimes overlap into each other's territory. It also reminds me that the "territorial imperative" of which Ardrey [33] spoke operates not only in the jungle but also in the university, the church and the clinic.

I work a great deal in intensive group processes, and some in individual counseling, but I never in the remotest sense think of myself as a psychotherapist, and have no desire to do psychotherapy. I am rather a minister, educator, ethicist and friend to my partners in the dialogue. I have spent sixteen years studying psychotherapy, both in clinical settings and as a cognitive exercise, and in fact I have spent many hours in one-to-one conversations which some might be inclined to describe as having some of the characteristics of therapeutic change, but I would prefer not to describe it for myself in that way. I think religious counselors are well-advised to disclaim the medical analogy altogether, and not to claim that their work is psychotherapeutic in any specific sense, since that ignites all sorts of defensive jurisdictional disputes with medical, psychological and clinical professionals. I would prefer to call this service in large part merely an attempt to improve human communication—interpersonal and intrapersonal communication, and in a larger sense transpersonal communication (i.e. addressing and being addressed by the unifying center of reality which transcends all human history and intentionality).

TRAINING FOR FACILITATING COMMUNICATION

WHO SHALL FACILITATE? What kinds of training are most functional in order to enable persons to facilitate intensive group experiencing? Clearly it is not medical training, nor psychological testing, nor diagnostic expertise, nor developmental psychology, nor psychoanalysis. In fact any or all of these may easily become a hindrance to facilitating group processes. Rather the focus must be upon enabling interpersonal communication, moral sensitivity, valuational awareness, self-understanding and intrapersonal congruence, upon spiritual formation and upon integrity of speech and life.

The answers that we get to the question of what sort of training is necessary in order to facilitate intensive group experiencing depend entirely upon the persons, or more accurately, the vocation of the persons, to whom that question is addressed. If addressed to a clinical psychologist, the answer will probably be a Ph.D. in clinical psychology with special work in group psychotherapy. If addressed to a psychiatrist, the answer (with few exceptions) will be medical training plus a psychiatric specialization, and anything short of that would be suspect. Ask the same question to a National Training Laboratories associate and the answer will be enough laboratory experience and knowledge of group processes to qualify as an NTL associate.

The same question when asked to the transactional analysts and the growth center people, however, yields a different answer. Although TA has been shaped by psychiatrists, they say ordinary people without advanced education of any sort can do effective TA. The Esalen-oriented growth center people often find their best material for group facilitators among dancers, mystics, freaked out psychologists, former body builders and health faddists, actors, priests, advertising executives, and physicists (I list here only those I have known personally who work as facilitators in such places). There is a tendency among these people to assume that professional therapeutic experience mostly has to be unlearned if one is really to function effectively with groups.[34]

Training in facilitating group communication should be part of seminary education, or should at least be available within the framework of theological education, much as clinical pastoral education (CPE) is available in institutional settings under supervision, although I would hope that group facilitating would not pattern itself after an already overprofessionalized CPE. I do not think that the seminaries should look toward the National Training Laboratories or Esalen or the psychiatric profession or the present structures of clinical pastoral education for such training settings. Rather the church must develop her own resources, rooted into her own history, shaped by her own theological understanding and responsive to the koinonla needs of local congregations.

I would like to see more deliberately built into the process of seminary education the task of learning what it means to be responsible to persons in a group, to be with others in a relationship of mutual accountability, to come close to others, and be accountable for one's words, body, deeds, thoughts, and gestures in the presence of others. Seminary education, as of now, is trapped in verbal linear modes of communication. It has yet to feel the full impact of the encounter culture.

In this connection, we need to distinguish between a "professional attitude" and professionalism. Ministers need a professional attitude without becoming trapped in the self-defensive maneuvers of professionalism.[35]

Facilitating in the parish. Recent conversations with ministerial candidates who are turned on to the intensive group experience as an arena for ministry have made me especially excited about the possibilities for such a ministry within the local parish. They ask me about credentials and I tell them to forget about credentials, and go ahead and do their own experimentation within the parish. Do not always look to psychologists to legitimize what you are doing as an expression of Christ's ministry in and for the world. While the minister does have certain role expectations, most people do not think of him as an expert in psychology or an expert in group processes. Therefore they are not intimidated by a person whom they do not perceive as an expert.

The parish is an appropriate context, and has been for over 200 years in Protestantism, for experimentation in the intensive group process. We do not need the medical and clinical people to authenticate and provide surveillance for what we do there.

The theological education which most American theological students are getting today in Protestant seminaries is probably no more debilitating, and often not as debilitating, as the restricted, reductionistic, clinical and medical training which most psychotherapists are burdened by and much of which they have to unlearn if they are to become effective therapeutic agents.

Work with intensive group processes is no more out of place in the local parish today than was the religious education movement out of place in the local parish of two generations ago. It is no more impertinent than it was in the days of Spener, Zinzendorf, Wesley and Israel ben Eliezer. By long historical precedent, the intensive group experience is the legitimate concern of the religious community, and in fact the psychologists should be prepared properly to be identified as late comers.[36]

PLACEBO THERAPY

A placebo is a pharmacologically inert substance which a physician may give to relieve discomfort when he does not wish to use active medications. It is now known that most medications in the history of medicine have been placebos, i.e. the medications have been pharmacologically relatively innocuous, even though they were regarded by both patient and doctor as powerful. The effectiveness of the placebo rests strictly in its ability to "mobilize the patient's expectancy of help."[37]

Placebo in psychotherapy. If the placebo effect has been an important factor in the entire history of medicine, it has surely been a decisive element in the history of psychotherapy, where the mobilization of the expectation of help is so essential. In five studies of psychiatric outpatients, 55 percent showed symptomatic improvement from pure placebos.[38] They help combat anxiety. Placebos even have been known to heal tissue damage, and to have such powerful effects as to reverse the pharmacological actions of a drug.

Psychotherapy, as its very name indicates, has modeled itself after medical therapy, with diagnosis and treatment procedures, and the language of illness and health. The medical doctor has behind him the momentum of the long tradition of unquestioned authority. Is it the expectation of healing which results from that authoritarian tradition which actually accounts for much so-called psychotherapeutic change? We have good reason to think so.

In placebo healing, one gets better because he is doing what the authority says will be helpful. Without a reservoir of awe for this authority, healing by placebo would be totally ineffective. Pastors, e.g. cannot utilize the placebo as well as physicians, unless the person comes to the pastor with the expectation of his power to heal.

Reduction of placebo. With the laicization of psychotherapy in small group encounter, the placebo effect is being reduced severely although probably still not eliminated. In small group encounter and leaderless groups, constructive change occurs without the particular authoritative nuances found in medicine. Therapy-like outcomes are mediated through ordinary people.

One might argue that the placebo effect should be continued. If it has had a positive effect, why not employ it? Maybe it is better than nothing, but finally it must be viewed as an authoritarian dissimulation. More and more people are refusing to take seriously the sugar-coated pill. As the awareness of the clientele of its own therapeutic resources increases, the effectiveness of placebo therapy decreases.

HISTORY AND SOCIOLOGY OF PROFESSIONALIZATION

Sociology of professionalization. Professionalization and laicization are complementary, and in some ways contradictory, dialectical tendencies. Sociologists have described the professionalization process as a useful means of generating upward social mobility in a vocational area where high standards of competence are needed for specialized services.[39]

Laicization, on the other hand, is the process by which a practice formerly done by professionals or elites is suddenly or slowly taken over by groups of ordinary persons. The function which was once done by professionals is appropriated by lay persons with or without direct professional oversight.

The usual aims of professionalism are (a) to improve prices for services, (b) to make work easier, (c) to build confidence in clientele, and (d) to elicit a corresponding distrust of nonprofessional services.

The sociology of professionalization may be described and identified in almost any vocational group in this country as a striving for secure acceptance, for the feeling of unchallenged competence, and for establishment status. Professionals sometimes play cruel games as they write their ethical principles, which not uncommonly serve the interests of the established professionals against peripheral challenges, more than they do the interests of clients. The sociology of professionalization may be observed just as much among funeral directors, insurance salesmen, labor organizers, truck drivers and quality ice cream distributors, as in physicians and psychologists. Each is searching for professional territorial definition. (An example of a profession which has moved *down* the social mobility ladder during the last fifty years is the chiropractic practicioner, where professionalization efforts have been tried and resisted, and have failed.) In each case there are built in rationalizations which intensify professionalism for reasons that are purely intrinsic to the persons who want to be professionalized.

Pietism as a role model of laicization. The most influential western historical model for therapeutic laicization is the Protestant tradition. The laicizing momentum began in the sixteenth-century Lutheran, Calvinist and Anabaptist developments, first primarily as a doctrinal statement, but later reaching greater concretion in the lay preaching, charismatic leadership and ecclesial voluntarism of eighteenth-century pietism.[40] Especially we see in American frontier pietism a radical laicization of the search for authentic community. Like the encounter culture, it was intuitively anti-professional. The laicization process was saying to the priestly professionals that the healing love of God could be proclaimed and communicated just as ably, and perhaps more so, apart from the elaborate mechanics of professional priesthood, apostolic succession, etc.

The whole apparatus and ideology of apostolic succession, incidentally, is also eminently applicable to our discussion of therapeutic history. I cannot avoid feeling on the one hand amused and on the other hand disturbed as I read the post Freudian literature, to note how seriously the line of succession from Freud to present practicioners is taken. How crucial it becomes, for example, for an Esalen to sport a Perls whose training analyst was Reich who knew Freud and rebelled against him. Admittedly there have been many schisms, heresies and revolts in the psychotherapeutic tradition, but practically all have been *within* the framework of some direct line of succession from Freud. I regard this as regrettable, for it seems to me that Freud did some almost irreparable damage to the practice of therapy, with his naturalistic reductionism, inverted messianism, sexual redemptionism, and misleading ideas of the inevitable repressiveness of society. It was bad seed that multiplied profusely, and we now must deal with an unweeded garden.

THE BURDEN OF MEDICAL PROFESSIONALISM. To return to our theme, the particular profession in American society which is most overburdened with professionalization is the tradition of medicine. Heavy constraints are placed upon persons both by law and custom not to meddle with medicine without license. There are many ways in which the medical profession sends messages to nonmedically trained people that their function is utterly without analogy, and that there must be a firm line between those who are competent and those who are not. Doctor and patient do not exist in a dialogical relationship, generally, but rather in an authoritarian stratification of admitted unequals. The patient awaits the expert's diagnosis and prescription, all strictly within an authoritarian framework. I am not protesting this as a pattern for the treatment of physical illness, but for its awkward imposition on the care of souls and interpersonal growth.

It was a regrettable and fate-laden decision when the practice of psychotherapy took its essential shape under the image of professionalization which unhappily prevails in the American medical profession and which, despite many achievements, represents an absolutely rigorist, authoritarian style of professionalization which works against the sensitive kinds of dialogical relationships which are most promising for therapeutic behavioral outcome.[41] I agree fully with psychiatrist Thomas Szasz's overall criticism of the use of the medical model in psychotherapy. The irony is that the style of professionalization which has been bequeathed to and chosen by psychotherapy, and reinforced for two generations, is clearly counterproductive to the deeper intent of psychotherapy.

DANGERS OF LAICIZATION

Admittedly certain practices need to be professionalized in order that structures can be developed for policing, training, self-criticizing, and for establishing professional continuity from one generation to another through accrediting institutions and established procedures of professional ethics. Professionalization is needed in order to develop theories of practice and criticism of alternative theories. This is one of the reasons why the universities have often stood on the side of professionalization rather than against it. There are certain functions such as in the dental profession, the legal profession, the practice of medicine, and so forth, where great expertise is required. You do not want to entrust your teeth, your legal defense or your heart to just anybody. That is the reason we have laws in most states which say that persons cannot practice law, television repairing, real estate, or plumbing unless they are accredited through professional procedures. I would like, for my own part, to underscore the desirability of such laws and such processes of accreditation.

What troubles me, however, is the danger of exploitation and abuse in the professionalization process against which the client is defenseless. We see this most clearly in those professions that have become the most acutely and uncritically professionalized. American medicine is at the apex of that process enjoying the greatest prestige without checking its own temptations to abuse.

It should also be clear that there are dangers and limitations in the laicization process. Both in the pietistic and the encounter group patterns of laicization, we can see tendencies toward fanaticism, which often are lacking in the kind of self-critical, self-policing functions which are found in the better forms of professionalization. Among other dangers of the laicization of therapy are (1) anti-intellectualism, which fails to develop adequate theoretical groundwork for its operational practices; (2) anti-establishment paranoia, which assumes that anything which has any precedent of historic continuity must by definition be ill-advised; (3) utopianism which imagines that completely new reconstructions of human environments are indeed possible, demanded and easily achieved; and (4) lack of realism about human and institutional limitations. Although there may be many admirable features in the laicization process, it almost always has its foolhardy edge which imagines that the professional role can be easily, and in fact happily, dismissed without harmful consequences.

CONCLUSION: THE END OF PSYCHOTHERAPY

I will conclude by briefly wrapping up the several dimensions of the theme, "The End of Psychotherapy." The English term *end* is rich in nuances, and may be taken in at least *four different senses,* each of which has significance for our discussion:

a) First we can apply the concept of *end as termination* to psychotherapy. Psychotherapy, as we have become most familiar with it as a one-to-one client relationship following a professionalized medical model, is now exhibiting many aspects of senility. The earlier hegemony or unquestioned preeminence of psychotherapeutic professionalism seems to be waning. This is not to suggest that the therapeutic tradition as such will not continue in some form, but the bequeathed tradition of one-to-one counseling under the medical model of diagnosis and treatment appears to be coming to an end.

Of course in a general sense everything is ultimately coming to an end including all theological, as well as psychological, conceptualities and languages. Psychotherapy finds it as painful as any other finite reality to be reminded that it too is not infinite, that it has no more guaranteed permanence in long-range historical processes than did medieval sacramentalism or Roman law or feudal economics or gnostic mythology, each of which thought of itself as permanent and absolute, and failed to deal with the relativities of history, which is another way of talking about that which the Bible calls the judgment of God upon our gods, and the awareness that all flesh is grass, (Ps. 90).

b) Secondly, we can apply the concept of the *end as goal* to psychotherapy. For end in English means not only death and deterioration, but also telos and final cause. We need to press much further the question of the teleology of psychotherapy. Toward what end is the process finally pointing?[42]

The end as telos refers also to the ultimate state or condition of therapy. That meaning may well be applied to the further probing of the implicit ontological assumptions of humanistic psychotherapy.

c) Thirdly, we can apply the concept of *end as boundary* to psychotherapy. End connotes also an edge or outermost part. Etymologically, end is cognate with the Sanscrit *anta,* which means boundary or limit. So we must ask: What are the limits of therapy? We need further to delineate the boundaries of the psychotherapeutic task, especially as they relate to universal history, which is the proper concern of theology.

d) Finally, we can most meaningfully apply the concept of *end as eschaton* to psychotherapy. This concept comes from the apocalyptic view of history's end. The dialogue with therapy and theology has not been probed deeply enough along eschatological lines. What is the relationship of therapy to universal history and to the end of history?

This leads to our final hypothesis: the dialogue between theology and therapy has suffered by being de-eschatologized. Laicizing psychotherapy would benefit positively by being placed within the frame of reference of universal history, which involves dealing with the future and finally the end of history. At this point, contemporary theology can make a significant contribution to an understanding of therapeutic change, in its approach to the future and the final end of universal history, which is the context in which all therapeutic change finally exists.

NOTES

1. *Kerygma and Counseling: Toward a Covenant Ontology for Secular Psychotherapy.* Philadelphia, Westminster, 1966. *Contemporary Theology and Psychotherapy.* Philadelphia, Westminster, 1967. This present chapter presuposes also the historical argument which I have developed in *The Intensive Group Experience: The New Pietism,* (Philadelhia, Westminster, 1972) especially in Chapter 2. Portions of *The Structure of Awareness,*

(Nashville, Abingdon, 1969) especially Sections 2, 6, 11, 12, 16, 17, 23-27, 34, 40-44 stand as background to the thesis of this essay.

2. Fromm, Erich: *The Crisis of Psychoanalysis.* New York, Fawcett World, 1970, pp. 12, 16, 40.
3. *Kerygma and Counseling, op. cit.,* chapters 2-4; *Contemporary Theology and Psychotherapy,* pp. 128 ff.
4. *Ibid.,* pp. 136 ff.
5. Carl Rogers' discussions of "A Theory of Therapy, Personality and Interpersonal Relationship," in Koch, S. (Ed.): *Psychology: A Study of a Science,* (New York, McGraw, 1959, vol. III) and an article, "The necessary and sufficient conditions of therapeutic personality change," *J Consult Psychol, 21:*95 ff, 1957, have identified the essential conditions of therapeutic agency as empathy, congruence and unconditional positive regard. These conditions may be found in nonprofessionals, and sometimes regrettably are not found among professionals. On the basis of Rogers' identification of the conditions for therapeutic change, C. B. Truax has shown that persons receiving high levels of these therapeutic conditions, whether from lay or professional agents, showed improvement in psychological functioning (see The Use of Supportive Personnel in Rehabilitation Counseling: Process and Outcome, Symposium sponsored by the Association of Rehabilitation Centers, Inc., and Arkansas Rehabilitation Research and Training Center, Colorado Springs, 1967). Further, Truax showed that when counselor aids were trained in empathy, congruence and unconditional positive regard, they showed higher therapeutic proficiency than professionals who did not exhibit these qualities. Elsewhere, R. R. Carkhuff and C. B. Truax have shown that persons without any knowledge of developmental psychology or therapeutic categories could be trained effectively in these necessary conditions of therapeutic agency (see Lay mental health counseling, *J Consult Psychol, 29:*426 ff., 1965). For developing studies of lay counselor training in religious settings, see Oraker, J. R.: A Study of Prediction in a Short-term Counselor Training Program for Religiously Oriented Nonprofessionals, unpublished doctoral dissertation, Fuller Theological Seminary, 1970; and Penner, C. L.: The Effectiveness of a Theologically Oriented Approach to Lay Counselor Training, unpublished doctoral dissertation, Fuller Theological Seminary, 1971. See also Truax, C. B. and Carkhuff, R. R.: "For Better or for Worse: The Process of Psychotherapeutic Personality Change," in *Recent Advances in the Study of Behavior Change,* McGill U Pr, 1963, pp. 118-163; and *Toward Effective Counseling and Psychotherapy,* Chicago, Aldine, 1967.
6. Jackson, M. P.: Their Brothers' Keeper—A Directory of Therapeutic Self-Help Groups, Intentional Communities, and Lay Training Centers. Unpublished, Department of Psychology, University of Illinois. Mowrer, O. H.: *The New Group Therapy.* New York, Van N-Rein, 1964. Burton, Arthur (Ed.): *Encounter,* San Francisco, Jossey-Bass, 1969. Mill, C. R.: *Selections for Human Relations Training News,* National Training Laboratories, 1969.

For a fuller bibliography see the Notes of *The Intensive Group Experience, op. cit.*

7. Jourard, Sidney: *The Transparent Self.* New York, Van N-Rein, 1968. Rogers, Carl R.: A Theory of Therapy, Personality and Interpersonal Relationships. In Koch, Sigmund (Ed.): *Psychology: A Study of a Science,* New York, McGraw, 1959, vol. III. Maslow, Abraham: *Toward a Psychology of Being.* New York, Van N-Rein, 1962.
8. Severin, Frank T. (ed.): *Humanistic Viewpoints in Psychology.* New York, McGraw, 1965. Stevens, John O.: *Awareness: Exploring, Experimenting, Experiencing.* Lafeyette, Real People, 1971. Berne, Eric: *Transactional Analysis in Psychotherapy,* New York, Grove, 1961.
9. Szasz, Thomas: *The Myth of Mental Illness.* New York, Dell, 1961, pp. 296-297.
10. Eysenck, H. J.: The effects of psychotherapy: An evaluation, *J Consult Psychol, 16:*319 ff., 1952. Shaffer, L. F., and Shoben, E. J. Jr.: *The Psychology of Adjustment,* 2nd ed. Boston, HM, 1956, 545 f. For another view see Luborsky, L.: A note on Eysenck. *Br J Psychol, 45:*129-131, 1954.
11. Skinner, B. F.: *Science and Human Behavior.* New York, Macmillan, 1953. *Verbal Behavior.* New York, Appleton, 1957. Mowrer, O. H.: *The Crisis in Psychiatry and Religion.* New York, Van N-Rein, 1961. London, Perry: *The Modes and Morals of Psychotherapy.* New York, HR&W, 1964; Wolpe, Joseph and Lazarus, Arnold: *Behavior Therapy Techniques.* Elsmford, Pergamon, 1966.
12. Mowrer, O. H.: *The Crisis in Psychiatry and Religion.* New York, Van N-Rein, 1961, pp. 81 ff.
13. Laing, R. D.: *The Politics of Experience.* New York, Ballantine, 1967.
14. Cooper, David: *Psychiatry and Anti-Psychiatry.* New York, Ballantine, 1967, p. 17.
15. LaPiere, Richard: *The Freudian Ethic: An Analysis of the Subversion of Western Character.* London. George Allen And Unwin, 1960, p. 280.
16. *The Last Whole Earth Catalogue.* Portola Institute, New York, Random 1971. Cf. Hicks, J. D.: *The Populist Revolt.* Minneapolis, U of Minn Pr, 1931.
17. Schofield, William: *Psychotherapy: The Purchase of Friendship.* Englewood Cliffs, P-H, 1964. Cf. London, Perry, *op. cit.,* pp. 240 ff.
18. Rogers, Carl R.: *Carl Rogers on Encounter Groups.* New York, Har-Row 1970, p. 33.
19. Gibb, J. R.: The Effects of Human Relations Training. In Bergin, A. E. and Garfield, S. L. (Eds.): *Handbook of Psychotherapy and Behavior Change.* Wiley, 1970, Chapter 22, pp. 839-862.
20. Rogers, *op. cit.,* p. 119.
21. Gibb, *op. cit.*
22. *Ibid.*
23. Rogers, *ibid.,* p. 121.
24. Rabinowicz, Harry M.: *A Guide to Hassidism.* New York, Yoseloff, 1960;

Wiesel, Elie: *Souls on Fire*. New York, Random, 1972. Stoeffler, F. E.: *The Rise of Evangelical Pietism,* Leiden, E. J. Brill, 1965.

25. Congar, Yves: *Lay People in the Church*. London, Chapman, 1957. Kraemer, Hendrick: *A Theology of the Laity*. Philadelphia, Westminster, 1958. For a broader review of recent literature on this subject see my discussion of Articles on current literature: Laity and vocation. *Encounter,* vol. 26, no. 4.
26. Ephesians *4:*11 ff. Ayres, Francis O.: *The Ministry of the Laity: A Biblical Exposition*. Philadelphia, Westminster, 1962.
27. Mowrer, O. H.: Is the Small-groups Movement a religious revolution? *Voices,* Fall, Vol. 7, No. 3, pp. 17 ff, 1971.
28. de Guibert, Joseph, S. J.: *The Theology of the Spiritual Life*. New York, Sheed, 1954. Congar, *op. cit.*
29. I am indebted personally to Phil Hanson and Walter O'Connell of the Human Relations Training Laboratory of the V. A. Hospital in Houston, with whom I worked as cotrainer in 1970, for much of my reflection on leaderless groups. For a major series of significant studies of autonomous groups see studies of the Houston Human Relations Training Laboratory: Hanson, Philip G., Rothaus, Paul, Johnson, Dale L., Lyle, Francis A.: Autonomous groups in human relations training for psychiatric patients. *J Appl Behav Sci,* July-September, 1966, vol. 2, no. 3. Hanson, Philip, Rothaus, Paul, and Lyle, Francis: Basic Concepts and Underlying Processes in a Human Relations Training Laboratory for Psychiatric Patients. Unpublished. Hanson, Philip G., Rothaus, Paul, Johnson, Dale L., Lyle, Francis A., and Moyer, Raymond: The anticipation and management of crisis on an open psychiatric ward. *J Appl Behav Sci,* vol. 2, no. 4, 1966. Hanson, Philip G. and Rothaus, Paul: A Philosophy of Treatment, unpublished. Also, Ansbacher, H. L.: The history of the leaderless group discussion technique. *Psychol Bull, 48:*383 ff, 1951. White, R. K. and Lippit, R.: *Autocracy and Democracy*. New York, Har-Row, 1960. O'Connell, Walter: The equal collaborator. *Individual Psychol Newsletter,* vol. 19, no. 2, November-December, 1961. Demorcacy in Human Relations. *Desert Call,* 5:8-9, no. 2, 1970. Bass, B. M. and Klubeck, S.: Effects of seating arrangement on leaderless group discussions. *J Abnormal Soc Psychol, 47:*724-727, 1952. O'Connell, Walter: The failure of psychotherapy in mental hospitals. *Individ Psychologist,* vol. II, no. 1, March, 1964. Two recent studies, widely regarded as significant contributions are Guerney, B. G. (Ed.): *Psychotherapeutic Agents: New Roles for Non-Professionals, Parents and Teachers.* New York, HR&W, 1969, and Goodman, Gerald: *Companionship Therapy*. San Francisco, Jossey-Bass, 1972.
30. Gibb, *op. cit.* Cf. Hanson and Rothaus, A Philosophy of Treatment, *op. cit.*
31. Rogers, *op. cit.,* p. 119.
32. Winn, See
32. See Winn, A.: Training Groups and Therapy Groups. *Hum Relations Train*

News, Fall, p. 7, 1963. Cf. O'Connell, Walter: Psychotherapy for Everyman. *J Existentialism,* vol. VII, no. 25, Fall, 1966.

33. Ardrey, Robert: *The Territorial Imperative.* New York, Dell, 1966.
34. Schutz, William: *Here Comes Everybody.* New York, Har-Row, 1971.
35. Cf. Glasse, James: *Profession Minister.* Nashville, Abingdon, 1968.
36. Schmidt, M. and Jannasch, W. (Eds.): *Das Zeitalter des Pietismus.* C. Schunemann, 1965. McLoughlin, W. G.: Pietism and the American character. *Am Q, 17:*163 ff., no. 2, Summer, 1965. Buber, Martin: *Hasidism.* New York, Philos Lib, 1948.
37. Frank, Jerome D.: *Persuasion and Healing.* New York, Schocken, 1963, pp. 65 ff.
38. Gliedman, L. H.: et al.: Reduction of symptoms by pharmacologically inert substances and by short-term psychotherapy. *AMA Arch Neurol Psychiatry, 79:*345-351, 1958. Cf. Fish, Jefferson M.: *Placebo Therapy*: A Practical Guide to Social Influence in Psychotherapy. San Francisco, Jossey-Bass, 1973.
39. For sociological discussions of the professionalizing process see Carr-Saunders, A. M. and Wilson, P. A.: *The Professions.* Clarendon, Clarendon Pr, 1933. Ross, E. A.: *Principles of Sociology.* New York, Appleton, 1930, Chapter 44. Brady, R. A.: *Business as a System of Power.* New York, Columbia U Pr, 1943. Whyte, W. H. Jr.: *The Organization Man.* Garden City, Doubleday, 1956. Fox, R. C.: An American Sociologist in the Land of Belgian Medical Research. In Hammond, P. E. (Ed.): *Sociologists at Work.* Garden City, Doubleday, 1967, pp. 399 ff.
40. For a fuller statement of this argument, see my discussion of *The Intensive Group Experience, op. cit.,* Chapter 2.
41. It is to Freud's credit that he recognized some of the dangers of medical professionalization; see his letters to Oscar Pfister. In Meng, H. and Freud, E. (Eds.): *Psychoanalysis and Faith.* New York, Basic, 1964.
42. Don Browning's recent stunningly brilliant book, *Generative Man,* Philadelphia: The Westminster Press, 1973, probes the moral assumptions and teleology of recent psychoanalytic anthropologies.

CHAPTER 2

THE POST-THERAPEUTIC SITUATION

WHERE THE PATH OF psychology and the search for religious understanding meet, it is always a significant, even if sometimes confusing, crossroad. As one who has stood at the crossroad and perused the cryptic signs and studied the available maps, I have often been astonished that the two roads run more parallel to each other than it is often assumed. Although kergyma and counseling are two distinguishable paths, they often move in the same direction and cross each other at many significant junctures.

When I have sought previously to discern these cryptic signs and tried to remap this territory, I have never had any illusions that my rough sketches were definitive, or unable to be further refined. From the outset I have sought to clarify and acknowledge the radically contingent, historically relative character of all constructive thinking today, and to affirm my hope that it would not take long for someone more adequately to state my deeper intention. I have always thought of my effort as a heuristic, anticipatory proposal, and its criticism as a part of my own education. Thus I have been grateful when a wide variety of respondents, including European critics such as Father Godin[1] and Professor Stollberg,[2] and such valuable American partners in dialogue as Don Browning, Wayne Oates, Charles Stinnette, Sam Keen, Seward Hiltner, Edgar Jackson, Edward Thornton, Russell Becker, James Lapsley, William Rogers and others, have developed my concerns further and sought to identify their weak points and unanswered questions.[3]

Much change has occurred in the intervening years, both in therapeutic theory and practice, and of course in theology where the situation is especially confusing. According to widespread reports,

God is dead but Jesus is alive. The church is ailing but the Spirit is very lively.

To some it may seem an unworthy task to be required to update theology in the light of new cultural dilemmas, yet that seems to be what is demanded of theological reflection, when it is true to its source and ground, not only in our time, but in other times. Contemporary man keeps on being contemporary in new ways, although his continuity with previous histories is always more powerful than he perceives. The good news deserves to be addressed in the ever changing languages of man. One such language which has unprecedented power and plausibility in our time is the language of psychology.

And yet psychology, as we have seen, is now caught in its own identity crisis. Religious communities are well advised to look squarely into the depths of this crisis, and to revise some of their expectations of therapeutic change. This section explores several dimensions of the new situation of the dialogue between psychology and religion, asking whether we now in fact exist in what might be called a "post-therapeutic situation," and whether our options have accordingly shifted.

AFTER THERAPY WHAT?

Until now we have left unclarified the specific intent of our overarching question: "After therapy what?" By focusing on this single theme, I hope to probe at least three broadly related issues: (1) What happens to *therapy* after its laicization? (2) After therapy, what happens to *persons?* and (3) After therapy, what happens to theology? Broadly gauged, these three questions constitute the three essential questions raised in these first three chapters.

To be more specific: first, under the theme "After Therapy What?" we have explored the question of the alleged incipient death of psychotherapy. We have asked in Chapter one whether Erich Fromm's foreboding talk about the general malaise of therapy is realistic.[4] If so, what comes *after therapy* as we have known it (essentially as a one-to-one relationship following the medical analogy)? In this connection we have spoken of the laicization of therapeutic agency, its problems and consequences, and will continue to explore its implications.

The second level on which we are raising the question, "After therapy what?" concerns *what happens to persons* after they have experienced effective psychotherapy. After it has done its proper work, what remains yet to be done in fulfilling the human possibility? This issue concerns us presently in this chapter.

A third level of the question, "After therapy what?" focuses on the future of religious thought and community following the impact of psychotherapy. After therapy, what directions might religious experience and reflection take? How is Christian thought specifically to develop amid and beyond that dialogue? Although we will address aspects of this question in this chapter, this third level will be the essential concern of the third chapter.

So the single theme, "After therapy what?" embraces a variety of related issues, each of which are crucial to the definition of the post-therapeutic situation, and each of which are urgent for current religious reflection. Similarly we are exploring the construction of the *post-therapeutic situation* on three related levels: (1) the characteristic shape of the post-therapeutic situation is determined by the laicization of therapy, challenging both the prevailing medical model and the professionalization process which have characterized the prior history of psychotherapy; (2) more personally we are interested in the post-therapeutic situation as it is being felt and experienced by individuals who have come through and moved beyond the best and worst that psychotherapy can give them; (3) finally we are interested in the post-therapeutic situation which exists in theology and in religious communities.

Now to turn to the question of this current section: how would the human task and possibility be redefined if psychotherapy turned out to be widely recognized as a wholly incomplete solution to the human quandary, pointing explicitly beyond itself? How are we to care for persons who have experienced seriously all that effective psychotherapy can do, and still find themselves struggling with the primordial questions of life, death, meaning, honesty, spiritual awareness and moral courage?

After the therapeutic process has run its course, what are the tasks that remain to be done, what understandings remain to be achieved? *After* therapy, where lie the meanings yet to be grasped, the excellent forms of behavior to be solidified? After therapy has achieved its aims, must we still speak of virtues that need to be realized, intellectual integrations which remain to be formulated? What forms of moral compassion remain to be embodied, what historical perspectives need yet to be grasped, and above all, what religious vision is yet to be beheld? After therapy has fulfilled its proper function, what yet remains humanly to be done?[5] This properly is the arena for the religious philosopher, theologian and moralist.

The question arises concretely out of the history of therapeutic care. The momentum of the history of psychotherapy itself demands that we face such a question.[6] It is a question being raised not only by the moralist but by the therapist himself out of his painful experience of dealing with a new kind of client whom he is now unexpectedly confronting. The therapist is often unprepared to deal with the profundity of the trip his client is taking.[7] The new client is being described by therapists such as Rollo May as one who has already experimented widely with erotic sexuality and is now searching for the deeper meaning of interpersonal intimacy.[8] The new client is described by Jourard as one for whom the strategems of previous therapies are often inappropriate because they presuppose that there is no desire for self-disclosure. Yet people increasingly are coming to the therapeutic process with a manifest desire for disclosure, and a willingness to behold and respond to the disclosure of the therapist.[9] The question is being raised by such therapists as Van den Berg, Boss and Caruso who find their clients struggling at the most elementary levels with questions of meaning and death.[10] Therapists today are struggling daily with fundamental moral and philosophical questions which two or three decades ago most would have ruled out of their range. They are being required to struggle with these questions by their clientele who are often demanding of them a greater philosophical, moral, religious and historical sophistication than they are prepared to deliver.

You have heard the intriguing popular ballad, the lyrics of which center on the theme *"Is that all there is?"* The song describes the watching of a fire in childhood, and several events which culminate finally in falling in love. In each case there is a high expectation of what that event would bring, followed by a corresponding disillusionment—"Is that all there is?" The song ends with a statement about death, namely, that on the basis of the accumulated disillusionments of life, death also doubtless will be overburdened with expectation, and thus the anticipated response: "Is that all there is?" An increasingly critical clientele today is coming to respond similarly to much psychotherapeutic treatment. Their expectations for human growth, personal meaning and redemptive change have been so expansive that when persons have actually been through the mill of psychotherapy, a significant number come away with the plaintive question, "Is that all there is?"

The best therapists have always tried to protect themselves against such exaggerated expectations. They have always tried to state modestly the limits of what they are able to do. But somehow the momentum of exaggerated hope concerning the therapeutic process has placed upon them a series of role expectations which they find themselves increasingly hard put to fulfill. They are expected to be the moral experts of society, the all-purpose healers, the medicine men, the avatars, the high priests. They are also expected to be the prophets and the social reformers, so much so that the burden of idealized role expectations upon the therapist, if he takes them at all seriously, is very weighty indeed.

It is relevant here to inquire into the history of what people want from therapy. That desire or expectation has itself had a mutating history. In the Freudian preiod. the focus of expectation from psychoanalysis, with its free association and dream interpretation, was principally on the alleviation of phobic and hysteric symptoms through the renegotiating of guilt feelings amid a father-dominated, sexually repressed, patriarchal society. That society no longer exists. The cultural peculiarities which it produced, and which asked of it a certain kind of therapy, no longer prevail.

When psychoanalysis came to make its home in America, it had to adjust its methods of cure to the new American clientele which in fact had an entirely different nosology and symptomology than European clients. Although the story of the symptomology is much too complex to try to tell now, we must at least note that American men and women, nurtured by the special features of American equalitarianism, idealism and pietism, asked and expected of psychotherapy something substantially different than their European counterparts. During the 1930's and 1940's American psychotherapists began to shift to this new set of expectations. Horney, Fromm and Sullivan began to talk about the cultural determinants of neurosis.[11] So there developed a second stage in the history of what people expect from therapy, shaped decisively by the American experience. One might accurately speak of the Americanization of the therapeutic process with a consequent thoroughgoing reformulation of the Freudian categories even though much of the language remained the same. Up to this point, the focus still remained on the cure of neurotic symptoms.

Toward the end of the 1950's, however, a new period begins to take shape in which increasingly people have begun to ask broader and more profound philosophical and moral questions of the therapeutic process. I am not speaking here about changes in the techniques of psychotherapy, but a change in the cultural situation in which the therapy itself exists. Clients began asking more and more about the meaning of their existence, searching for intimacy, for community, looking for some way of putting it all together, of integrating their lives into some cohesive unity.

As a result of those demands, psychotherapy moved toward a new set of strategies, techniques and approaches sometimes called the "third force" in psychotherapy, or humanistic psychology, namely a third option beyond psychoanalysis and behaviorism. In this phase we saw the development of existential psychotherapy, with its concern for meaning, intentionality, decision-making and freedom, a psychology of being, a broader definition of health, and the new search for peak experiencing.[12]

Finally, as this history of expectation comes full circle, we are now seeing the development of the encounter group process, which in some sense is an extension and proliferation of the therapeutic tradition, although it eclectically combines many other traditions.[13] But it is a most significant development in the history of therapeutic expectations that through group encounter, persons expect to enter into a community of persons searching for wholeness and to achieve relationships of intimacy and trust which they find missing in ordinary social transactions.

Oddly enough, the encounter group is both an extension of the psychotherapeutic tradition, on the one hand, and also a disintegrating factor in the psychotherapeutic tradition, on the other. It is an extension in as much as most of its leadership has come from the left wing of humanistic psychology, existential psychology and Gestalt therapy with an emphasis on the I-Thou encounter model, self-disclosure and intimacy. It stands as a disintegrating factor, however, since the therapeutic process is becoming increasingly laicized, i.e. shaped by and for laymen. It is increasingly seen as a lay function done in groups of ordinary human beings, sometimes led by professional therapists, but not always.

So the question "After therapy what?" has partly to do with the special history of what people expect from psychotherapy. If therapy is defined in a modest way as the cure of neurotic symptoms, then what comes after therapy is what has actually happened in the proliferation of psychotherapy's tasks and orientations in the last twenty years. If therapy is defined in a specific way as the cure of symptoms, then something must come beyond that cure, namely the quest for meaning, intimacy, moral depth and understanding.[14] If therapy should be defined as inclusive of those quests, however, then the question "After therapy what?" is much more diffuse and difficult, since therapy broadly defined would embrace the continuing quest for meaning and depth in religion, moral experiencing, social and political responsibility.

When I speak of *the post-therapeutic situation,* I think more personally of specific individuals whom I have come to know well, who have been engaged in long periods of reasonably effective psychotherapy. They have come to me *after* that experience and put to me a tough genre of post-therapeutic questions. They know the depth of the quest they are on. Some are very young people who nonetheless have already accumulated a large fund of human experience. They have taken all sorts of trips. They now understand themselves in the light of those trips as only now beginning a quest for a deeper spiritual identity. They continue to experience anxiety, guilt and boredom. They do not see their existential situation as fundamentally changed by the psychotherapy which is effectively behind them.

What special relevance does the Christian community, with its witness to God's occurring love, have for these post-therapeutics? Most of them regard therapy as a part of their spiritual growth, even if it has come in the form of the torment which Luther called *Anfechtung,* an intense period of suffering and temptation.[15] Even there amid the arduousness of therapy, many have beheld the providence of God in their lives or have found some alternative language to point to the same reality. They have been able to frame the human question more profoundly, to address the human predicament more clearly, and in that sense they understand themselves to be closer to salvation than they were before.

Later we will show how certain new developments in Christian thought provide a richer, more plausible frame of orientation for understanding therapeutic change. Contrary to its deeper interests, psychotherapy has neglected the concept of universal history. Yet it always in fact exists within a larger universal history, a vision of which might liberate it in unexpected ways, both to be more modest about its own mission, and to be more operationally effective. The recent stodies of apocalypticism, of eschatology and of a philosophy of the future (especially as they have been brilliantly brought together by Wolfhart Pannenberg) address the post-therapeutic situation with fundamental questions which must no longer be avoided.

In order to move the conversation further, however, it is useful at this point to review the present state of dialogue between psychology and religion, to probe its awkwardness and resistance, asking why it has failed, and how the dialogical agenda might be expanded.

THE TASK OF DIALOGUE IN THE POST-THERAPEUTIC SITUATION

THE NEGLECT OF PASSION. There is a curious irony in the fact that Protestant worship today is still desperately trying to divest itself of emotive display, just at the time when current therapeutic practice, notably Gestalt therapy, is seeking to intensify emotive self-disclosure. For several generations now, Protestantism has suffered under the onus of nineteenth-century memories of unbridled charismatic emotionalism. Wild, frontier, emotive fanaticism has given pietism a bad name, and elicited a strong backlash against any expression of emotion whatever. Consequently, in the staid churches which have slowly emerged out of the pietistic tradition, which are the main stream of American Protestant denominations, there has been for many decades a steady polemic against emotive self-disclosure. There is a corresponding drive toward propriety, respectability, order and consistency.

There was a time, of course, when that polemic was a relevant corrective. It becomes ironic, however, when the churches confront youth culture and the encounter culture. For what is happening in youth culture and the encounter culture is precisely a search for charisma, peak experiencing, self-disclosure and emotive expression. This is often accompanied by an impatient put-down of verbal or cognitive expressions which are divorced from experience, often rejecting any "head trip" which does not exhibit personal wholeness and emotive congruence.[16] Youth culture and the encounter culture are very busy rejecting the traditions of scholasticism and abstract thinking which do not express psychic depth and emotive wholeness, while religious worship is still busy rejecting fanatical emotionalism. Consequently when young people come into churches whose ethos has been irreversibly shaped by that reaction against emotionalism, they find themselves totally out of place and alienated. It constitutes a certain moral crisis, since emotive congruence has in fact become a central moral value, and incongruence a widely recognized hypocrisy. So just at the time the churches are holding tight reign against emotive expression, youth culture is going far out ahead in the honest expression of eros, spontaneity, intuition and joy. Therapeutic and group processes are concerned with deep feeling levels in personal encounters here and now. A good deal of what used to be an *avant-garde* religious corrective is thus extremely contrary to the pivotal values and virtues of the encounter culture and youth culture. It appears that the pietistic concerns for charisma and interpersonal community which were present 100 to 200 years ago in American Protestantism are now, oddly enough, in the process of being secularly reappropriated in the encounter culture, with its personal testimonials, small group mutual care, tough feedback procedures and revival-like marathons.[17]

Doggedly, however, the church continues to prepare for crises which have just passed. Consequently it tends not to want to look at the hunger in popular culture for the charismatic, the ecstatic gifts of the spirit, the wholistic and the personal. Contemporary theology has good reason to be in dialogue with these cultural developments, however, since they are secular by-products of the Judeo-Christian tradition in an extended and implicit sense. They could never have happened without the achievements of Protestant personalism and individualism.

Often when I teach seminars in contemporary theology and psychotherapy, I am astonished when students level with me about how difficult it is for them really to dig the language of the church, and yet how easy and native it is for them to get inside the language frame of psychology. I remind you that these are theological students, preparing for ordained ministries. One might understandably assume, as I had earlier assumed, that these ministerial candidates would come to such an inquiry steeped in the language of biblical witness and the Christian tradition, and that consequently it might be difficult for them to move themselves out of that world and into the categories of developmental psychology and psychotherapy. Yet student feedback inlicates that psychology is their home territory linguistically.

RESISTANCE TO DIALOGUE. In order for the dialogue between psychotherapy and theology to be sustained and significant, it must be a dialogue between two equal and equally respected partners. If there is no mutual respect and confidence in the other's competence, there can be little real dialogue. Dialogue means intent listening and candid self-disclosure.

In my opinion, the dialogue between theology and therapy has regretably failed in the last decade, in large part, because theology has been unwilling to risk full self-disclosure about its own distinctive evangelical presuppositions. It has tried too hard to gain false credentials of acceptability, credentials which from the beginning have been inconsistent with its deepest self-understanding. Theology has often been too shaky, defensive and uncertain of itself to engage in genuine two-way dialogue.

There has been another reason for the failure of dialogue. Quite frankly, the therapeutic tradition has not been noticeably eager or motivated really to listen to what an evangelical theology most uniquely and profoundly has to say. Much psychology has remained too deeply indebted to an empiricist conception of verification to be able to listen to the most profound questions which a congruent Christian theology might otherwise have been able to put.[18]

So it has been an aborted and truncated conversation. I am speaking out of ten years of involvement myself, and out of personal acquaintance with many of the therapists and theologians who have attempted to engage in the dialogue. Theologians have a difficult time getting psychologists, with a few remarkable exceptions (Erikson, Mowrer, Maslow, Jourard),[19] to take their presuppositions seriously. Few clinicians are reading such important works as those of Frank Lake, Wolfhart Pannenberg, Herbert Richardson or Arend van Leeuwen,[20] and if their colleagues knew that they were reading theology, they might feel a diffuse sense of guilt and might need to make apologies to legitimize such an inquiry.

THREE LEVELS OF DIALOGUE. The dialogue between therapy and theology may be viewed at varying levels of intensity. I would note especially three possible levels of dialogue:

(1) There is a type of minimal, *pragmatic,* professional dialogue, which exists only because it is necessary within institutional contexts, because ministers and counselors happen to be working with the same people, or in the same institutional setting. There is a practical need for consultation, but it is understood by both parties as relatively dispensable and limited.

(2) There is another level which might be called an *exploratory* dialogue, which wonders if possibly the two parties might have something significant to give to each other. But this is tentative, exploratory, and the possibility of mutual benefit is only vaguely conceived.

Then there is a form of nondialogue which only looks like dialogue, which exists where one party is ready for dialogue but the other is not. That usually turns out to be a monologue, with one party speaking and the other listening, but no mutual listening to each other. That is how so-called dialogue between theology and therapy often develops, with therapy doing most of the contributing.

(3) Finally there is a level of dialogue which might be called existentially *significant* dialogue, in which something essential is at stake in one's self-understanding, in which both parties are willing to take substantial risks in listening to each other. Even this level may be divided into two types: the significant dialogue which is occasional but not sustained, and the dialogue which is significant and sustained.

We have not arrived at the point where we are yet in a significant dialogue, with the exception of a few isolated individuals who are out of the main stream of their own traditions of theology and therapy. The dialogue is still merely exploratory and tentative, and in some cases only practical and professional. It will require a more substantial commitment from both sides to develop sustained significant dialogue.

COMPARING THE WORST WITH THE BEST. There is one particularly disastrous tendency in discussions of religion and psychology[21] to commit a common error in logic and discourse which need to be better understood. The error is to take the most admirable aspects of one viewpoint, e.g. psychology, and compare those with the worst aspects of religious communities or vice versa.

One popular form of this game is to take the worst, most deteriorated, unconscionable forms of religion as examples of religion generally, and compare these to idealized approaches to therapy. It is understandable that the therapist might occasionally be prone to think in these terms, since many of his clients are persons whose lives have been so horridly messed up by religion or by pseudoreligious distortions. He sees in his therapeutic practice the dysfunctional character of religion so vividly, and often much deeper than the growth-producing dimensions. Further, he himself may operate with a highly idealized self-image of his own therapeutic work. So it is understandable, even if not reasonable, for him to compare his therapeutic ideal-type with the anxious, guilty, life-rejecting pseudoreligious distortions so evident in his practice.

Likewise, it is also unfair, however understandable, for religiously oriented persons to see only the worst aspects of psychotherapy, and compare those with the best aspects of religious community. That game is played something like this: Psychotherapy is seen as a very costly professional relationship with an unimpressive rate of cure. The process is often extremely long, arduous, and ineffective. While religion seeks to see man wholistically, psychotherapy is often based upon a limited understanding of man as a naturalistic instinctual mechanism, or as an idealized rational creature, or as an individual without a social context, or as a creature with no relationship to his ultimate source and end. Whereas religions are ordinarily supportive of social morality, there allegedly remains a general immaturity on moral issues among many therapists. Many are trapped, e.g. in a misleading estimate of the high importance of coitus and orgasm as the basic sexual definition of man, rather than a more polymorphous understanding of man as embodying an existence which is more profoundly erotic in ways transcending genital sex.[22]

The point is: If we should wish to play the game of looking only at the seamy side of therapy, and there is plenty of it, and plenty more than I have mentioned, we could easily take the most desirable aspects and achievements of religious communities and compare them very favorably.

It serves little purpose to continue such a comparison, other than merely to point out that the procedure itself is prejudiced. Yet this is just the procedure which so often is applied. The dialogue will best proceed when both communities are aware of their limitations and achievements, when both parties are willing to see themselves in an historical continuum as interdependent, and when neither idealizes itself against the other. In such a dialogue both have a great deal to learn.

THE EXEGESIS OF THERAPY. It has been said that the Christ event is God's exegesis of the meaning of history and creation, if exegesis is understood to be the making known of what is there in the text. Exegesis brings to light the text, looks into its language and makes its meaning transparent. This is what the life, death and resurrection of Jesus does for the created order—provides an exegesis into the meaning of the cosmos. In this sense I am proposing that the history of Jesus is also thc exegesis of the inner meaning of the therapeutic process.[23]

Evangelical dialogue with mutating psychotherapy will best proceed in the spirit of Paul who wrote that he could not be "ashamed of the gospel: it is the power of God for salvation to everyone who has faith, to the Jew first and also to the Greek. For in it the righteousness of God is revealed through faith for faith" (Romans *1*:16-17). Our conversation will not be genuine dialogue, but merely listening or phony echoing or defensive withdrawal, if we are ashamed of the good news of the event of God's occurring love, or if we back off already at the most crucial point of speaking of the meaning of God's self-revelation for therapy.

In answering the charges of some who apparently thought that Paul was becoming captivated by worldly concerns and secular processes, the apostle spoke in a way which we can direct toward our concern with psychotherapy: "For though we live in the world, we are not carrying on a worldly war" (II Corinthians *10*:3) i.e. in our terms: although we are certainly addressing secular processes, we are not grappling with those processes strictly from within the limits of their own self-interpretation. For the struggle against man's despair and alienation is God's own struggle. "For the weapons of our warfare are not worldly but have divine power to destroy strongholds" (II Corinthians *10*:4). In summary, Paul calls upon his hearers to "take every thought captive to obey Christ." The theological method which with greatest promise shapes our approach to psychotherapy lives in the hope that we can take captive *omnem intellectum in obsequium Christi.* For some of us are convinced that the secular process of psychotherapy indeed has a right to know of the Christ event, God's occuring love, as its hidden center.

EMPATHY AS BEYOND THERAPY

One way of expanding the agenda of dialogue is to show that a central feature of both evangelical thought and therapeutic change is found in many other human concerns. My thesis in this section is: The process of empathy, so crucial to psychotherapy and so suggestive for a theology of Incarnation,[24] is also visible in a broader range of human experience, which if clearly grasped, yields some fresh illumination on therapeutic empathy. Psychotherapy is sometimes narrowly regarded as a wholly unique human relationship, without analogies to other forms of human transactions. There are, in fact, several strata of analogies, which I will spell out briefly, each of which amplify the meaning of empathy in the others and of God's empathy for all.

As a leading case in point, the empathetic process is at work in the study of (1) *archeology and anthropology.* When a competent archeologist, anthropologist, or paleontologist digs into a primitive culture, he must use whatever artifacts are available to help him project himself into the internal frame of reference of that culture, using broken shards, bones, architectural foundations, and all available remaining cultural fragments in order to reconstruct in his imagination the life of that culture, to behold it as it saw and felt itself, and to translate that internal perception into meaningful terms of himself and others.

When Malinowski sought to understand the Melanisian culture, Margaret Mead the Samoan, or Durkheim the Arunta culture, they in effect moved empathetically into the primitive context and *listened,* as it were, with unconditional positive regard, to it speaking in its own language. This process could rightly be termed purely phenomenological in its intention and method, in the sense that it attempts to enter into the phenomenon observed and to understand the phenomenon from the viewpoint of the phenomenon itself, without external judgments reshaping or silently twisting the perception.[25] It seeks to avoid imposing alien viewpoints upon the phenomenon, and tries to let an understanding of the phenomenon emerge exclusively out of the listening itself. In this way anthropological study expresses an act of historical empathy which is clearly analogous to psychotherapy, which enters into the frame of reference of another to allow that consciousness to achieve better definition. It is not fully analogous, of course, in that the therapeutic agent is attempting also to recommunicate to the individual his own self-understanding more perceptively and accurately than he has himself perceived, whereas anthropology has no healing intention.

(2) All *historical* study in fact has a similar empathetic focus. Whether one is studying the history of a civilization or a country or a period, the object is to get inside of that particular place, time or orientation, not to project one's own presuppositions upon it, but to let it speak with its own distinctive language. Whether studying the history of art, literature, politics, or philosophy, one must project oneself with empathetic imagination into the past through the study of documents.

As all historical study illustrates empathy, so also does (3) *sociological* study. When the sociologist studies a particular subculture, he often lives within it as a "participant observer" for a sufficient time for its own meanings to emerge and penetrate his own horizon and language.

Thus, what the therapist must learn with keen awareness about empathy must also be learned in a more general sense by all persons in the historical and social sciences. If so, it must also be the case that whatever the Christian faith knows about the empathy of God, the participation of God in human estrangement will stand as an overarching analogy before which the empathetic process in many other sciences and knowing processes may be finally understandable.

If the empathetic process can be seen in all historical and sociological study, it can be seen even more dramatically in the work of the (4) *artist*. A great *novel* like William Faulkner's *The Sound and the Fury* would serve as an example. Faulkner writes a profound internal phenomenology of four characters in the four chapters of the book: the demented Benjy, Quentin who finds himself unable to bear the burden of upholding a deteriarating society, Jason the cynical enterpreneur, and the sensitive, reconciling maid Dilsey. The gifted novelist is capable of showing his reader the internal frame of reference of his characters, revealing the symbols and values which bestow meaning in their interior experience. It takes a profound empathetic imagination and story-telling capacity for this to come alive. Thus the same empathetic, imaginative quality which makes for good therapy is also powerfully present in good literary creation.

(5) An effective *playwright* must develop even more remarkable empathetic powers of creation, if he is to bring to life out of his imagination a dramatic situation with which the audience will identify meaningfully. The playwright must be capable of making the internal orientation of his character transparent to the audience through dialogue and action, in order that the character becomes believable. In some ways that is a much more complex task than psychotherapy. A mediocre playwright will not really accomplish this process or make transparent the character's depth. The gifted playwright will create a character or situation in which the hearer will find himself intensely caught up in the man or his situation imaginatively. To do this takes empathy, the ability to project oneself imaginatively into the frame of reference of another, to live out the characters' lives in fantasy so that their story can be written down on paper, and brought to life by a director on stage.

Likewise the (6) *actor* constitutes one of the most transparent examples of empathetic participation. Despite the variety of schools of acting, each in its own way seeks means of enabling the actor to get inside the character he is portraying, so as not only to be able to feel imaginatively the experiences of his role. but to make these feelings transparent to others. It is the director's task to help him reach inside the character and to weave all the characters together into an integral dramatic situation. The actor's essential task is to project himself into the frame of reference of his character, not to be lost in it, but to allow others to sense the feelings of the dramatis personae. The analogy cannot be pressed further since the therapist continues in a live dialogue with a real person, not merely a fictional creation of the imagination of another.

What we have said about the arts in general can be amplified in connection with (7) *painting* in particular. The painter first must listen to the situation or object he is painting, attend to it carefully, even if it exists only in his consciousness. Regardless of content or media, or theoretical viewpoint, the artist must have his eyes open, visually experience reality, and in that visual encounter experience a kind of unconditional positive regard for what he sees. The artist's eye is capable of visually grasping the reality which the common eye often misses. The therapist's eye likewise should be capable of seeing levels of emotive response which ordinarily are not perceived.

In the history of contemporary art, *cubism* represents a significant development, following impressionism and expressionism, of this empathetic capacity to reach imaginatively into structures, to rebuild the object, to participate in the situation so profoundly that one can break it up in effect into its component parts, atomizing it, seeing its inner movements from several vantage points simultaneously, portraying its being through the use of color, design, texture and all the artist does to make the reality he sees transparent to the beholder. Picasso's "Young Girl at the Mirror" is an especially fitting example of this process of empathetic participation with another. When the viewer witnesses four different levels of a girl's vision of herself, one sees two dimensions of the unmirrored girl herself, and two other dimensions of her perception of herself in the mirror in a complex creation of Picasso's empathetic imagination. When Christian faith beholds in the empathetic participation of God in the world a means of understanding all empathetic action, it beholds with special amazement such achievements of artistic empathy.

The affinity between artistic creation and cathartic healing has long been noted by aesthetic philosophers. This is seen primordially in Greek theater and expressed philosophically in Aristotelian aesthetics, where *katharsis* is crucial to theater. Future research will hopefully determine more specific forms of analogy between the empathetic positive regard in therapy and empathetic transparency in the dramatic arts. Might all of the arts at some level be partners in the healing reality of divine, empathetic, accepting love, whether or not confessed or acknowledged? They enable empathetic participation through which the human situation is revealed and through which Christ takes form in history. Possibilities which in ordinary life would be blocked are opened up by the artist's ministry, as it participates in God's own ministry to the world. Stated as a formula: just as we see authentic psychotherapy participating implicitly in the accepting reality made known in the Christian kerygma, likewise artistic creation participates in that same reconciling reality through imaginative empathy, and that is why the artist's work lays hold of human existence at a depth similar to therapeutic work.

What we have said of history and art may also be said of (8) *hermeneutics* and the process of translating meanings from one language to another and discovering reliable patterns and principles for interpreting one frame of reference to another. When one learns another language, he must learn to project himself into another wholly different ethos, not only learning vocabulary and syntax but more so, if he grasps the meanings, projecting himself empathetically within the whole cultural frame of reference in which the language is spoken. He cannot merely translate word by word. He must translate experience by experience across cultural and historical barriers. It is the profound difficulty of this process with which a hermeneutic must deal.[26]

Often we put our own questions to ancient texts so as to fail to hear their questions to us. Whenever we authentically encounter the text, however, we move in a hermeneutical circle in which we are putting our human questions to it and it is putting the claim of its understandings to us. The text *addresses us with its own distinctive question in such a way that it transforms our question about it.*[27]

Something of the same hermeneutic circle exists in psychotherapy, for as the client addresses the therapist with his own orientation, twisted though it may be, the therapist's task becomes to allow that witness to speak to him in such a way that his consciousness is reshaped by that address. In producing an exegesis of, or making present, biblical texts, we are thus not doing something wholly different from therapeutic empathy.

Having discussed the empathetic process present in historical, humanistic studies, artistic creation and having not mentioned the (9) *natural sciences,* one might suppose that the range of our discussion of empathy would be limited only to the human and historical dimension and not to natural processes. But this is not the case. Alfred North Whitehead and Teilhard de Chardin have pointed the way for a panpsychic perception of the natural.

We shall take as a key example what might seem a most unlikely one, viz. a *geologist.* In what sense is the process of empathetic participation in natural history elemental to his task? If one looks carefully at geological thinking, the analogy with empathy is apparent, bizarre as it may sound. For the geologist, with his capacity for analysis and imagination, and his knowledge of the historical development of structures of rock and soil underneath the earth, projects himself back literally into millions of years, aeons of natural history. He tries to put the puzzle together. He sees a fault here, a hill there, a cleavage here. It takes a very imaginative act of sustained empathy with natural processes to project oneself into remote geological time and reconstruct layer for layer its developmental sequence. There is a level at which therapeutic empathy for the client's history may be strongly analogous to the geologist as he confronts the unique geological history of a particular earth structure. Likewise, we could speak similarly of the study of botany, zoology, physics and chemistry, and other natural sciences which involve an imaginative projective movement of the consciousness into various forms of natural processes.

What we see so sharply clarified in therapy has interesting possibilities also in (10) *political science* and international relations. For if we could develop in a democratic society a capacity to behold empathetically the orientations and values of others, it would be much easier for us to deal with power conflicts. Empathy has a profound implication for diplomacy. With deeper empathy, healing might emerge in political communities in ways similar to individual consciousness.

Our overall purpose has been to show that empathy is not circumscribed by the process of psychotherapy *per se,* although that process is still the place where empathy is probably best focused and clarified. The same features that make for health in psychotherapy also are requisite for other forms of human understanding.

Finally it seems clear that the freedom to engage in the frame of reference of another is only possible if one exists in the midst of an accepting reality which frees one from defensive self-righteousness. For *how can I enter your frame of reference or language or culture unless I understand myself to be understandable to and understood by a larger process of empathetic understanding which embraces us both?* This larger process of empathetic understanding is the unacknowledged ontological presupposition of all of these forms of empathy at work, regardless of where they occur. The kerygma speaks precisely at this point, saying that this empathetic love and unconditional positive regard, which is implicitly assumed in all these processes, has made itself known explicitly in the history of Jesus in such a way that it can be known and celebrated in all other forms of human knowing.[28] Thus all our means of employing this empathetic love become an implicit doxology, a hymn of praise to the One who has made himself known as sharing in our condition, providing an over-arching frame of orientation for all of our empathetic extensions and concerns.[29]

The Christian faith has no desire to limit or derogate these humanistic empathetic processes. It only wants to celebrate its good news that the ground of all unconditional positive regard in all these processes is revealed in the history of Israel and in the Christ. Therefore, Christian proclamation is not a denial or rejection of human empathy but an affirmation of its inner center as God's own initiative, and a witness to its constitution in the One who gives us life.

GOD-LANGUAGE AND THE LANGUAGE OF THERAPY

In pursuing an evangelical theology for a post-therapeutic situation, it is important to me that I not dodge one of the most difficult issues which has been put to me repeatedly by my critics. Is God language necessary to my hypothesis on the implicit ontological assumption of therapeutic agency?

I have continued deliberately to use God-language to speak of the internal meaning of the therapeutic process. After one discovers the incognito love of God mediated through another human being in a therapeutic relationship without God language, what function does God language have? Is it merely a burden which can be dispensed with? Is theology outmoded by therapy, in the sense that therapy mediates already the accepting reality to which theology points with God language? Why speak of God, in other words, *after* effective therapy? Isn't God language an unnecessary burden? Why not jettison the excess baggage? If a person in a secularized, demythologized relationship can mediate in fact the incognito love of God, why talk of God at all?

"A humanistic therapist may well ask why God has to be introduced at all," writes Belford of my proposals. Dietrich Stollberg of Erlangen, Germany, asks: What is the purpose of kerygma, if the love of God is nonverbally expressed?[30] In order to respond to these questions, I wish first to discuss the alternatives to God language in the developing Judeo-Christian tradition, and then to further distinguish between the verbal and transverbal address of God, and finally indicate the uniqueness of the church's *therapeia.*

A) The surprising recentness of God language. It is astonishing to many modern users of God language to reflect on the simple historical fact that God language is relatively new in the Judeo-Christian tradition, if we review the overall historic time line of that tradition.

The Hebraic tradition functioned reasonably well for some 1,000 years before God language, or Gott language, which is a distinctively Teutonic development. Hebraic religion used the symbol YHWH to point cryptically to that to which one cannot adequately point with words. There was no good word for a reality beyond words, so they made one up with an artificial language, a symbol so holy that for many centuries it was not even pronounced. We translate that into artificial English words like Yahweh or Jehovah. Other names like Adonai and El attempted to point to the reality which unifies, embraces and completes history, which Jesus knew as the father of Abraham, Isaac and Jacob. We do not know that reality except as we behold the history of its self-presentation. That is why the biblical witness takes the form of narratives of the self-vindication of Yahweh.

Yahweh language then became translated into alternative world views. The Greek and Roman mythologies of Theos and Deus already had a set of meanings attached to them which the church bought into and transmuted. Yahweh language is a developing tradition. We pick up this tradition far, far down the line, after it has undergone many mutations which we need better to understand. The reason why I refer to the recentness of God language is that talk about Yahweh, El, Adonai, Theos, and Deus was a very old tradition, centuries old, when it finally ran into a dialogue with the Teutonic language from which our word Gott or God is derived. It was far into the fourth century before any Christian ever thought of using God language. The New Testament canon was entirely completed before God language entered the Christian tradition. So even though we are now accustomed by 1500 years of experience to the use of God language in the West, from a larger historical perspective God language is a relatively recent introduction into the Judeo-Christian tradition. The church got along quite well in its first several centuries without God language.

We could continue that story, if we wanted to turn this series into a historical theology exercise, and trace in detail the changing mutations and developments of God language through the medieval period, reformation, the enlightenment, the nineteenth century and to the present, through idealism, empiricism, naturalism, etc. We are the heir to all these already actualized mutations. We do not come to this conversation afresh. It is only reasonable to ask of Christians that if we are to use God language, we are called to have some sense of the historically mutating character of that language.

B) The verbal and transverbal address of Yahweh. With that preliminary observation, I now move on to the question of God language in therapy. Ralph Wilburn [31] makes a useful distinction between the disbelieving and believing therapist which I would like to develop. Already it is clear from what I have said in previous essays that the disbelieving therapist, who may have no direct awareness of the Christian kerygma, nevertheless may (not necessarily does, but may) function as an effective therapist with congruence and empathy.[32] If he communicates to his client that the ground of acceptance is accessible, and that he too stands on that ground, and that the acceptance mediated through him is also rooted in that ground, and beyond the mere interpersonal relationship, then that relationship will tend toward growth.[33]

The therapist can work operationally with more or less full adequacy without announcing the kerygma. But the question is pressed upon me: could he not work with fuller adequacy if he understood and declared the kerygma? The distinction I am struggling for is difficult. Operationally he is capable of functioning reasonably well, but the ground upon which he actually stands is made known in a particular history which is remembered by the church.

If we look then at the functioning of the believing therapist, there is one major difference: he is more deliberately aware of the ground upon which his acceptance stands, because he has heard the witness of a community about the revelation of that reality in history. Suppose he has heard and believed that Word, and knows that the self-disclosure of the accepting reality is made known in history so as to transform his knowledge about that reality. It is precisely in his meeting with the event of divine acceptance that his knowledge of the accepting reality has been reshaped and augmented. Would such knowledge affect his therapy? It seems more reasonable to suppose it would.

This is a fine line to walk: even though the disbelieving therapist may function operationally much in the same way, the believing therapist has a broader and more realistic perception of the ground upon which he stands, and of the larger dynamics of the relationship. This is not grounds for boasting, but for simple thanksgiving for the self-disclosure in history of the accepting reality.

Why, if the love of God is nonverbally expressed, is it necessary to express it verbally in the form of proclamation? Clearly the love of God does not need to be verbally expressed in order to be present. But from time to time this community which has emerged in response to the Christ event does in fact declare and celebrate its origin and ground. The believing therapist hears about the accepting event always only in a community of proclamation. He did not just intuit it. No act of Christian faith exists completely apart from a community of witness which points to the event to which the Christian life refers itself.

Why is the verbal proclamation necessary at all? It is clear that the nonverbal is prior to the verbal, and the event is prior to language about the event. But that should not imply that the act of proclamation is thereby innocuous or irrelevant. It is simply not primary. The prior reality is the event of God's love. It is rooted in history.[34] If we are addressed and grasped by that story, we must find human language, frail though it always is, to try to express in understandable ways God's own language to us. Theology's only importance is that fragile task. Theology has a modest task: to put to work human language in an attempt to point to how God has made himself known in history. So Christianity is not primarily a religion of law or ideas, but of event. That does not mean that preaching is innocuous or meaningless, but that it is secondary to the event itself.

Although we may theoretically speak of unconditional positive regard as an element of therapeutic change, we know that finally no human love is unconditional. The believing therapist may be more aware of his finiteness and the limitation of his loves, if he has internalized the Christian understanding of man's alienation and deliverance. No finite person will ever comprehensively express eternal love. Neither the believing or disbelieving therapist fully expresses the range and depth of unconditional acceptance, but the believing therapist knows of a community which has a story to tell about the historical event of unconditional acceptance, which his behavior reflects hopefully even if in fragmented ways.

C) PASTORAL THERAPEIA. A corollary question: Does the thesis of the implicit ontological assumption of therapy tend to lose sight of the unique subject matter of pastoral care? Doesn't pastoral care lose its distinctiveness if secular psychotherapy also mediates God's love and care and perhaps even better than pastoral care? Can the uniqueness of the pastor as counselor be clarified and redefined in the light of my previous proposals?

Edgar Jackson has rightly pointed out that I have said little about the uniqueness of the pastor as counselor. Dietrich Stollberg also wonders, in response to my proposals, whether the proprium, the proper subject of pastoral care, can any longer be identified.[35]

Part of what I am trying to accomplish is the preservation of the autonomy of the therapeutic task, without theological imperialism, yet also with an attempt at theological interpretation of the so-called secular healing process. What becomes of the church's pastoral care in all this? We need better to distinguish between the lay believing therapist and the ordained therapeutic agent, including the pastor as counselor, who has taken upon himself as a matter of vocational commitment the maintaining of continuity with the apostolic tradition. The reason why apostolic succession was important in the early centuries was to maintain deliberate continuity with the earliest witnesses to the originative event of Christianity.

The committed lay Christian therapist certainly may have in his consciousness all that the pastoral office hopes to celebrate, protect and develop. The pastoral office functions in order to insure the deliverance of that message and the nurture of that consciousness.

What is the nature of the pastoral office, without which the church's particular *therapeia* is not fully understood? We cannot answer that without pointing to the flow of historic continuity between the originative event and now. We would not be anywhere at a given moment if it were not for that complex historic continuity, with all its embarrassing ambiguities and inadequacies. We exist because that tradition exists. We may be accountable or unaccountable to it, but we do not exist without that real historical process occurring. The pastoral office exists to mediate that history to the now.

The ordained minister differs from the believing lay therapist, not in the respect that he is more therapeutic, or less therapeutic, but in that he is set aside by the church in a visible way for a specific purpose: to be deliberately prepared to be accountable to that tradition. There have been and will continue to be many alternative views of how that accountability is to be understood. But no one is well-advised to set his foot on the path of ordination unless he is ready to take that task upon himself seriously. It is not something to be taken on routinely, nor a matter of answering someone else's questions. It is a matter of discovering what your identity is, whether you are called to that vocational commitment, the yoke of which is experienced (by those who are called to it) as easy, and the burden light.

The model of the professional therapist with his office, receptionist, records, fifty minute sessions, and fee-basis relationship, is a very tempting professional model which has gotten transferred to parish relationships. That model is not best for Christian pastoral care. Ordained ministry has many advantages and freedoms which the professional therapist does not have. It has the freedom to witness. It enjoys greater durations of interpersonal continuity. Ordained ministry is organic to a community. This is not true of professional psychotherapy which ordinarily has no sustained relationships which are not paid for. The ordained ministry shares consciously in Christ's own prophetic, priestly and worldly ministry, whereas professional therapeutic care often shares in it only unselfconsciously and implicitly.

SOME EXPERIENCE-BASED CONCLUSIONS FOR POST-THERAPEUTICS

The central thread of this chapter has been the attempt to understand the post-therapeutic situation as a new context for dialogue between theology and psychotherapy I began by asking what happens to persons *after* they have experienced effective psychotherapy. I explored the curious, mutating history of expectations of what psychotherapy ought to be able to do, a history which has tended constantly to extend itself in the direction of religious and philosophical questions. I then asked why the dialogue between therapy and theology has become so immobilized in this situation, pointing out abuses of previous patterns of dialogue and stating the norm of significant, sustained dialogue. Then as a means of expanding the agenda of that dialogue I examined the wide-ranging implications of the concept of empathy, so crucial to both therapy and theology, for understanding the human condition. In such crucial questions as the viability of God-language, the relation of verbal and non-verbal expression, and the uniqueness of the church's own *therapeia,* I have attempted to approach anew some of the thorniest unsolved problems of the dialogue.

I conclude with some hard-won inferences which have been hammered out of my experience as a teacher, counselor, sufferer and fellow human being. As we proceed, I hope the reader will find that some of these conclusions correspond with his own experience and intuitive insight:

(1) *Every moment of human history is meaningful in the light of the whole.* Each part of the historical process points always beyond itself as being put together, wholistically, even though you and I may not have it put together. Despite its evident distortions, evils, and alienation, the whole is whole. There is no part finally unreconciled. That is an intuitive judgment which I cannot prove, but of which I am as experientially certain as I am that tomorrow will follow today.

(2) *A great deal may be learned from the sufferings and distortions about the whole.* It is precisely at the points where the whole does not seem to be whole that the greatest learnings about its wholeness are capable of being perceived. The cross is, for Christians, the key example of such learning. It is just at those points at which the wholeness of the historical process seems least credible, where its internal stresses are greatest, that the most profound insights are offered into the wholeness. The wholeness often surprises our expectations, revealing itself at the points of greatest stress. The wholeness is not something which we project or impose upon reality, or even something which we must discover in order for it to be there. It is a wholeness in cosmic reality itself, there whether we discover it or not, yet always inviting its discovery, and in fact seeking to make itself known. Its presence is not dependent upon our recognition, but searches for our recognition.[36]

I speak from my experience, not assuming its credibility to your experience, but hoping that something in your experience correlates with mine. At precisely those points where I have experienced the most guilt about my past and anxiety about my future, when I have faced myself seriously to explore the dynamics of that guilt and anxiety, I have come furthest toward learning about the wholeness of the historical process beyond my limitations and suffering.

(3) *Growth toward free, responsible love and self-determination develops and grows best in the atmosphere of free, responsible love and self-determination.* If we are serious about ideas of freedom, love and responsibility, we are being called to create and nurture communities of freedom, love and responsibility. Until free, authentic, accountable love is shared in community, it is an abstract idea lacking credibility. In fact it cannot be adequately conceptualized until it is first actualized (despite persistent claims of abstract idealists who assume that the biggest step to the actualization of an idea is first of all its proper conceptualization). Without a forgiving community, forgiveness is not understandable, and in fact is somewhat implausible as an idea. We may need a partial moratorium on talk about freedom, love and community, while we search for its concretion in actual communities of free, responsible love.[37]

(4) *Freedom emerges best not in an atmosphere of uncritical acceptance, but in a deeper acceptance which is able to solicit and receive negative feedback and caring judgment.* Genuine love is not sheer permissiveness, despite what I have said earlier[38] about the analogies between grace and permissiveness, which probably were overstated. Cheap, uncritical (and therefore unloving) acceptance of the phoniness of another is not *agape,* but carelessness.[39]

(5) *Every person I have met has some healing resources in him,* some element or image of authenticity, from which we can detect some unmistakable clue pointing to the wholeness of history, and to potential human wholeness. Reflecting on my own experience, I can think of no person whom I have come to know well, from whom it was impossible to receive strength, grace, insight, and witness to the "light that enlightens every man" (John *1*:9). I do not know anyone intimately who has finally failed to address me with some hopeful possibility or perception of the wholeness of history.

When I have from time to time functioned as a growth-producing agent, I have discovered that each of those persons whose lives were seeking correction in my presence, have given something growth-producing to me. It has never failed. If this is true, it has a broader implication for the study of therapeutic change. The therapeutic task is not a task which is reserved only for professionals trained in diagnostic skills and therapeutic stratagems. No professional group has a corner on *therapeia*. God's spirit is bestowed upon all flesh. Every man has gifts to be shared.

That which Paul says concerning spiritual gifts among the body of Christ, I also find among all persons whom I have encountered at a level of depth: There are a variety of gifts, but the same Spirit, and there are varieties of service, but the same Lord. There are varieties of working, but it is the same God who inspires them all in everyone. "In each of us the Spirit is manifested in one particular way, for some useful purpose" (I Corinthians *12:7*). It is something like that which I experience in effective group encounter. Each person, however distorted, has his own gift to give or word to address to me, even in a relationship in which I presumably am the one expected to facilitate constructive change. I discover that my troubled colleague is mediating to me growth and healing, even amid his anxious or frustrated quest, so that in surprising ways "all things work together for good for those who love God."

Joel prophesied that God would pour out his spirit on all flesh (Joel *2:28*). According to Jeremiah, God enters into judgment not with some but with all flesh (Jeremiah *25:31*). The biblical notion of *all flesh* as recipient of God's spirit remains profoundly relevant to a theological interpretation of therapeutic agency. Even in the so-called secular counseling situation in which man is presenting his brokenness and alienation, laying bare his estrangement, in hope that through that process growth will occur, even and precisely there amid alienation, one finds the Spirit's address and witness beyond itself to the wholeness of history amid the alienation. In this way the Spirit's presence in the wholeness of history is being fulfilled in the *saeculum,* to be sure in a fragmented, but nonetheless real, way in secular therapeutic change.

The New Testament shared the hope of Isaiah: "The glory of the Lord shall be revealed and all flesh shall see the salvation of God." That glory, once for all revealed in the history of Jesus, is being revealed anew at those points in the secularizating process where wholeness is being enabled, where persons are moving toward authenticity, freedom and love.

NOTES

1. Godin, Andre: Psychologie religieuse et pastorale. *Lumen Vitae, 4:*243, no. 2. Chapelle, A., Godin, A., and Oden, T.: *Experience Therapeutique et Revelation: Un Symposium. Nouvelle Revue Theologique,* 97, Tome 87, no. 8, September-October, 1965, pp. 785-835. See *Continuum,* vol. II, nos. 2 and 4, and vol. III, nos. 2 and 2, 1964-1965.
2. Strollberg, Dietrich: *Therapeutische Seelsorge.* Munich, Chr. Kaiser Verlag, 1969, pp. 342-380. *Seelsorge Praktisch,* Vandenhoeck and Ruprecht. Gottingen, 1970. Was ist Pastoralpsychologie? *Wege zum Menschen,* Heft 6, 1968, pp. 210 ff.
3. Among American responses and criticisms of my proposals on theology and therapy are the following: Stinnette, Charles R. Jr.: Reflections and Transformation, Knowing and Change in Psychotherapy and in Religious Faith. In Homans, P. (Ed.): Essays in Divinity, III, *The Dialogue Between Theology and Psychology.* Chicago, U of Chicago Pr, 1968. Stinnette, Charles R. Jr.: Review of contemporary theology and psychotherapy. *Psychology Today,* February, 1968. Lapsley, James N.: Kerygma and Counseling: A Review Article. *J Pastoral Care, 21:*100-103, no. 2, June, 1967. Lapsley, James N.: Contemporary theology and psychotherapy (Review). *Theology Today, 24:*547, January, 1968. Browning, Don S.: *Atonement and Psychotherapy.* Philadelphia, Westminster Pr, 1966, review in *Religion in Life,* Summer, 1968. Keen, Sam: No Mortar. Christian Century, January 28, 1970. Jackson, Edgar N.: *Pastoral Psychol, 17:*60, June, 1966. Thornton, Edward E.: *Pastoral Psychol, 18:*58, May, 1967. Becker, Russell: Kerygma and counseling (Review). *Religious Education,* March-April, 1967. Oates, Wayne: *Review and Expositor,* Winter, 1969, Fall 1970. Kemp, C. F.: *Encounter, 28:*387-388, August, 1967. Hartshorne, M. H.: *Interp, 22:*368, July, 1968. Rogers, William F.: *J Pastoral Care, 21:*186-187, September 1967. Belford, Lee: Review. *Witness,* November, 1966. Wilburn, Ralph G.: Psychotherapeutic theory. *Christian,* October, 1966. Malony, H. Newton: Differences in Pastoral Care as a Function of Religious Beliefs and Attitudes Toward Human Nature. Paper presented at the American Association of Pastoral Counselors, San Francisco, November, 1971. Bower, Robert K.: Review. *Reformed J* October, 1967. Other reviews by David K. Switzer, *Christian Advocate,* November 30, 1967; Carl M. Davidson, *Church Management,* September, 1967; Herbert Anderson, *Princeton Seminary Bull,* 1967; Elton Einegenberg, *Reformed Rev,* December, 1968; C. W. Brister, *Southwest J Theol,* 1967.
4. Fromm, Erich: *The Crisis of Psychoanalysis.* New York, Fawcett World, 1970, pp. 11 ff.

5. Albert Outler is the first one, to my knowledge, to raise this question with unmistakable clarity, in *Psychotherapy and the Christian Message.* New York, Har-Row, 1954.
6. Szasz, Thomas S.: *Ideology and Insanity.* Garden City, Doubleday, 1970, pp. 218 ff.
7. Keniston, Kenneth: *The Uncommitted.* New York, Delta. Dell, 1969. Egenter, Richard and Matussek, Paul: *Ideologie, Glaube und Gewissen: Diskussion an der Granze Zwischen Moraltheologie und Psychotherapie.* Droemer Knaur, 1965.
8. May, Rollo: *Love and Will.* New York, Norton, 1969.
9. Jourard, Sidney: *The Transparent Self.* New York, Van N-Rein, 1964.
10. van den Berg, J. H.: *The Changing Nature of Man.* New York, Dell, 1964; Boss, Medard: *Psychoanalysis and Daseinsanalysis.* New York, Basic, 1963. Caruso, I. A.: *Existential Psychology.* New York, Herder & Herder, 1964. Cf. Binswanger, L.: *Ausgewählte Vorträge und Aufsätze.* Bern, Francke, 1955. Frankl, V.: *Man's Search for Meaning.* New York, Washington Square Pr, 1963.
11. See Mullahy, Patrick: *Oedipus: Myth and Complex.* New York, Grove, 1955, and Sykes, Gerald: *The Hidden Remnant.* New York, Har-Row, 1962, for accounts of this history.
12. Goble, Frank: *The Third Force.* New York, Pocket Books, 1971. For an excellent bibliography see Maslow, A.: *Toward a Psychology of Being.* New York, Van N-Rein, 1962, pp. 206 ff; also see Bugental, J. F. T.: The Third Force in Psychology. *J Humanistic Psychol, 4:*19-26, no. 1, 1964.
13. Burton, Arthur (ed.): *Encounter.* San Francisco, Jossey-Bass, 1969. Ruitenbeek, H. M.: *The New Group Therapies.* New York, Avon, 1970.
14. Bugental, J. F. T.: *The Search for Authenticity.* New York, HR&W, 1965.
15. Luther, Martin: *Sermons on the Catechism, 1528.* Luther's Works, Muhlenberg Pr, 1959, vol. 51, pp. 162 ff.
16. Snyder, Ross: *Young People and their Culture.* Nashville, Abingdon, 1969, pp. 27 ff.
17. Although this line of historical development cannot be elaborated upon here, the reader is referred to Chapter two of my book, *The Intensive Group Experience: The New Pietism.* Philadelphia, Westminster, 1972, for a fuller clarification of it.
18. For a significant alternative, see Polanyi, Michael: *Personal Knowledge.* New York, Har-Row, 1958, and *The Tacit Dimension.* Garden City, Doubleday, 1966.
19. Erikson, Erik: *Young Man Luther.* New York, Norton, 1962. Mowrer, O. H.: *The Crisis in Psychiatry and Religion, op. cit.* Maslow, A.: *Religions, Values and Peak-Experiences.* New York, Viking Pr, 1970. Jourard, S. M.: *Disclosing Man to Himself.* New York, Van N-Rein, 1968.
20. Lake, Frank: *Clinical Theology.* Darton, Longman and Todd, 1966. Pannenberg, W.: *Basic Questions in Theology.* Philadelphia, Fortress, 1970-1971, vols. I and II. Richardson, Herbert: *Nun, Witch, Playmate.* New York,

Har-Row, 1971. van Leeuwen, Arend Theodoor: *Christianity and World History.* New York, Scribner, 1964.

21. Thurneysen, W.: *Seelsorge und Psychotherapie.* Munich, Chr. Kaiser Verlag, 1950. Fromm, E.: *The Fear of Freedom.* Boston, Routledge and Kegan Paul, 1960.
22. Reich, Wilhelm: *The Sexual Revolution.* New York, FS&G, 1969. Lowen, Alexander: *Pleasure.* New York, Lancer, 1970. Ellis, Albert: *Sex Without Guilt.* New York, Grove, 1965. Cf. Richardson, Herbert, *op. cit.*
23. Bonhoeffer, Dietrich: *No Rusty Swords.* New York, Har-Row, 1965, pp. 308 ff.
24. Katz, Robert L.: *Empathy: Its Nature and Uses.* New York, Crowell Collier, 1963. Rogers, C. R.: *On Becoming a Person.* New York, HM, 1961.
25. Husserl, Edmund: *Ideas.* New York, Crowell Collier, 1962.
26. Gadamer, H. G.: *Wahrheit und Methode: Grundzüge einer Philosophischen Hermeneutik.* Tübingen, 1960.
27. Bultmann, Rudolf: *Essays, Philosophical and Theological.* SCM Pr, 1955.
28. Bultmann, Rudolf: The New Testament and Mythology. In Bartsch, H. W. (Ed.): *Kerygma and Myth,* SPCK, 1957.
29. Pannenberg, W.: *Basic Questions in Theology,* vol. I, pp. 212 ff.
30. Belford, *op. cit.* Stollberg, Dietrich: Doctoral dissertation, Erlangen Universität, 1968.
31. Wilburn, Ralph: Psychotherapeutic Theory. *Christian,* October, 1966.
32. *Kerygma and Counseling, op. cit.,* Chapter 1. *Contemporary Theology and Psychotherapy, op. cit.,* Chapter VII.
33. Rogers, C. R.: *On Becoming a Person, op. cit.*
34. Pannenberg, W.: *Basic Questions in Theology, op. cit.,* pp. 15 ff.
35. Along with Jackson, *op. cit.* and Stollberg, *op. cit.,* Becker, Hiltner and Stinnette have raised similar questions.
36. Among writers who have decisively shaped these conclusions are Teilhard de Chardin, *Le Milieu Divin.* Brooklyn, John M. Fontana, 1964. Richardson, Herbert: *Toward an American Theology.* New York, Har-Row, 1968. Kierkegaard, Soren: *The Gospel of Suffering.*
37. For a statement of the relevance of this idea for pedagogy, see Optimal conditions for learning: Toward a clarification of the learning contract. *Religious Education,* March-April, 1972.
38. *Kerygma and Counseling,* op. cit., pp. 66 ff.
39. Has Protestant theology diluted its understanding of the human predicament by accommodating itself to the language of humanistic psychology? To the extent that we depend upon psychologists like Maslow, Rogers and Horney who speak of a reliable "self-actualizing tendency" in man, and of the neonate experiencing as a model of psychic health, do we not to that extent undercut the Christian understanding of man's fallenness and alienation? Among my critics no one has pointed to this issue more clearly than Russell Becker, *op. cit.,* who wonders whether I have minimized Rogers' excessive optimism. He has expressed his doubts that the glimpses of sin in

humanistic psychology are really the stubborn facts of sin as seen in the Biblical account of man. "Incongruity . . . distortion . . . and anxiety," he says, "are not the full equivalents of pride and falling short of the mark in Christian theology." Becker wonders if I have missed "something of the difficult problem which God has with humanity . . . symbolized in the expulsion from Eden, the crucifixion, and the Second Coming as expressions of the unrealized fulfillment of God's kingdom this side of the eschaton."

The answer is not an easy one to give. On the one hand, one of the most forceful contributions the evangelical tradition can make to contemporary analyses of man is its continuing awareness of the temptations of man to self-assertiveness and pride, and even further to man's unawareness of his own self-assertiveness. Precisely the predicament of man is that he is blind to his own blindness, unaware of his own lack of awareness. My critics correctly remind me that this is a word which deserves to be addressed by the Christian tradition to humanistic psychology.

On the other hand, however, there is a sense in which effective therapeutic transactions probe more deeply into the actual dynamics of human estrangement and self-deception than do theological schemas about them. So we are left with the question of who takes more seriously the human predicament, the theologians' doctrinal statements or the therapists' actual conversation with existing forms of human estrangement. Neither should be neglected. At least I do not feel accurately described by my critics as an "ontological optimist." Optimism implies a lack of willingness to look with full determination at the depths and power of the human predicament. To fail to recognize this willingness is to misunderstand the intent of my proposals, however inadequately that intent might have come to expression.

CHAPTER 3

THE HUMAN POTENTIAL AND EVANGELICAL HOPE

THE HUMAN POTENTIAL MOVEMENT constitutes a current restatement of a recurring theme in Western civilization. It is a hope for radical human transformation, found both within and outside the Christian tradition. Our imaginations have again been stirred by the bold spirit of Marcion, the Montanists, the monastic visionaries, Renaissance men like Pico and Alberti. reformers like Calvin, Menno Simons and Roger Williams, and a long tradition of utopian visionaries from Joachim to Ernst Bloch.[1] We have been challenged anew to think of the human possibility.

I want to acknowledge the periodic need to hear this challenge, and to make clear that its most recent exponents, leaders of the human potential movement such as Maslow, Perls, Rogers, Schutz, Watts, Gunther, Otto and others,[2] have been important to me as I reflect upon my own human possibility, and upon the potentials of religious community in the emergent society. Again we find a prevailing mood of hope for a radical transformation of the human condition transcending what are viewed as the contemporary limits of man.

We need not here repeat these hopes, but it does seem relevant to ask how the evangelical hope relates to them, or in what sense it stands in tension with them.

In this chapter I would like to formulate that response at several different levels. (1) I will begin by describing the predicament of current theology as a mismanagement of the concept of hope. I will delineate a series of disastrous miscalculations of recent theology hinging on its misunderstanding of the future. (2) Then I will turn to a more constructive alternative, asking whether recent apocalyptic and eschatological studies bear more profound relevance to psychotherapy and pastoral counseling than at first might seem to be the case, and surely more than the faddish changes of pop theology. (3) I am also concerned about how we shall develop a theological method which will help us relate tradition and renewal in a time of rapid social change, and I am going to propose that Newman's idea of doctrinal development is an exceptionally useful model for Protestant theology in times such as these. (4) Finally I shall be trying to clarify for myself and for you a subtle direction in my own thinking which I can only identify as a stabilizing and conserving direction,[3] which in some sense stands in tension with the exaggerated, freaked-out edge of the human potential movement.

THE MISCALCULATIONS OF THE FUTURE

The history of popular American theology in the last decade is a history of some ludicrous miscalculations of the future. Pop theology is a particular expression of lay-oriented theology on the American scene after World War II, fortified by the phenomena of mass religiosity during the 1950's. Many laymen were reading theology. Church membership was increasing. During the mid-1960's, when pop theology peaked and began to fade, the publishers searched incessantly for that next great trend in theology, assuming that if they could just identify that trend in advance, they could supply books for that emerging market.

My thesis is: the central tradition of recent avante-garde American theology has been almost consistently wrong in its estimate of the future. For that reason we have experienced an incredibly varied and demoralizing sequence of fads. Pop theology is by definition faddish, and it is not surprising that astute laymen and pastors would soon cease to take such theology seriously, and for good reason, because they have come to expect it to be ostentatious and shallow. When theology is accommodative and faddish and phony and compulsively relevant and simply trying to get on board with the current movements of society, it does not deserve the attention of the laity.[4]

I would briefly like to demonstrate that each of the key moments of pop theology of the 1960's miscalculated disastrously the situation to which they addressed themselves. We will see that each one of the achievements of pop theology soon encountered a stiff reactive backlash against the very movement predicted to be the wave of the future.

(a) Harvey Cox's *Secular City* is an evident example. Cox idealized the city as a context for humanization and freedom just at the moment when the greatest American cities were beginning to fall to pieces. In 1964, Cox, only months before the burning of Watts, idealized the urban environment, coupled with a withering polemic against all pastoral, rural and village images of ministry, mission and community.

Much in that argument remains useful. But what was decisively miscalculated was the unprecedented predicament of the city, and the fact that the city is not being judged by many who live in it these days as the great humanizing context of freedom of which Cox so romantically spoke. As Cox wrote, in fact, youth culture was just beginning to search for new, small-scale village images of community and developing communes, and was trying to learn to scratch the ground and produce its own food, reading the *Whole Earth Catalogue,* shaping for itself a new social model on a very modest scale reminiscent of frontier America.

(b) Another crest of pop theology focused on the alleged promise of high level ecumenism, reaching its height in the early sixties just about the time when the ecumenical movement was beginning to disintegrate. Among young people today there is a pervasive boredom toward even the desire or necessity to build ecumenical institutions, and yet only a few years ago pop ecumenical theology was pointing to Council on Church Union as the impending wave of the future.

(c) The apex of pop theology was probably around 1965 when the death of God movement hit the fan, and for about two or three years, due to the exceptional interest of the media, the death of God was on the lips of everyone, as if it were something important generally to American religious communities. The irony in all this was that just at the time the death of God was being reported, there was beginning to take shape quietly in youth culture and the counterculture a profound God trip, a serious quest for religious experience, a new concern to nurture the meditative disciplines. It is odd that the death of God movement would predict the future of theology without God just at the time when there was a substantial new movement taking shape in culture earnestly seeking personal religious experience.

(d) Another high point in pop theology was William Hamilton's famous essay on "The New Optimism"[5] in which he spoke of the new affirmation of man beyond existential despair, and of the new joyful spirit at work in culture. The publication of that essay coincided almost exactly with the escalation of our involvement in Vietnam. For the last half-decade we have been swimming in a deepening morass of demoralization which has undercut the expected new optimism. It would be hard to find a social moment in the history of the west which has been more demoralized and pessimistic than the moment that immediately followed Hamilton's essay.

(e) After the death of God and the new optimism came the brief fading moment for a "theology of revolution," when Richard Shaull, Stephen Rose [6] and others did some relaxed armchair talk about bloody revolution as an option for Christian political ethics. What its proponents did not anticipate was that the American context which they were addressing was at that very time moving in a conserving, stabilizing direction, as is shown by the careful public opinion research of Scammon and Wattenberg.[7] Just at the moment of the most volatile talk of revolutionary politics, American labor, management and politics were taking a decisively conservative turn, making the volatile, revolutionary language look dated and out of place even among youth culture, where privatism and the values of personal self-actualization were again becoming predominant.

Charles Reich's *The Greening of America* [8] presented another picture of an alleged wave of the future which was no sooner published than it was thoroughly out of date. It depended upon fads which were changing already while the galleys were being read.

Bandwagon "wave-of-the-future" theology has proven to be a very hazardous occupation in an era of accelerating change, especially when the continuities of history are not as evident as its discontinuities, and when the media focus the public eye upon society's distortions rather than its solidities. As a theologian who has lived through all this, and in fact ridden on his share of bandwagons, I am now generally dissatisfied with, and often feel abhorrence toward, the pop theologies in the last decade. They have been truly objectionable in many ways. What is the appropriate response to such developments? The same as before. The task of theology still remains to make self-consistent and intelligible the life of faith in the Christian community. It does not do this by simply trying to anticipate the direction of the future and flow with it. It may have to go against the stream. Its final accountability is not to the world, or to the particular current of national or regional history, but to the reality to which scripture points, viz. the judgment, grace and care of God in and for the whole of history.

The antidote to pop theology is thus classical, orthodox, holistic Christian thinking in dialogue with the emerging realities of history. It is a conserving radicalism, an attempt to reappropriate the Christian tradition in terms of the current situations and languages of man amid his predicament and possibility. Theology cannot be fully accountable to the present without being accountable at the same time to scripture and tradition. A theology which is most accountable to scripture and tradition will prove to be most accountable to the future.[9] One such contribution, which appears to me to be far more enduring than the ebbs and flows of pop theology, is the work of Wolfhart Pannenberg.

ESCHATON AS BEGINNING

In this section I will ask whether the theology of Wolfhart Pannenberg [10] offers a viable theological model for a new form of significant dialogue with psychotherapy. Ordinarily one does not think of Pannenberg's work as associated in any way with psychotherapy, the healing arts or pastoral care. His work has focused on the study of apocalyptic, universal history, eschatology and philosophical theology—concerns which appear to be quite distant from therapeutic transactions. I will explore some of Pannenberg's ideas in a way that will hopefully initiate some fresh perceptions of possibilities for dialogue between evangelical theology and humanistic therapy. Pannenberg is not only presenting theology with a significant set of new options, but I think he is presenting theology's dialogue with therapy with a new set of decisive questions, even though he himself has not formulated his project in such a way that it has yet made firm contact with psychotherapy.

A) Limitations. I begin by stating in advance some of the quarrels I have with Pannenberg, in order that this will not be viewed as another simplistic, programmatic statement by an uncritical disciple. Since much of what will follow later will be an exposition of Pannenberg's views, I would like for you to know from the outset that I find some serious limitations and problems in Pannenberg's proposals:

(1) He belongs to an exaggerated polemical tradition of abrasive Teutonic theology which sometimes sees nothing except the tragic flaws of one's opponents, who in this case are thinkers such as Barth

and Bultmann, who have much more to offer than their occasional edges of overstatement and inconsistency.

(2) Much of Pannenberg's effort lives out of the spirit of Protestant scholasticism with its tendency to dogmatism, objectivism, and the cerebral neglect of charisma, intuition and individual experience. This involves the loss of certain subjective dimensions which I value as a part of the Protestant pietistic tradition which he rejects.

(3) When Pannenberg says that "There is a restricted, yet important, way in which God does not yet exist,"[11] I respect his intention, but I do not trust his way of stating it.

(4) When he speaks of the Resurrection as historically demonstrable from the viewpoint of objective, scientific history as a historical fact,[12] this may be a meaningful corrective to existentialist tendencies to dehistoricize the Resurrection, but I do not think that the church's faith finally rests upon the objective historical demonstrability of events.

(5) Hope is a virtue which easily turns into vice.[13] Theologies of hope generally are not sufficiently aware of the temptation of hope to despise the now, to idealize the future, to lack realism. Pannenberg's theology of hope, though less vulnerable to this charge than Moltmann's and others, nonetheless tends to become so futuristic at points as to neglect the present.

B) POSSIBILITIES. An entirely personal way of indicating the importance of Pannenberg's proposal is to state why he is important to me. There are several ways in which he has been increasingly shaping my way of framing questions, and in all of these respects, there are important implications for the dialogue with therapy:

(1) Pannenberg is taking us a significant step past existentialism. Most of my training has been in the existentialist theologians: Bultmann, Niebuhr, Tillich, Kierkegaard, Michaelson, Buber, Gogarten. Pannenberg is leading us in the direction of a new cosmic objectivism and rationalism in theology, and away from the introverted, subjectivized momentum of existentialism. If Kierkegaard emphasized inwardness as opposed to Hegel's stress on an objective, universal historical perspective, Pannenberg's thrust may correctly be seen as neo-Hegelian. For Pannenberg is speaking about universal history as the arena of theological reflection.[14]

(2) Pannenberg wants to lead us beyond a realized eschatology

which would understand the end time as exclusively now (a Bultmannian theme which is deeply imprinted into my book on *Radical Obedience*[15]). Pannenberg's work has required me to rethink and affirm the need for a futuristic eschatology, without denying the impingement of the end time upon the now.[16]

(3) In Pannenberg I experience a refreshing new affirmation of reason in theology after a long period of distrust of the rational competence of the natural man. He is leading us toward a new apologetics which may have some important consequences for theology's willingness to dialogue with therapy. My own conversation with therapy has been deeply indebted to Barth with his strictures against apologetics and emphasis upon a confessional approach to dialogue. Without denying the validity of that, I am persuaded with Pannenberg that kerygmatic theology must get out of a ghetto mentality and learn to use alternative language models which are available to us in the modern world, so that the inner meaning of the whole historical process, which is the real content of theology's reflection, will be evident to any man who looks at it clearly and without bias.[17]

(4) Above all, Pannenberg is pressing me toward a new way of framing the task of theology as concerned fundamentally with the wholeness of history.[18] That has profound consequences for the dialogue with therapy, which has viewed its task almost exclusively in a subjectivist and personalist frame. Seldom do therapists talk about an integral perception of universal history, but I am convinced that such a perception, even at an intuitive level, is crucial to therapy both operationally and theoretically.

(5) Pannenberg has helped me to appreciate (in such a forceful way that I am surprised how personally significant it is to me) the apocalyptic understanding of history.[19] Until recently any eschatology other than a eschatology of the here and now has turned me off. Pannenberg has helped me to see how apocalyptic thought is the most decisive way of talking about hope, which I think is an important ingredient in therapeutic agency, and a category whose depths we have yet to plumb.

(6) Utilizing rigorous historical-critical scholarship, Pannenberg has grasped a new understanding of Jesus' ministry as a prolepsis or anticipation of the end time. He is pressing a whole new set of questions on the resurrection, stressing that the resurrection makes little

sense without an apocalyptic understanding of history, but within that tradition it makes evident historical sense.[20]

(7) Pannenberg gives us a much needed role model for building a systematic theology on the basis of solid biblical exegesis.[21] If that is hard to come by in theology generally, it is even more so in American theology, whose systematicians have never been conspicuously adept at biblical exegesis, and have recently assumed that they can pursue their theological inquiries quite apart from solid historical-critical study of the Bible.

C) An eschatological perspective for therapy? Again we frame our general question: Is Pannenberg giving us a new option for a more significant sustained dialogue with psychotherapy, the healing arts and pastoral care?

Psychotherapy and counseling in America have basically accommodated themselves to a subjectivist approach to reality, and a limited personal definition of self. Only what is real to me is real. History begins with my history. In fact there is a deep strain of antihistorical narrowness in much psychotherapy, with its inveterate focus on here and now transactions.

Psychotherapy and pastoral care profoundly lack a historical, and even more pointedly, an eschatological perspective. This is central to what therapy might learn from the trajectory of the Pannenberg circle.

Such a point seems to some so antiquated and quaint and archaic. How can it be argued freshly that such a question is decisive? Because therapeutic transactions are so enmeshed in human limitation, despair, guilt and anxiety. Therapy must deal with those limits. I would hypothesize that if the individual in therapy, or even in the post-therapeutic situation, could envision and experience a larger embrace of the historical process, and an affirmation of its end, then that would be immensely constructive in achieving the desired outcome of psychotherapy. Historical awareness is therapeutic, I believe, not only with respect to digging into one's past, but with respect to affirming and embracing the future, the whole future, and finally its end. Such an affirmation is what the apocalyptic hope, the ministry and the resurrection of Jesus are all about.

D) Universal History. History is for Pannenberg the most comprehensive horizon of Christian theology. "All theological questions and answers are meaningful only within the framework of the history which God has with humanity and through humanity with his whole creation—the history moving toward a future still hidden from the world but already revealed in Jesus Christ."[22]

The task of theology is not the rehearsal of only a part of history, but the attempt to grasp, and be grasped by its totality. The earliest church's proclamation (kerygma) is not merely a recital of past events, but an interpretation of the whole of history, from beginning to end. The total sweep of history is thus the subject matter of Christian theology.

Apocalyptic speaks of the incompleteness of the contemporary. Only at the end is the meaning of history fully revealed. History's meaning is not derived from some particular segment, but from the whole. Existentialist interpretations of history so focused on the here and now that the meaning of history has been truncated, and the eschaton wholly existentialized in the now. Pannenberg rejects this as out of touch with the radically futuristic hope of the New Testament. It is only at the end of history that the meaning of history will be apparent. The resurrection is the key for grasping the meaning of the whole of history.[23]

American pastoral care is alienated from both exegesis and systematic theology. Pannenberg is trying to bring them together, rejecting either one which would see itself as independent of the other. Pannenberg is struggling against an introverted inwardness which dissolves history into subjective existence. He is also in tension with a dogmatics which would see salvation history as something over and above universal history. Thus two of the major schools of twentieth century theology, Bultmann's existentialist exegesis, and Barth's *Urgeschichte,* both stand as exaggerated tendencies, in Pannenberg's view. Despite the many differences between Barth and Bultmann, Pannenberg argues that there is one way in which they are very similar, and that is precisely where he rejects both of them, viz. both reject the category of universal history.[24]

Eschatology is the beginning, not the end, of all theological reflection, according to Pannenberg. My question for further exploration is: Might eschatology be a significant, and generally neglected, key to pastoral care, to practical theology and to the dialogue between theology and therapy?

In Jesus' message the apocalyptic understanding of history was most sharply stated and transmuted. Jesus did not invent apocalypticism, but he modified it decisively both in his proclamation and his resurrection. He intensified apocalypticism by speaking of the future of God as at hand, and by speaking of himself as the sign of its imminent coming.

Why should pastoral care and counseling take seriously a viewpoint so odd and antiquated as apocalypticism? It seems so fatalistic, so skewed with fanaticism, so deterministic, so peculiar. And yet, as Pannenberg argues, it is impossible to understand history without the end of history, which is the subject of apocalypticism. It is in fact impossible to understand the particular dynamics of the western intellectual tradition, which depends upon historical consciousness, without dealing with apocalypticism. For western historicism has been shaped by apocalypticism.[25] Furthermore, Jesus' resurrection does not make full sense unless it is seen in the context of the overarching anticipation of the resurrection of all believers at the end time, which was an apocalyptic hope.

A prolepsis is an anticipation, a preactualization. Jesus shared the apocalyptic perception of history. Jesus came proclaiming a particular understanding of the end time, namely, that it was at hand, and that it was present proleptically, by way of a kind of preactualization, in his own ministry, life, death and resurrection. That was an audacious claim for anyone who shared the apocalyptic hope. The end is already anticipated in the present. The end time occurs, in a sense, ahead of time, as a foretaste of the age to come. It is as if the future, or a glimpse of the future, were taken into the present.

The presence of the kingdom does not mean that it does not still have a future. Here Pannenberg differs with the realized eschatologists who say that the end time is now and thus no longer future. Pannenberg insists that there is a lot of the historical process left, yet unknown. Yet the center of the Christian hope is that in Jesus' resurrection we meet an anticipation of the future, the last word of history. For anyone who shared the apocalyptic hope in a general resurrection, the resurrection of Jesus thus revealed the meaning of universal history. It glimpsed the end of history. The meaning of the whole was made known through the lens of this one end time event. Jesus was regarded as the first born from the dead, the one through whom the believing community learned to look for the final coming of the kingdom of God, and the general fulfillment of the apocalyptic hope.

All that may sound very antiquarian. The main point is that persons who shared in that hope felt themselves to be grasped by the end time in the presence of Jesus. They acquired an incredibly confident and otherwise implausibly courageous attitude toward history, suffering and life's ambiguities. Why is the New Testament community so confident about the historical process? Because in Jesus' resurrection, the end is already beheld. The resurrection is thus the clue to the whole of history, through which God makes himself known.

What is the relevance of this radical affirmation for pastoral care and counseling? Pathologies and therapies generally only have a limited interest in the larger historical process. They look toward single events and transactions, or at most toward one single life history. But they need the completion of a larger historical perspective which seeks to behold current suffering and alienation from the vantage point of its completion in the whole. In the light of the resurrection, if properly understood, all suffering and human bondage, which are the concern of psychotherapy, are glimpsed from the vantage point of their end, and the now is faced as if its incompletions were completed, as if the worst possible things were already in effect past and transcended. Psychology, like history, is full of inner contradictions, but when each part is beheld in the light of the anticipated whole, all contradictions are in principle, even if not in fact, resolved and viewed wholistically.[26]

IDENTITY AND DEVELOPMENT

In the concluding pages of this series I would like to inquire into an available theological model which might help us deal with radical historical change in the post-therapeutic situation without denying the historic continuity of the Christian hope.

How are we to understand the development and task of Christian theology following its encounter with psychotherapy? Again, after therapy what? An exceptionally useful model, I believe, is provided by the remarkable nineteenth century Roman Catholic scholar, John Henry Newman, in terms of which we can understand the reshaping of doctrine in the light of new encounters and challenges such as those of psychotherapy.[27] For Newman achieved what few have: A strong emphasis on the apostolic continuity of the church, without denying the necessity of change.

The heart of Newman's thought is that it is only through change that the church can remain the same. It is only through the living process of constant rebirth and reappropriation of the Christian tradition that the church can be the same, i.e. be itself, authentically renew the ancient tradition. Put differently, it is only through mutation that the church can maintain its apostolicity. This is why Newman is intriguing. He provides an original formula for the solution of the dilemma between archaism and sheer novelty.

A leading spirit in the Oxford Movement, Newman was profoundly awed by the capacity of the Christian tradition to negotiate new historical circumstances. Preeminent amoung conserving radicals of the nineteenth century, Newman's imaginative formulations on the relation of tradition and renewal still have remarkable relevance for our situation today, and in fact for the entire issue of doctrinal mutation in the light of psychotherapy.

Newman's idea of development. In 1845, in his famous essay on *The Development of Christian Doctrine,* Newman tried to show that the mutability of the church does not deny its continuity with the primitive church. This essay was an intellectual exercise which he undertook during the period of his conversion in order to satisfy his own conscience that the Catholic tradition as it appears in the modern world is substantially the same as it was in its originative period. It is particularly his notion of development [28] that interests us.

The concept of development may be readily grasped by recalling its usage in photography where a chemical process is at work on film. The development of the Christian tradition is like chemical emulsion developing a film. History (analogous to the chemicals) works on the film so as to expose only what is already there. If an image was not imprinted on the negative originally by light rays, then it cannot be developed by the chemical processes. But as those processes go to work, suddenly it appears. In time one sees what was already there waiting to be revealed. So analogously, in the whole process of doctrinal development, everything that unfolds is implicit in the process from the very beginning. We are in the midst, and not yet at the end, of this development. The *depositum* of faith is fully given—the negative is there, waiting historical circumstances to bring it into view.

HISTORICAL CHANGE AND THE DEPOSITION OF FAITH. Although Newman was concerned with the question of relevance and of the reasonableness of the faith to the modern world, he is certainly not to be viewed as an accommodative modernist bent upon simply *adapting* Christian faith to the conditions of the modern world. Against such adaptive chameleon Christianity he conducted a vigorous polemic. Nonetheless, his concept of development coalesced from a sophisticated set of ingredients which post-liberal historicism was never quite able to achieve; viz. it showed the consistency between historical change and the essential unchangeability of the evangelical faith.[29]

Is it possible, Newman asked, for doctrine to change and still remain the same? His essential thesis is: "The differences and additions in doctrinal teaching observable in the history of the Church are only apparent, being necessary for the development of her ideas."[30] The uniqueness of his formulation is the sensitivity with which he shows that the Christian tradition is both changing and unchanging. There is a constant changing of historical conditions, but the depositum of faith out of which doctrine is constantly reformulated is unchanging. This is a very imaginative formulation. It rejects, on the one hand, a simple archaism which alleges a fixed deposit of faith but without the possibility of relating that deposit to new historical circumstances. It rejects, on the other hand, a simple cultural absorption which identifies Christian doctrine with whatever happens to be going on in the world. It affirms that the Christian tradition can relate itself relevantly to each new historical circumstance and that in each new occurrence something new is to be learned in dialogue with the *depositum,* and that these new insights are all in a deeper sense implicit in the original faith once for all delivered to the apostles.

Newman is especially valuable to us at the point of grasping the continuity of the tradition precisely in the midst of its changeability. He does not offer us an unacceptable alternative between either fixed doctrine or mutability. He does not limit our choices, as they often are limited, between defensive archaism and normless change. He is proposing that it is normative for the Christian tradition precisely to be mutating, and that in each new appropriation, something new is discovered about what was there in the beginning implicitly, namely the apostolic faith.

Post-therapeutic theology will again have to face the task of holding together tradition and renewal without neglecting either. In Newman's view, each requires the other in order to be authentically itself. Like man and woman as sexual beings, neither is fulfilled without the other.

ORGANIC CHANGE. Newman compared doctrinal development to the growth of an organism. Referring to the parable of the mustard seed, he argues that evangelical development does not take shape as "an effect of wishing and resolving, or of forced enthusiasm, or of any mechanism of reasoning, or of any mere subtlety of intellect; but comes of its own innate power of expansion within the mind in its season, though with the use of reflection and argument and original thought. . . ."[31]

The real shaping force of development is the organism itself, not its environment. No matter how much we may cultivate the seed, or wish it to grow, if it does not have within itself the innate possibility to become a mature plant through growing in its seasons, then you cannot make it grow. This has crucial bearing on a post-therapeutic theology. In such an environment, catholicity cannot grow without the seed, the originative apostolic witness, fully imprinted on its consciousness, no matter how wise it may be in other ways.

With remarkable originality, Newman speaks of a "particular superiority (so to speak) of later times over apostolic,"[32] by which he means that later historic situations call for and enable more explicit definitions of faith than were called for or needed in apostolic times. The post-therapeutic situation constitutes just such a challenge.

The church was given the fullness of the depositum of faith from the very beginning, but

> What then is meant by the *depositum?* Is it a list of articles that can be numbered? No, it is a large philosophy; all parts of which are connected together, and in a certain sense correlative together, so that he who really knows one part, may be said to know all, as *ex pede* Herculem.[33]

The tradition, however, only becomes open and active at moments when it is challenged, when it is called deliberately to articulate itself as in the presence of mutating psychotherapy. It is like a river that encounters resistance; the only time you see a ripple in it is when it encounters a rock. Similarly tradition is a process that moves, is passed on and remembered, taught, developed and embodied in life. Therefore it is necessarily unwritten. It is too vital, too active, too existential, too responsive to become fully committed to writing. Tradition in this sense refers to the whole life of the church, the actual embodiment of Christian community. That is the experienced reality to which Newman is appealing. In that sense scripture is the product of tradition, of a living, experiencing community.

It is interesting to note that in each period of Christian experience, it is impossible to anticipate with certainty what is coming next. If Luther could have predicted the later Melanchthon, he surely would have been horrified, yet Melanchthon, it can be argued, expanded and deepened the Lutheran tradition. Augustine quite likely would have been horrified by a Wycliffe or Calvin, but each one deepened and enriched the Augustinian tradition. Each stage of the tradition builds upon the wholeness and complexity of the past. Although any given century might tend to judge each subsequent century as inadequate, nonetheless from an antecedent view, each later century of the Christian tradition is given the possibility of a richer perception of God's revelation in history than the previous one.

CHRISTIANITY MUST CHANGE IN ORDER TO REMAIN THE SAME. The notion of a mutating theology at first seems to imply the very accommodative tendency against which Newman inveighed in his polemic against modernism and liberalism. But on deeper examination, the mutation image is precisely what Newman himself implied when he argued that the Christian witness must "change . . . in order to remain the same!" "In the other world it is otherwise, but here below to live is to change and to be perfect is to have changed often."[34]

Although there is much about Newman to which I am not particularly attracted—his polemic against Protestantism, his tendency to become defensive toward the modern mentality, his temptation to idealize the Christian tradition—nonetheless his idea of development is a brilliant, methodological contribution which impinges directly upon dialogue with the sciences of behavioral change. The profundity of this idea is that what is currently being unfolded in the emerging Christian tradition is precisely what is implicit in it from the beginning.

What is now happening in theology's dialogue with therapy, I would suggest on the basis of Newman's thinking, is a new development of what has been there from the beginning, a new imprint, a new expression of the primitive Christian faith. It is new because the emerging world is new. It is the same because the love of God is the same. That sameness can only be known if it is constantly renewed.

THE CONSERVING-STABILIZING DIRECTION

CONFESSION OF A DISPLACED PERSON. For years I have thought of the more progressive, liberalizing alternatives as requiring the greater courage, foresight and determination. Now I experience a strange irony. No road now could be more bland and challengeless than being a so-called revolutionary in a sea of so-called revolutionaries, moving with the current at a time when no one cares enough to give it self-critical resistance. The irony is that it now seems to me to take much more courage to move in a stabilizing-conserving direction, to go against the stream and struggle for order, continuity, due process and a vision of organic change. Strangely enough, contrary to previous periods in my consciousness, it now requires more sagacity, patience and commitment to *conserve* the valid aspects of our social processes than to revolutionize them.

One reason I find myself to be something of a displaced person among the intellectual currents of our time is that the predominating academic mood has acquired the peculiar habit of applauding any sort of hope for any radical transformation of man without asking further about its problematic second and third order consequences. Some look forward eagerly to the complete change of everything. They grind out endless, marvelous blueprints of the new order of futurity. Amid all this I find myself searching for stability, continuity, internal integrity, congruence, a sustained viable self-understanding, political reality, and an understanding of the wholeness of the historical process without the rejection of any one part of it. That is simply where I am as a human being. I do not want to suggest or imply that others should be there. I am merely trying to confess why I find myself in an awkward position in academia today, and in some sense within the therapeutic community as well, which wants so desperately to believe in the romanticist, idealist statement of the human possibility.

CONSERVING RADICALISM. It is regrettable that so much American academic theology is blocked from and out of touch with the conservative religious critique. In *Beyond Revolution,* I developed more fully the construction of conserving radicalism, i.e. being committed to conserving the tradition of western radicalism which at its heart is the biblical witness. It is difficult to express how impoverished I feel as an heir of American Protestant liberalism, with the continued lack of sustained dialogue with the Roman tradition on the one hand, and the conservative evangelical tradition on the other.

Imaginative aspects of the conservative evangelical tradition which I admire are:

(1) Despite its reputation for being resistant to change, this tradition has shown itself more capable of swinging with a changing environment, moving with people, being where they actually are, as illustrated by the "Jesus movement" in youth culture.

(2) It does not have quite the tragic depth of alienation between laity and clergy that liberal ecclesiology experiences, of which Jeffery Hadden has spoken so powerfully in *The Gathering Storm in the Churches.* There may be tensions, but as a whole there is a greater internal homogeneity and consistency in the communities of the conservative evangelical tradition which I value and would hope could be preserved, not detached from the pluralistic environment, but not acquiescent to it either. I passionately hope that evangelical Protestantism will not too easily accommodate itself to its environment in the name of vague appeals to relevance and accommodation.

(3) The conservative tradition has been able to break into youth culture at an emotive and charismatic depth much more than the churches of the liberal establishment, where the parent generation seems almost totally estranged from its youth.

THE ESCALATION OF RESISTANCE. I would like to shift these observations toward the dialogue with psychotherapy, or lack of it. What is now required is a certain *escalation of resistance to the psychologizing of man.* There remains a powerful psychic reductionism at work in our midst attempting to reduce man to an instinctual mechanism, to deny moral responsibility, to reduce hope to this worldly expectation, to reduce faith to interpersonal confidence, to reduce God's acceptance to human acceptance, to reduce all transcendence to this worldly language. Unchecked, the tendency aims toward a total psychologizing of history, of man, of society. On behalf of man, Jewish Christian theology is called to resist the excessive psychologizing of man.

It is not really to the interest of humanistic psychology, either, to be deprived of its long-term critical partner, the evangelical hope. I would hope that the resistance would now harden at strategic points, without making the dialogue any less cordial, so that the accommodating tendencies of the liberal establishment would not be the only voices which psychologists would take to be available as serious partners in dialogue.

The Judeo-Christian hope has wisely been wary of instant claims of total change of man and society which would fail to recognize the tenaciousness of man's sin and self-deception. This tradition's hope in God's coming kingdom does not diminish the clarity of its perception of the human will as prone to self-assertiveness, pride, self-righteousness and sloth which is easily beguiled precisely by its own idealistic, utopian expectations. Biblical realism beholds psychotherapy and therapeutic social goals within the framework of such hopes, hopes which deserve to be affirmed in their deeper intention, but not taken with surface seriousness.[38] There comes a time when we must face honestly the enigmatic human dilemmas which lie beyond the disillusionment of all humanistic expectations. These dilemmas, however, are never the last word for those who are nurtured by the evangelical hope. For each newly grasped dilemma points beyond itself to the future toward which the God of the Bible is drawing human history.

NOTES

1. For a brief sampling of bibliography, see Cohn, N.: *The Pursuit of the Millennium*. New York, Har-Row, 1961. Mumford, L.: *The Story of Utopias*. New York, Viking Pr, 1962. Bloch, E.: *Das Prinzip Hoffnung*. Berlin, Aufbau Verlag, 1959-1960. Shklar, Judith N.: *After Utopia*. Princeton, Princeton U Pr, 1957. Becker, C.: *The Heavenly City of the Eighteenth Century Philosophers*. New Haven, Yale U Pr, 1950. Note especially Giovanni Pico della Mirandola: *Oration on the Dignity of Man*. Chicago, Regnery, 1956.
2. Maslow, *op. cit.* Perls, F.: *Gestalt Therapy Verbatim*. Lafeyette, Real People, 1969. Watts, A.: *The Book on the Taboo Against Knowing Who You Are*. New York, Crowell Collier, 1966. Otto, H. (ed.): *Human Potentialities*. St. Louis, Green, 1968. Otto, H. (ed.): *Explorations in Human Potentialities*. Springfield, Thomas, 1966. Rogers, *op. cit.* Gunther, B.: *Sense Relaxation*. New York, Crowell Collier, 1968. Schutz, William C.: *Here Comes Everybody*. New York, Har-Row, 1971. For an overview see Howard, Jane: *Please Touch*. New York, McGraw, 1970.

3. The tradition of social thought which is shaping this direction includes Hobbes, Burke, Hegel, Tocqueville, Coleridge and Newman.
4. Cf. Dittes, James: *The Church in the Way.* New York, Scribner, 1967, The Relevance of Being Irrelevant, pp. 3 ff.
5. Hamilton, Wm. H.: The new optimism. *Theology Today, 22:*479, January, 1966.
6. Shaull, R.: "Revolutionary Change in Theological Perspective", in Bennett, J. (ed.): *Christian Social Ethics in a Changing World.* New York, Assn Pr, 1966. pp. 23-44. Rose, Stephen: *The Grass Roots Church.* Nashville, Abingdon, 1966. Novak, M.: The Revolution of 1976. *Commonweal, 86:*441 ff., no. 16, 1967.
7. Scammon, Richard and Wattenberg, Ben: *This USA.* Garden City, Doubleday, 1965. *The Real Majority.* New York, Coward, 1970.
8. Published in 1970 by Random House, this book describes some features of the situation in certain aspects of youth culture which prevailed about 1968, but which have shifted dramatically since then.
9. Oden, T.: *Beyond Revolution.* Philadelphia, Westminster, 1970, Chapters 6 and 8.
10. The heart of Pannenberg's contribution is to be found in *Basic Questions in Theology* (hereafter *BQT*) vols. I and II. Philadelphia, Fortress, 1970-1971.
11. "The Question of God," *BQT*, II, pp. 201 ff.
12. *Jesus—God and Man.* Philadelphia, Westminster, 1968, pp. 53 ff.
13. Cf. Hoffer, E.: *The True Believer.* New York, New American Library, 1958.
14. *BQT,* I, pp. 96 ff.
15. Philadelphia, Westminster, 1964, with a response by Rudolf Bultmann.
16. Pannenberg, W.: *Theology and the Kingdom of God.* Philadelphia, Westminster, 1969, pp. 51 ff.
17. Pannenberg, W. (ed.): *Revelation as History.* New York, Crowell Collier, 1968, pp. 135 ff.
18. *BQT,* I, pp. 81 ff.
19. *Revelation as History, op. cit.,* pp. 139 ff. *Jesus—God and Man, op. cit.* passim.
20. *BQT,* I, pp. 15 ff.
21. *BQT,* I, pp. 1 ff., pp. 137 ff.
22. *BQT,* I, pp. 15.
23. *Theology and the Kingdom of God, op. cit.,* pp. 51 ff.
24. *Revelation as History, op. cit.,* pp. 3 ff.
25. *BQT,* II, pp. 2 ff.
26. Cf. Richardson, H.: *Toward an American Theology.* New York, Har-Row, 1967, Chater 4.
27. In tension with Charles Davis' conclusion in *A Question of Conscience* (New York, Har-Row, 1967), that "the Roman Church lacked credibility as the Church of Christ," as a Protestant I would reverse the issue to state my impression that the credibility of the Roman Church is clearly increas-

ing among non-Romans. There can be no doubt of a newly emerging Protestant reassessment of the plausibility of the post-Tridentine Roman Church and its distinguished theological tradition. Even though I do not always fully grasp its subtleties, I find myself increasingly drawn toward attempting seriously to follow the Roman *magisterium* as directly as possible as I see it addressing our situation. Surprising as it is to myself as much as anyone, I find myself trying to be accountable to that *magisterium* even to the point of affirming the Petrine authority, as it has been conceived by the Vatican II Council fathers, provided it does not constitute a serious denial of conscience, which in most cases it does not. Without implying any denial of my own Protestant heritage, I no longer wish to withhold my dawning awareness that non-Roman Christians have good reason to study, venerate and carefully consider the resources and wisdom of the *magisterium*. Catholics have eagerly recognized their need for dialogue with Protestant renewal, but we have not learned from them as rapidly as we might of the wellsprings for relevant reconstruction *within* the rich historic Christian tradition, and within the framework of the unity, holiness and catholicity of the church.

28. Newman, John Henry: *An Essay on the Development of Christian Doctrine.* New York, Sheed, 1960, pp. 25 ff.

29. It may seem curious that when Protestants plunge toward what they think is a catholicizing direction, they are often drawn not to recent Vatican II conciliar thinking, but rather to preconciliar concerns and attitudes which so many Roman Catholics are struggling to leave behind. Why should Protestants rehearse this older Catholic history and theology prior to Vatican II? Why not just swing with current conciliar developments? Why return to the period from Trent to Newman? The answer is important to understand. Protestants know the core of most post-conciliar ideas and attitudes like the back of their hand. Many of the leading ideas of Congar, Thils, Baum and Küng, which were so decisive in recent conciliar reformulation, are already, for good or ill, familiar to the history of Protestant theology. This is what Protestants have lived with for four ambiguous and problematic centuries: the lay apostolate, experimental liturgies, ecclesial voluntarism, etc. Some Protestant theologians are convinced that the Roman Church is trying to repeat in twenty years every mistake made by Protestants in the last 400 years.

    Now for another heuristic thesis: *What is really new and most fresh in Roman Catholic thinking is what Protestants least need,* because we have had it already up to our ears. Protestants have long overstressed the very points now being discovered as new by post-conciliar Catholicism, as the story of Protestant thought from Kant through Schleiermacher, Hegel, Ritschl, Harnack, and to the present should make evident. What Protestants need instead is to rediscover some of the rich Catholic themes which have been misplaced for four centuries (apostolicity, authority, the presence of

Christ in the sacrament), even though it may look to conciliar Catholics like a retreat into a past era.

The irony is that in an age of ecumenism, Protestants and Catholics seem to be moving in entirely different, even opposite directions, although they in fact are moving toward one another. At some points in history it is ironic that two parties must move in opposite directions in order to come together. So it is today that many Protestants sense the need of moving in a conservative, sacramental, catholicizing, institutional direction, while Roman Catholics are quite clear that they need to be moving in an experimental, secularizing, liberalizing direction. Roman Catholics are learning what we have learned over these four experimental centuries of ambiguous division, while we must now learn what is to be legitimately learned from their counter-reformational experiment in authority, unity and continuity. If the Protestant pendulum has swung so far away from catholicity that it needs the earlier counter-reformational themes to correct it, then the Catholic pendulum may have swung so far away from charismatic individual voluntarism that it needs the earlier Protestant themes as correctives, even though these themes are now being abandoned by Protestants who are disillusioned with some of their inadequacies.

30. "Letter to Mr. Joulkes," quoted in Walgrave, J. H.: *Newman the Theologian*. New York, Sheed and Ward, 1960, 51.
31. Newman, *Essay, op. cit.*, p. 54.
32. Newman, An unpublished paper on the development of doctrine, 15 February, 1868. In Barry, C. J. (Ed.): *Readings in Church History*. Newman Press, 1965, vol. III, p. 195.
33. *Ibid.*, p. 195.
34. Newman, *Essay, op. cit.*, p. 30.

CHAPTER 4

A PSYCHOLOGIST'S PERSPECTIVE ON ODEN'S INTEGRATION

NEIL CLARK WARREN

PERHAPS NO AMERICAN THINKER has contributed more effectively during the last ten years to the dialogue between theology and psychotherapy than has Thomas Oden. However sharply his conclusions may be criticized, the fact remains that he has stimulated a remarkable amount of discussion and has served as a point of reference for the formulation of alternative approaches to the task of integration. His recent visit to Fuller Seminary triggered considerable interest and exploration. His strategy seemed to be based on his belief that every person must discover truth for himself if he is ever really to possess it. If this was indeed his strategy, the result was decidedly successful.

It is my purpose in this paper to respond briefly to Professor Oden's lectures. My objective is to understand his position at critical points and then to challenge or support it, finally offering suggestions as to how it might be clarified, amplified, or modified. I trust that the spirit of my undertaking may be one of cooperation in a process designed to discover truth, and humility in the recognition that Dr. Oden has made through his writings and person a significant and unique contribution to our understanding of the interrelationship of theology and psychotherapy.

Especially in his discussion of "The Post-Therapeutic Situation"[1] and certainly throughout his three books, Oden becomes deeply involved in the question of how psychotherapy offered by Christian and non-Christian, or believeing and nonbelieving, therapists differs. He develops his position as follows:

> Already it is clear from what I have said in previous essays that the disbelieving therapist who may have no direct awareness of the Christian kerygma, nevertheless may (not necessarily does, but may) function as an effective therapist with congruence and empathy.[2]
>
> The therapist can work operationally with more or less full adequacy without the kerygma.[3]
>
> If we look then at the functioning of the believing therapist, there is one major difference: He is more deliberately aware of the ground upon which his acceptance stands, because he has heard the witness of a community about the revelation of that reality in history.[4]
>
> This is a fine line to walk: Even though the disbelieving therapist may function operationally much the same, the believing therapist has a broader and more realistic perception of the ground upon which he stands, and of the larger dynamics of the relationship. This is not grounds for boasting, but for simple thanksgiving for the self-disclosure in history of the accepting reality.[5]

It occurs to me that this is a highly complex question to which Oden has addressed himself. It is perhaps the single most important theme in his writings, receiving the majority of his attention. It closely borders on the question of the relationship of kerygma and counseling which itself is the subject for one of his entire books.[6]

But I am of the opinion that any answer to this question is at this time premature. At the very least, it seems to me, we should recognize that any answer will be necessarily heavy with speculation. To engage in serious dialogue about the differences between Christian and non-Christian therapists without defining terms in a highly specific way is sure to lead to considerable controversy and confusion. Furthermore, when one fails to examine carefully relevant empirical findings, one becomes guilty of armchair theorizing, and the likelihood of progress is minimal. The very task of defining terms is a perilous one with a low order reliability quotient at this stage of development. Let me offer some examples of this:

(1) Who is a *Christian* or believing therapist? Perhaps Oden would say, "one who has received and understood the kerygma." But on what level? Some theologians argue, for instance, that to have a conceptual understanding of the good news on a purely cognitive level is sufficient for one to be termed a "Christian." It is quite possible in this case that the person could be inadequately organized, out of touch with his affect, bound up in neurotic defenses, and incapable of

a deep relationship with another, or even with himself. Yet, according to these scholars, he is a Christian because he can cognize about the Christ event and has even developed, perhaps, a logical mastery of the faith. Imagine the destructiveness of this type of person in a therapy situation where delicacy, intimacy and finely tuned sensitivity are so critical.

Are there not many levels on which one may receive the kerygma? On what level or at what stage does a person have to be to receive the label "Chrtsian?" To whom are we referring when we use the term "believing therapist?" If the conceptual level is often reached last and thought to be a relatively superficial stage of awareness, what stage is profound enough to establish the identity of "believer?"

(2) Who is a *therapist?* This seems an especially important question in light of Oden's first chapter on the "diffusion of therapeutic agency." Must he be licensed, e.g. by the Board of Medical Examiners, such as is the case in the state of California? Is a therapist different from a counselor? Or is a therapist anyone who listens attentively to another while they both hold, however tentatively and informally, that gain will occur for one or both because of the encounter? Is the term limited to those of an existential orientation? Can someone who is behaviorally oriented be a believer? Do we include medicine men, Christian Science Practitioners, or even psychoanalysts?

(3) Who is an *effective* therapist? Oden has said that "a disbelieving therapist . . . may . . . function as an *effective therapist. . . .*" One might assume because of Oden's heavy reliance on Carl Rogers throughout his writings that he would define an effective therapist as one who scores high on the therapeutic triad. But there are many pitfalls here. For instance, therapists fluctuate in their scores on the dimensions of the triad. Is a therapist only effective when he scores at a certain point on one or all of these dimensions? Is it impossible then to even think of a therapist as effective in some static, unchanging way?

(4) What is *effective therapy?* This is a closely related and perhaps even more complex issue. It occurs to me that much of the difficulty in Oden's theorizing results from a lack of crispness at this point. What is his dependent variable, i.e. his criterion for effective therapy, and how does he propose to measure it? When he says "effective therapy," does he mean the same thing that his hearers do? A recent issue of

the *Journal of Consulting and Clinical Psychology* [7] devotes considerable attention to the evaluation of outcome in psychotherapy. It becomes readily apparent that agreement between separate observers of change in therapy is very limited. In what type of change are we interested? And when do we think it is reasonable to assess the effects of therapy? There is disagreement and confusion across the therapeutic orientations about criteria appropriate for evaluating therapy. Furthermore, studies almost never include follow-up assessments completed six months, one year, or two years after therapy terminates. When they do, the results must be interpreted from one's own perspective. It seems to me that what some theologians might call effective therapy, and how they would suggest measuring it, could be very different from what therapy participants or therapy theorizers might propose.

The point is that the question about differing effects of believing and nonbelieving therapists may well be an impossible one to answer. When our definitions are so imprecise that we are unsure about the identity of believers and nonbelievers and puzzled about the type of therapy outcome we will focus on and the method we will use to measure it, we may need to abandon the effort or at least approach it from a new angle.

A second point is that Oden seriously neglects research findings which bear on his argument. (We will see this even more clearly when we discuss his laicization theme.) Furthermore, he seems to imply that he is uninterested in any contribution to an answer that might be made by a careful, empirical approach to the question. For instance, might it not be possible to develop definitions of believing and disbelieving therapists and of effective therapy, decide on appropriate times for measurement, and then develop a study which would throw light on whether one type of therapist is more effective than another and, if so, with whom, and for how long? Is there any need for us to fault the empirical approach to the point of ignoring it?

Engaging myself in the speculative game, let me mention six points which seem to bear directly on the question of whether believing and non-believing therapists produce differential effects.

(1) First of all, therapy on any level is an intensely dangerous enterprise. Rogers and Truax [8] developed this insight in the mid-1960's and Bergin [9] has called it the most significant breakthrough in psy-

chotherapeutic research to this point. It has yet to influence the training and practice of psychotherapists in the profound way it must, but the fact is well established that persons can be seriously damaged by therapy. Some of those who become involved in therapy show decided gain, but others show deterioration, whereas control subjects receiving no therapy tend to show neither growth nor deterioration. Involvement in psychotherapy involves significant risk, and it seems to me that Oden throughout his writings fails to come to grips with the reality of this finding.

(2) Whether therapy eventuates in constructive growth or deterioration seems to largely depend on the person of the therapist.[10] When he has become whole to the point of being able to relate empathetically and respond warmly to the client even when the client seems not to be progressing, and able throughout the process to remain aware of his own feelings at a deep level, constructive growth usually takes place. When these qualities are not present in the therapist, deterioration is likely to occur for the client.

(3) How does a therapist develop these therapeutic qualities in himself? I strongly suspect that a life-changing encounter with the Kerygma is involved in the process. But, I have no confidence at all in a superficial kerygmatic encounter which leaves one unable to love God, man or self. As a matter of fact, a therapist who has had this experience may be least characterized as therapeutic. A kerygmatic encounter resulting in no fundamental life changes sets up a tension which may result in counter-productive attitudes. The New Testament irrevocably fuses the authentic conversion experience and the changed life. The work of the Holy Spirit in every believer changes radically the relationship of that believer with himself, with creation as a whole, and certainly with other believers in the life of the church. To assume that conversion in the New Testament was a purely conceptual event which left the new believer as defensive and miserable as before, as unable to relate meaningfully, as impaired psychologically and emotionally is to miss the whole thrust of the New Testament. These were changed people—both in terms of their cognition and their affect.

(4) The role of the Christ event in the redemptive process needs further consideration. Oden seems to argue that the Christ event only pointed toward the reality that has always been true. Is it not dan-

gerous to build an entire argument on only a single strand of Scripture? I suggest that to be true to the New Testament account it is necessary to consider the Christ event as more than sign or symbol. It participates in causality. And to the degree that this is true, the client who is to be changed by the reality of this acceptance needs to encounter fully the Christ of history who established the new reality.

(5) More than just encountering a therapist who serves as an agent of grace, my experience has convinced me that a client is best served when he is introduced in a meaningful way to the source of empathy and warmth, and is assisted in establishing an understanding of his relationship to that source. When this does not happen, it has been my experience that clients often fail to maintain the growth they experience during their actual therapy involvement. The goal is to help them to gradually understand and appropriate the kerygma in such a way that it can continually renew them. When this happens in the context of a valid church experience, it happens I think as it was designed to happen in the beginning.

(6) Finally, Oden himself strongly suggests the importance of every psychotherapeutic relationship being informed by a Christian world view including the wholeness of history idea and the importance of an end-time which is beyond the here and now. If a therapist's timing is right, it seems to me that his task is incomplete until he has helped the client understand the total context of his existence.

A related point from Oden's second chapter involves the tendency on the part of the church to emphasize cognition and relegate affect to a decidedly peripheral position:

> There is curious irony in the fact that Protestant worship today is still desperately trying to divest itself of emotionalism, just at the time when current therapeutic practice, and notably Gestalt therapy, is seeking to intensify emotive self-disclosure.[11]

Oden's analysis seems accurate enough. There is indeed a strong tendency among those Protestant groups with which I am familiar to deny the importance of emotion. Deep feelings of any kind are immediately suspect, and man is taught to think but not feel. The focus of so much of the church is on the objective, and there is a strong dogmatic emphasis. The work of the Holy Spirit is trapped in the past tense so that it can be contained and appropriately controlled. Subjective dimensions are threatening because they can involve such risk to the church's staid consistency. That some of this is a reaction to "wild, frontier, emotive fanaticism," as Oden suggests, is undoubtedly true. But surely there are additional reasons, since Western culture in general seems so much more comfortable with cerebral events than with emotional experiences and Oden is correct in pointing to the opposite tendency among the youth and encounter culture.

There is a hunger for experience and a strong emphasis on feelings. The Rogerian focus, for instance, is entirely on feeling. The Gestalt movement concerns itself almost exclusively with the detection and differentiation of feelings and feeling states. The underlying theory has asserted that most Americans have repressed their affect so severely that it is distorted and twisted and badly in need of being set free through openness and self-disclosure.

Two problems remain, however. How can a balanced approach—i.e. treating cognition and affect as equal and necessary partners—be secured? And how can the church be assisted in its efforts to engage in self-correction and to discover those aspects of itself which it is currently ignoring?

With regard to the former, the historical tendency has been to swing radically from an emphasis on cognition to an overemphasis on affect. If some theologians seem "mind-dominated," the fact is that they are reacting to a heart-dominated period which was permeated with abuses as destructive as those pointed out in today's situation by youth who fail to find feeling in the church. What we need is a delicate balance between cognition and affect. Jesus said that we are "to love the Lord our God with all our heart, soul, strength, and mind." There is no single channel which is to be dominant.

The ideal, it seems to me, will involve a careful transmission of the faith on a conceptual, cognitive level. This will center on a full understanding of and emphasis on the Christ event. It will be historically integrious and cognitively consistent. At the same time there will be a corresponding and equal emphasis on the appropriation of that faith, the working out of one's salvation, the reception of the kerygma, and the radical response in faith. This will include going deeply into the virgin territory of one's inner world in an effort to allow the kergyma to extend its reign to the uttermost parts. And there will be a concentrated effort to keep either of these aspects from taking control in a way that causes a barren, logically consistent, dull-to-dead mind-dominated faith, or a fanatical, chaotic, inconsistent, out-of-control heart-dominated faith.

But the second problem remains. How can the church be assisted in its effort to discover itself on new levels—i.e. how can it come alive to its affect without threatening to overwhelm its more rational and insight-oriented emphasis? Oden speaks boldly about what must be done, but he chooses not to touch upon the herculean task of trying to help the church move away from rigid and secure, even if boring and ineffective, patterns.

I think he makes a sound point when he says, "We may need a partial moratorium on talk about freedom, love and community, while we search for its concretion in actual communities of free, responsible love."[12] It is the task of searching for its "concretion in actual communities" which requires attention. Is it enough to change the worship service and simply make it more relevant to present needs and concerns? Surely not. Is it enough to start a few small groups for those who are hungry for relationship? Obviously not. The church is in need of major correction from within. The discovery of freedom and love in community involves radical change. Is it enough to merely recognize the need for that change? No. In the context of biblical teaching about the quality of life the church is to enjoy and the type of process it is to facilitate in its members, it seems to me that it is now imperative that we apply to the problems of the church the best we know about social design procedures, and the most penetrating insights from sociology, psychology, and practical theology. If the church is to rise to the ideal of its designer, it must be encouraged to be uncomfortable with its single channel of response. It must be assured that it need not swing too radically to another single channel. It must be taught and instructed and motivated—to become alive to itself on new levels that are true to all that it knows it should be.

Oden's chapter on the diffusion of therapeutic agency is undoubtedly the most controversial of the three. He not only threatens the professional who has become comfortable with the tightly proscribed limits on therapy, but he also threatens the layman who feels inadequate to deal with the problems he encounters and who is currently released from responsibility for dealing with them by merely referring them to the nearest professional.

In this lecture Oden takes two positions: (1) Psychotherapy has become excessively professionalized, is now in a defensive posture, and as a profession is in the process of being transmuted; and (2) "the therapeutic agent need not be a professional," but may be a layman, and the layman may have certain important advantages which the professional does not have.[13]

At the outset I have some difficulty understanding Oden's purpose for pursuing this theme. At a point in time when laymen are interested in becoming more therapeutically involved but simply do not know where to look for structure or training, Oden is busy engaging in sharp criticism of the medical profession: "No vocation is more disastrously and excessively professionalized in our society than medicine."[14] In a period in which the church is talking about community more than ever before but finds the task of becoming one overwhelming, Oden seems somewhat irrelevantly intent on blaming the professional for claiming that "he can mediate a therapy or elicit constructive change which are entirely out of the range of the non-professional."[15] My point is this: who will deny that many psychotherapists have become artificially professional and thereby less effective, but how does an emotional attack on the medical model, the medical profession, and the overprofessionalization of therapy change anything? Who will deny that the layman has unusual abilities to effect constructive change in others, but what they need is *training* rather than bold comparisons of their abilities with those of the professional.

Oden seems to have no plan for using our knowledge about psychotherapy, most of which has been accumulated at a slow and painful rate through empirical research, in the development of training programs for ministers and laymen. His advice is to start over: "Rather the church must develop her own resources, rooted into her own history, shaped by her own theological understanding, and responsive to the koinonaic needs of local congregations."[16]

My most profound concern was aroused when Oden reported his instructions to ministerial candidates. "Go ahead and do your own experimentation within your parish," he tells them. "The parish is an appropriate context, and has been for over 200 years in Protestantism, for experimentation in the intensive group process."[17]

Certainly I would not dispute, but rather support, the fact that work with intensive group processes is the legitimate concern of religious communities. What I do not support is an excessive eagerness to throw off the restrictions of the psychotherapeutic profession, and an irresponsible, experimental approach within the church that refuses to recognize the possible dangers involved in individual and group psychotherapy.

By far the most complete review of therapeutic outcomes was written by Allen Bergin.[18] Referring to thirty published studies on therapeutic outcome, Bergin persuasively argues that in each one there is strong evidence that therapy led to positive growth for some clients and to deterioration for others. "Evidently there is something unique about psychotherapy," Bergin says, "which has the power to cause improvement beyond that occurring among controls, but equally evident is a contrary, deteriorating impact that makes some cases worse than they were to begin with." Psychotherapy was once thought capable of doing a client no harm, and on the basis of this assumption, Oden might be justified in advising ministerial candidates to "go ahead and do your experimentation. . . ." However, when it becomes obvious that therapy can also lead to disastrous results for some clients, experimentation becomes irresponsible.

Truax[19] has demonstrated that the deterioration effect can be largely explained on the basis of the therapist's attitudes. As reported earlier, a specific therapist's characteristics, on the basis of accumulating evidence, are associated with both deterioration and improvement. Is it sensible, then, to tell a group of ministerial candidates to simply experiment? When specific variables have been identified which relate directly to the likelihood of a therapist being helpful or hurtful, is it not more appropriate to encourage a candidate to determine his ability to be helpful, and then either assist him in reaching some minimal level of therapeutic skill or strongly encourage him to move into another area of ministry?

I have become more and more convinced from my study of the research findings and from my own experience that psychotherapy is particularly vulnerable to the demonic. When a therapist is not whole in a relationship, he can, in the name of being therapeutic, violate the person of the client in highly destructive and permanently damaging ways. To ignore the possibility of this process is to participate in its destructiveness.

First of all, I think it is important to recognize the realities with regard to laymen becoming involved in the task of therapy. We need to ask at least three questions: (1) Are they motivated to become involved? (2) Are they equipped to be therapeutic and, if not, can they become so after training? (3) Whom are they likely to help?

I have been directly involved in the training of a large number of nonprofessionals in the area of psychotherapy. When they seek this training, I have found them to be timid but eager. When they have not directly sought the training, but have seemed likely candidates, I have found them to be intrigued but, whether because of their fear or their time commitments, unlikely to become involved. At a stage in my own professional development when I have become committed to the training of every member of a large, local church to be part of a therapeutic community, I have become increasingly aware that the majority of laymen have little or no interest in becoming involved. So, we should be aware of what we are dealing with—a relatively few laymen who are quite highly motivated to seek training, a few more who might be persuaded but are not very eager, and the majority who are simply not interested.

Secondly, I have been struck by the fact that while many laymen seem to have therapeutic potential, this potential is not well developed. This is no different from what I discovered about myself when I reached graduate school, and what I have found to be true with other graduate students in our school of psychology. The point is that before training, laymen, like the professionals, have skills which must be developed. Whether it is due to the culture in which we grow up or to something else, it does seem to be true that few persons are naturally therapeutic across a very wide range of contacts or moments. What we have discovered, however, and what has been discovered in countless research studies on training is that laymen can become perfectly capable of being therapeutic, at least in relatively pressure-free dyads. But there must be training, and the training must be done by those who have been found to be therapeutic themselves and to understand the principles that are involved in that process.

Thirdly, it seems important that we face squarely the issue of who the laymen are likely to be helping after they are trained. What I find is that mothers are generally trained to help their children, group members are trained to help each other, and vocational counselors are trained to assist young people or others who desire work. It is important to note that no one is trained to assist the neurotic or psychotic who is unable to be of help to others. It may be that some of these persons who are in deep distress will end up in encounter groups and receive support and concern there. But I have been struck with the high percentage of encounter group participants who are growth-oriented rather than being concerned with nagging deficits. And in the presence of a fairly serious problem, laymen seem best equipped to give temporary assistance rather than long-term help. Both in terms of the amount of time they wish to give and the depth of skill they possess, their chief contribution seems to be in the area of prevention rather than remediation.

This is an important point to recognize. The layman can most effectively be involved with prevention. The practical goal should be to assist him in learning to relate to the primary figures in his life in such a way as to help prevent emotional breakdown. When emotional problems do develop, the layman can be trained to intervene in the early stages and help persons redirect their life process so that the number of more severe problems is decreased.

The established professionals, on the other hand, can devote their energies to the training of laymen and to the task of remediation when the emotional problem has become tightly knotted and snarled. The professional thus involves himself in both preventive and remedial phases of therapy. He strives to decrease the number of problems which get beyond the early stage, and he labors with the persons whose problems have become critical in an effort to help them get themselves back on course.

Through all of this it should be clearly recognized that professionals have no corner on effective psychotherapeutic process. Laymen can develop interpersonal skills and behavioral techniques which will make them remarkably effective in assisting with many types of difficulties. But neither should we engage in a kind of fantasy that makes ordinary people into better healers than professionals. The fact is that most professionals were once laymen who have spent four or more years of full time study attempting to become effective in their therapeutic efforts. We should quickly grant that parts of that training can usually be recognized as irrelevant to the therapeutic task. And it is likely that a certain percentage of the professionals will be no more effective as therapists than they were prior to training. But to fail to recognize the high value of much of the training would be an unfortunate oversight. And to assume that laymen are equally effective or even more effective because they have received no training is, in my experience, a faulty assumption.

The challenge to the training institution is to be continually aware of the results of research which relate to the importance of various aspects of therapy and be prepared to modify the training program on the basis of these results. Recent research has focused chiefly on two areas: (1) the process of therapy, its relation to therapeutic outcome, and the effect of the therapist's person and the client's style on that process; and (2) the value of behavioral techniques in helping persons who complain of any one of several behavioral problems. If we train future therapists in terms of their personal and interpersonal skills, and if we teach them to use various proven techniques, they will be able to offer considerable help.

But in the process we must not become overly professionalized. If we do, we will have violated our concern for developing genuine persons. And we must make sure that we do not fail to professionalize either. Our responsibility as a profession is to maintain high standards, provide adequate undergraduate, graduate, and extension education, and insure the quality of our service in every way possible. If we are able to do this, I assume we will have dealt adequately with Dr. Oden's concerns.

Dr. Oden's final chapter focuses on two points: (1) the potential relevance to psychotherapy and pastoral counseling of the "recent apocalyptic and eschatological studies," and (2) the unfolding of Newman's idea of doctrinal development as a method to "help us relate tradition and renewal."[20]

At the end of this lecture I detected within myself at least five strong impressions: (1) what a relief to be convinced that the plethora of current theological approaches which are less focused on scripture than on the current scene tend to inhibit our progress in understanding man and the world rather than facilitate it; (2) the psychotherapeutic approaches which I have been taught are indeed heavily loaded with existentialist thought and need the radical influence of the universal historical emphasis of Pannenberg; (3) Newman's method is indeed clever, perhaps even correct, but I question its contribution beyond justifying an effort which we will make in any case; (4) so much of Oden's writing seems conceptual-theoretical and so little of it provides *functional* assistance that I sometimes feel he is guilty of making our burden heavier rather than lighter; and (5) for example, how can I use the "wholeness of history" idea as a psychologist, and how can eschatology contribute meaningfully to my therapeutic efforts?

Out of these various impressions I find one continually recurring. It has become a strangely profound question for me. Can it be that I have been caught, as a psychologist and psychotherapist, both by the time in which I live and the training I have received, in a powerful existential surge which was born out of concern for the here and now and a recognition, from former days, that the future had for too long eclipsed the present? I struggle with the insight that the corrective which was so necessary a few decades ago became for me the truth, and the corrective that is necessary now is the one toward which Pannenberg and others like him would lead us. For me, I fear, a necessary pendulum swing became confused with a movement toward a new kind of absolute truth. And so my concern in therapy has been with *this* moment, and my view of the ideal process has been saturated with the subjective.

I have found it necessary to translate biblical teaching about the age to come into biblical teaching about the present age, assuming all the time they were either the same or so continuous they could not be separated. I have failed rather miserably, I think, to understand the radical nature of the resurrection as an anticipation of the end of time. The very idea of the wholeness of history as a concept having meaning for all of my clients and for me never crossed my mind. For I was born in an age that peaked in the "now generation," that frowned on meaning that was not in the present tense, and ignored the entire biblical thrust relating to the importance of the triumphal end of history.

The decisively Christian message does seem to relate first to the final historical triumph, and then to the present moment and the quality of life which is appropriate to the total context. My emphasis as a therapist has been on the present moment with an implicit assumption that to productively process this moment is to assure a positive outcome. Now I question the order of my emphases.

Eschatology is so foreign to my mind. I need help in understanding it and applying it to the psychotherapy process. Universal history has been such a puzzle for me that I have ignored it in order to escape it. It must be clarified and understood in light of biblical teaching. What this could mean to the integration of theology and psychology boggles the mind. Perhaps no other theme has excited me as much in terms of its potential value.

But even as I sense my own enthusiasm for this major and necessary corrective to my own thinking and practice, I perceive an urgency in my concern about swinging too far in the other direction. It does seem to be the case that the New Testament witnesses to the "impingement of the end time upon the now," and I sense that the power and purpose of this witness could be easily lost in our growing appreciation for the importance of beginning with the eschaton.

For is there not a need to become available to oneself in this moment, even if this moment is to be filled with a looking beyond this moment? My experience leads me to see psychotherapy lending itself to the organizing of life *to* this point so that life *at* this point can be lived on the basis of careful and conscious choice. The Christian view of the world is often rejected because it is misunderstood both in transmission and reception. Perhaps the ministry of the word must bear responsibility for the former, while psychology can contribute meaningfully to a reduction of the latter.

Furthermore, it seems to me, psychotherapy should become involved in the appropriation of the Christian message. What we too often fail to recognize is that every Christian is immersed *in* the world, and as such, he is plagued by helpful and hindering relationships and processes. For him to receive the kerygma in all of its contextual power, allow it to relate to the various aspects of his life, and remake him so that he is a new creature—and do all of this while he is *in* this world—steady, supportive, understanding, insightful assistance is a must. If psychotherapy is to be directed at the facilitation of this process, the therapist will himself need to have "worked the problems" and be aware of the issues. This is a task which should require a six-year program on top of the present six-year program.

But if indeed psychology, which involves a psychologist who is Christian, operates without a message of its own, but with techniques and artistry designed to implement a message which is from beyond itself, then perhaps our task is incomplete until we have fully grasped the message we believe to be essential and have developed specific skills for the transmission process necessary to that message. Perhaps we have attempted to shape the message in such a way as to make our task—both of understanding and transmission—less difficult. This is what we must consider.

The fact is that for me Oden's third chapter is by far the most profound and helpful. It is a clear call for a new assessment, and that assessment is overdue. It provides a theological approach which is at once lucid and thorough. In the lecture Professor Oden brilliantly summarizes two extensive positions and makes available those parts of each which contribute potential assistance to the understanding of psychotherapy in the light of biblical revelation.

It should be fully recognized that this final lecture on "The Human Potential and Evangelical Hope" is a radical shift for Oden in terms of his position on the integration of theology and therapy. The former questions which were so existentially oriented become dull in the face of his new awareness. It is no longer meaningful to talk about the kerygma as though it could possibly be transmitted by a warm, empathic, genuine therapist who does not verbally refer to the reality represented by the Christ event. The starting point now is the eschaton, and it is difficult to see how the idea of a triumphant approach to universal history could be fused with a secular psychotherapy.

This is an emphasis which is totally new for Oden. It makes therapeutic approaches which emphasize merely the full awareness of feelings in the now seem superficial and inconsequential. It adds an entirely new dimension to the healing task. It refuses to let man view himself as his context; it picks him up and positions him in a time and space arena whose limits are as wide as the wholeness of history and as precise for every man as God alone could create. The discovery of meaning in this moment apart from the power of the cross suddenly seems like wishful fantasy from a prior era when the context was narrowly drawn and this moment was all that counted.

In light of this radical shift I see not one but two necessary spheres. There is, of course, that sphere which relates only to the here and now, in which a man sees his existence as bound up in the time and space of his finite frame. This sphere, while often so similar in momentary process and outcome to the other sphere, is ultimately limited and hollow. The second sphere begins with the awareness of a context which is historically responsible and futuristically defined by the radical, world changing event of the cross. To receive the message of this event is to be aware of God's final victory in history and of our participation in that victory. The relevance of that outcome has incredible consequences in the now. This life must be lived in the recognition that the final struggle has been decided already by the resurrection.

It is as though we are called to live each moment in celebration of an outcome that is certain. It is the very certainty of that outcome that gives life in this moment and changes existence from a moment-to-moment monotony into a process infused with meaning because of an eventuality which is beyond imagination. And that eventuality becomes the key to similar but lesser struggles and victories in this age.

Now, we should not become anesthetized to the struggle. The very heart of the matter has to do with the ultimate attraction of the moment. There is a subtle, magnetic quality which these moments have which would encourage us to believe that we can find our fullness in them. But the fact is that if we begin to believe that the here and now is ultimate, if we fail to recognize that we are part of a vastly larger piece, we will be swallowed up in isolation, in a fight for a few superficial relationships, and in danger of evaporating in subjectivity. When we are caught up in a new context, on the other hand, we are brought together into a community, for we share a common historical oneness. History has relevance to both of us then, and we are celebrants of the same certain event. But the point is that participating in this process requires the best we can give to each other in insight and care and support. For the moment in which we live and the thirty or forty years in which we have been alive have been largely given over to an emphasis on the immediate and a continual repudiation of the over-arching historical context.

What further can be said about the contribution which Dr. Oden has made? It is almost as though Chapter one was a response to the confusion and emptiness of that sphere which lives for today and fails to recognize the full, historical context. Chapter two was a struggle in the night, a search for understanding, a speculative effort which by itself dies in process and finally, Chapter three is a brilliant return to the witness of scripture which calls us afresh to a recognition of the impact of the cross and resurrection on creation as a whole and to lives that are drawn together in a faith that is supported by hope and eventuates in love. Dr. Oden has himself been an agent of grace and has contributed significantly to the process in which we have chosen to engage.

NOTES

1. Oden, Thomas C.: The Post Therapeutic Situation, *op. cit.*, p. 28.
2. *Ibid.*, p. 57.
3. *Ibid.*
4. *Ibid.*
5. *Ibid.*
6. Oden, Thomas C.: *Kerygma and Counseling.* Philadelphia, Westminster, 1966.
7. *J Consult Clin Psychol,* vol. 37, no. 3, December, 1971.
8. Rogers, Carl R., Gendlin, Eugene T., Kiesler, Donald J. and Truax, Charles B. (Eds.): *The Therapeutic Relationship and Its Impact.* Madison, U of Wis Pr, 1967.
9. Bergin, Allen E.: Negative Results Revisited. In Berenson, B. G. and Carkhuff, R. R. (Eds.): *Sources of Gain in Counseling and Psychotherapy.* New York, HR&W, 1967.
10. Traux, Charles B. and Carkhuff, Robert R.: *Toward Effective Counseling and Psychotherapy.* Chicago, Aldine, 1967.
11. Oden, Thomas C.: The Post Therapeutic Situation, *op. cit.* p. 37 f.
12. *Ibid.*, p. 63.
13. Oden, Thomas C.: The Diffusion of Therapeutic Agency, p. 3.
14. *Ibid.*, p. 2.
15. *Ibid.*, p. 4.
16. *Ibid.*, p. 17.
17. *Ibid.*, p. 18.
18. Bergin, Allen E.: The Evaluation of Therapeutic Outcomes. In Bergin, A. E. and Garfield, S. L. (Eds.): *Handbook of Psychotherapy and Behavior Change: An Empirical Analysis.* Somerset, Wiley, 1971.
19. Truax, Charles B. and Carkhuff, Robert R.: *Toward Effective Counseling and Psychotherapy.* Chicago, Aldine, 1967.
20. Oden, Thomas C.: The Human Potential and Evangelical Hope, p. 67.

CHAPTER 5

THERAPY AND THEOLOGY: BIBLICAL AND PASTORAL CONCERNS—A RESPONSE TO THE ODEN LECTURES

KENNETH B. MULHOLLAND

THE LECTURE SERIES presented by Dr. Thomas C. Oden, in conjunction with the reading of three of his books, proved to be personally stimulating and helpful. This paper is intended to constitute a rather selective personal response to those points in his three lectures which addressed me in my existential struggles, rather than a systematic critique of his thought.

I found the first lecture the most exciting of the series. Perhaps it was because it confirmed so many of my own prejudices and maybe it was because I see so many possibilities (as well as dangers) in the whole process of the laicization of therapy. In this lecture Oden's thesis was clearly stated from the beginning:

> The excessive professionalization of psychotherapy following the medical model has allowed therapeutic agents to see themselves trapped in a style of professional transactions which are counter-productive to the task of being a liberating, growth producing therapeutic agent.[1]

Although this thesis sounds very negative indeed, as the title "The Diffusion of Therapeutic Agency" indicated, the lecture contained many positive features. I can testify that I was personally helped with two concerns which I brought to the lecture: First, how can the Church best be a therapeutic community? Second, what are some clues toward a valid model for pastoral care? One interesting facet of this lecture was that it stirred student feeling. Although I am not sufficiently involved in the school of psychology to make an objective judgment, my own feeling was that the extreme defensiveness on the part of the psychology students—their picking at minor points, their refusal to let seemingly settled issues alone, and the irritated tone of voices, aggrieved facial expressions, and forced speech—communicated to me that Oden had indeed touched a tender nerve with his questioning of the medical model and his critique of professionalism, despite the fact that he critiqued *excessive* professionalism rather than professionalism *per se* as being counterproductive and pointed out extremely productive channels for the exercise of professional skills. For me, the highlight of this lecture centered around two foci: a helpful historical framework for understanding the contemporary church and some alternative models for ministry other than the medical model.

Helpful Historical Framework

In *The Structure of Awareness* [2] Oden presents a very helpful historical sketch on that strange conflation of moralistic pietism and liberal optimism which entered into a clandestine alliance to produce popular, acculturated Protestant Christianity. Despite their differences, both streams united in rejecting serious confession as unnecessary since the church is seen as the good community. A helpful footnote mentioned in the lectures, but developed more fully during the discussion periods, was Oden's distinction between a living pietism, such as that encountered in the evangelical revivals, and an ossified pietism blended with legalism. The latter is often found in contemporary evangelical congregations. He traced how the counterculture has rushed in to fill the vacuum created by the absence of a living pietistic tradition in conventional churches. At that very moment in history when persons in our culture, having experienced proximity without community, are seeking both intense interpersonal relationships and a mystical awareness of God, most conventional churches are turned off to gut level interpersonal encounter and/or the expression of emotion in worship. They are dedicated to keeping on an even keel in these turbulent times in which we live. Indeed, Dr. Oden's brief historical sketch has genuine diagnostic value for assessing a local church situation.

Models for Ministry

I also find myself in substantial agreement with Oden's critique of the medical model as the *sine qua non* for the role of the professional. I believe that the friend model is, as Oden suggested, more helpful than the medical model for the pastoral ministry. After all, Jesus was called a "friend of sinners" and said to his disciples, "You are my friends" (John 15:14). I trust that Dr. Oden will develop this model more fully in subsequent writings. The role of friend fits much better a relationship in which the parishioner is known in a variety of situations over a long period of time—and usually in his family, social, and vocational context. Even if the parishioner takes the initiative in seeking help, he is already known when he arrives. On the other hand the pastor may take the initiative in confronting the parishioner with his own perceptions. In a relationship of dialogue he can expect to receive as well as give, and questions of meaning and ethics can be freely dealt with. The resources of the church with their therapeutic potential cannot be ignored either.

Despite his diatribes directed at the medical model, I do not perceive Oden as calling for the abolition of the professional model. Rather, I see him pleading for a broadening of the base of therapy with the professionals fulfilling the role of trainers of para-therapists much as some sectors of the medical profession have trained paramedical personnel in order to improve their delivery systems. Not only are paraprofessionals able to handle the less serious situations; they also serve as a filter assuring that the professional's time is devoted to handling those situations which are beyond the capabilities of the paraprofessional and actually require those skills which the professional is uniquely trained to provide. In this vein, the establishment of the Church Consultation Service at Fuller Theological Seminary is a decided step in the right direction. Church counseling centers staffed by gifted pastors and laymen under the supervision of a clinical psychologist, who would also handle the more severe cases as well as carry on a teaching ministry in the churches of the community, are another possibility.

The need for laicization as an expansion of, rather than as competition with, the work of the professional became very clear to me in my own cross-cultural ministry as a missionary. We discovered that churches were strengthened far more by the steady ministry of gifted, but relatively untrained, local laymen than by occasional visits by highly trained pastors. In the same nation, medicine was highly professionalized. Top flight medical care was available to the handful who could afford it, and there was a constant push to upgrade the sophistication of modern medical practice. In the same country the life expectancy was fifty, and the number one cause of death was intestinal parasites. It was not that the majority lacked quality medical care. It was that they lacked *any* medical care. Seventy per cent of all deaths were judged preventable. Yet they took place in large part because of the refusal of the medical profession to expand their delivery systems through supervised paramedical personnel.

Although I feel that it is possible to nail Dr. Oden for his lack of concreteness at many points, he does seem to be sufficiently aware of the dangers of laicization and seems to have provided enough middle axioms in this lecture that those genuinely interested in widespread mental health will both welcome and facilitate the movement toward the laicization of therapy.

My reaction to the second lecture was a mixture of disappointment, appreciation, and stimulation.

Disappointment

This lecture originally bore the title "Beyond Normalcy," and the very title raised my expectations. I see myself more as a Pastor than as a therapist. As such, I relate to emotionally troubled persons before, after, as well as during therapy, even if I am not their counselor. I encounter them in everyday life—at the store, in church, at home—as well as during the therapy hour. I am concerned about the therapeutic quality of the climate which pervades the fellowship of believers of which I am pastor. As a pastor I am concerned about wholeness—intrapersonal, interpersonal, and transpersonal. What is the role of the Church in the promotion of mental health (which I loosely equate with normalcy)? What light does psychotherapy shed upon the role of the Church as classically defined beyond the promotion of mental health? I expected Dr. Oden to deal concretely with such questions in the light of the title of his second lecture.

My mind raced back to my parish experience. I remembered how I labored using the Rogerian approach I had gleaned through Hiltner in seeking to help persons work out their problems in living. Sometimes the knots were untied and the anxiety-producing symptoms which brought them to me receded. But while these persons were better adjusted to life, they remained both selfish and godless. In other cases there were dramatic incidences where God seemingly remade broken or shoddy lives through the proclamation of the Gospel. And there were still other times when genuinely devout persons failed to find relief from either psychotherapy or those religious resources upon which they relied. Whereas I wrestled with the relationship between counseling and proclamation in the parish, I found myself struggling with the issue of service and evangelism on the mission field. The approach of our mission was comprehensive—meeting the needs of the total man. But under that umbrella we were forced to establish priorities. If the immediate need was hunger, or education, or medicine, did we provide that and then proclaim the Gospel once the other needs were met? If so, were we not guilty of covert motivation in our service rather than providing genuine caring love regardless of the response? On the other hand, to evangelize along with service often resulted in the simple incapacity to hear the Gospel because of the preoccupation with physical needs.

Prior to the lectures I read Dr. Oden's books. In *Kerygma and Counseling* [3] and *Contemporary Theology and Psychotherapy* [4] he elaborates his central thesis that proclamation makes explicit the Christ event which illuminates the accepting reality mediated in effective psychotherapy. This excited me. It raised expectations—and questions. Since he seemed to allow little room for proclamation within the framework of the therapeutic process, would he now call for the illumination of that process through the presentation of the Gospel once the pathological symptomatology had abated? If that is not what lies beyond the removal of symptomatology, what is it that does? And just exactly what does he mean by normalcy? These were other questions with which I expected Dr. Oden to grapple under the rubric of "Beyond Normalcy."

Here I was disappointed. Dr. Oden never even arrived at a consideration of his announced theme until the conclusion of the lecture. True, he phrased the issue beautifully:

> How are we to care for persons who have experienced seriously all that effective psychotherapy can do, and still find themselves struggling with

> the primordial questions of life, death, meaning, honesty, spiritual awareness, moral courage?[5]
>
> After the therapeutic process has run its course, what are the tasks which remain to be done . . . ?[6]

But after all that buildup, he never really answers the questions that he raises except to state obliquely that this is a valid field for the religious philosopher, the theologian, and the moralist. He does hint that psychotherapy needs to protect itself against exaggerated expectations and that the alleviation of symptomatology is not sufficient. Also, he evidently feels that the Church does have a role to play by helping people to "frame human questions more profoundly, address the human predicaments more clearly and in that sense they are closer to human fulfillment than before."[7] I already knew that what happens beyond the removal of neurotic symptoms was a legitimate field of pastoral care, but Oden's almost total lack of concreteness amounted to a near betrayal of my expectations for this lecture.

Appreciation

I feel indebted to Dr. Oden for his discussion of empathy. I believe he has performed a real service by helping us to see the process of empathy at work in a wide range of human experiences. This section of the lecture freed the word "empathy" from becoming just more jargonistic verbiage in my expanding vocabulary. He helped me to formulate clearly what it means to establish empathetic relationships.

Stimulation

The first point of stimulation which this lecture afforded me was Dr. Oden's wrestling with the issue of language. In various parts of the lecture he seems to be saying that psychological language is the mother tongue of contemporary civilization and the theological language of the Western, historical-philosophical tradition is the foreign language. He implies that we should go with the former. I sense that this is too simplistic. I would favor the use of the technical jargon of theology for talking among the Christian community, along with a radical translation of traditional terminology into the contemporary language of the day. Contemporary terminology is necessary to keep in touch with each new generation. Classical terminology is necessary to keep in touch with our own historic roots. Contemporary terminology changes along with changing thought forms in this age of ever-accelerating change. So if you translate the historic concepts, they must be retranslated anew in each generation—and with each translation they get further away from the original connotation—unless continuity is maintained with the classical forms.

As an example, in missionary work in Latin America, the line of scripture translation may go something like this: Greek and Hebrew to English to Spanish to Indian Dialect. By the time the translation has reached the Indian dialect, it has already passed through the cultural grid of Anglo and Hispanic cultures and may have strayed far from the intention of the original writers of Scripture. It makes sense, therefore, to be continually translating the Scripture into new languages, but using the original as a basis for translation. This position I feel is more in keeping with the conserving, radical tradition in which Oden claims he wishes to stand than with his apparent stance of theology abandoning its classical framework to go native in regard to the thought forms of the day.

The second point of stimulation was Oden's elucidation of his central thesis which undergirds his effort at dialogue between theology and psychotherapy. While this thesis is assumed and mentioned in his lecture, it is elaborated and developed in detail in his books under the rubric of a covenant ontology. Briefly stated, this means that the accepting reality implicit in psychotherapy is rooted in being itself. The therapist, though he be an atheist, mediates that unmerited acceptance to the patient and in so doing is an agent of God. The role of the Christian Gospel is to make clear the event in which the reality of acceptance is rooted.

Dr. Oden's thesis was of special interest to me because my basic theological training combined a Rogerian (via Hiltner) approach to pastoral care with a Barthian theological stance. I sensed a tension, but knowing no other path, operated out of this framework in the parish which I served. At times I felt like I was living a double life as I felt a deep incongruity in my role as a preacher and as a counselor. Although I was able to correlate Barth's doctrine of justification and Roger's understanding of acceptance by telling myself they really meant the same thing and by comparing this with Tillich's famous dictum, "You are accepted. All you need to do is accept your acceptance," their assumptions regarding human nature seemed contradictory. Nevertheless, I plodded on with evangelical preaching and nondirective counseling. Sometimes, I comforted myself with the fact that I was certainly not doing anyone any harm and there were signs that I was doing some good. At certain points I was even grateful for the good side effects of my Rogerian-Hiltnerian orientation. It taught me how to listen and helped me to develop an awareness of my own feeling. Being conceptually oriented, this did not come easily. Also, many of the insights of my pastoral experiences enriched my pulpit and classroom ministry. Still, I was bothered by persons coming through the counseling process with problems resolved and symptoms relieved so that they were able to pursue comfortably a self-centered way of life and remain indifferent to God. I felt pushed toward a two-sphere style of thinking: Counseling is a human service I can offer out of Christian love to help persons adjust to themselves and their environment, but it has nothing to do with the communication of the Gospel—that is the task of proclamation.

This is why Oden's work is so stimulating to me. I cannot help but regard it as the finest attempt to reconcile and integrate the work of Barth and Rogers that I have yet encountered. I feel that Oden's use of the analogy of faith was a valid and creative contribution to the dialogue between religion and psychotherapy. His analysis of the limitations and contributions of Hiltner in *Contemporary Theology and Psychotherapy* admirably expressed my own deep, unuttered feelings which had grown out of experience.

However, with one aspect of his effort to integrate religion and psychotherapy, I found myself extremely uncomfortable. This centered around his use of the Bible in general, particularly his exegesis aimed at establishing the role of the Incognito Christ in therapy.

The Problem of Authority

Although Oden calls for a return to classical orthodox Christianity, he himself does not stand in the evangelical Protestant tradition of *Sola Scriptura.* This needs to be very clear for it is basic to the stance from which he enters the dialogue with psychotherapy. This is clearly stated in *The Structure of Awareness:*

> My authority for speaking of God therefore is fourfold: *scriptural* truth,

> *experienced* in life, made intelligible and self-consistent through *reasoning* and mediated through the historic Christian *tradition*.[8]

Oden argues that to focus upon a single one of these four resources in such a way as to exclude the others is to subvert a wholesome and logical method. Yet, by refusing to assign priority to any one resource, Oden has no criteria for validity when the resources contradict each other—as they have during the history of the Church.

The Principles of Hermeneutics

Violating sound principles of Biblical interpretation, Oden builds his doctrine for the Incognito Christ upon a single Biblical passage—and that a parable without parallel in the other synoptic Gospels. He brings no other scripture to bear upon his conclusion. Generally, evangelical hermeneutics has proceeded on the conviction that "the Epistles interpret the Gospels . . . systematic passages interpret the incidental . . . universal passages interpret the local . . . didactic passages interpret the Symbolic."[9] Matthew 25:31-46 is not only found in the Gospels, but is incidental, local, and symbolic. Despite his sophistication, at this point Dr. Oden appears to me strangely akin to those who build eschatological schemes on isolated, difficult, and symbolic passages found in the apocalyptic literature.

The Method of Exegesis

Though the word picture of the Last Judgment found in Matthew 25:31-46 is set within the framework of conventional Jewish apocalyptic, it has no parallel in the other gospels. Oden, however, sees this passage as crucial to his discussion with psychotherapy. He views it as impinging powerfully upon his understanding of the presence of Christ amid the therapeutic process and

> it focuses upon the inconspicuous service to the troubled neighbor as a ministry in which Christ himself is a hidden participant.[10]

Three questions are fundamental to any exegetical consideration of this passage. Who are "the least of these my brethren" in verse 40? Who is included in the gathering of the nations in verse 32? What and the criteria of judgment?

To the first question Oden answers that the King's brethren are suffering humanity:

> The least of these refers to those who share most deeply and inconspicuously in the estrangement, thirst, and sickness of the world, even as God himself shares in it. To the extent that we have served the least of these we have served Christ.[11]

Dr. Oden is somewhat vague as to the identity of the nations. He seems to equate them with all mankind. The blessed

> are those who have rendered concrete, sensitive, caring service to the expected Deliverer, the Son of man whom they have met in need.[12]

The cursed are those who by failing to exercise love have

> already chosen in their relation to the neighbor a certain form of self-damnation which misses the opportunity for authentic life which is at hand and being offered already in the least of these.[13]

In response to the third question, Dr. Oden affirms that the criteria of judgment

> focuses directly upon the relationship of the messianic King to the process of charitable, sensitive, helping service. . . . To the extent we have served the least of these we have served Christ . . . not merely that we ought to do good deeds in order to receive God's favor, but that the needy already bear God's favor to us in the most real sense.[14]

On the basis of his exegesis Dr. Oden affirms that Christ Himself is the hidden participant in the encounter of the helper and the helped —though he is the hidden or Incognito Christ.

I question the above conclusion because I believe that it is based upon faulty exegesis. Following T. W. Manson, I would hold, first, that the King's brethren are not suffering humanity in general, but rather Jesus' own persecuted followers. I checked out every single use of the word brother, brothers, and brethren, in the synoptic gospels. In every case the term refers to either a blood relationship or to Jesus' followers. There is not a single case of it referring to humanity in general or needy humanity in particular. In keeping with this theme, Christ is identified with his Body. He is the Head and the Body is the Church, not the world in general. Not only is this taught in I Corinthians *12*, but it can also be seen in Acts *9:*4-6 where the risen Lord asks, "Saul, Saul, why do you persecute me?" and identifies Himself as "Jesus, whom you are persecuting." As far as we know, Paul never persecuted the earthly Jesus. He persecuted the Church. In persecuting the Church, he persecuted Jesus. The King's brethren in the parable are not even numbered among the nations, for they are participating in the judgment as was hinted at by Jesus when he spoke of the twelve thrones delineated by Paul in I Corinthians *6:*2 as well as alluded to in such passages which see Christ's followers reigning with him (Cf. II Timothy *2:*12, Revelation *5:*10).

Second, rather than identifying the nations of verse 32 with all humanity as Dr. Oden seems to do, I would maintain that this word refers to all mankind except the disciples of Jesus. The brethren are not included among the nations. The word "nations" can be translated either as nations or gentiles. Its usage in the rest of Matthew would indicate that "all the gentiles" may be the best rendering. What about the Jews? Those who have rejected the Kingdom as preached to them by Jesus and his disciples are reckoned among the gentiles.[15] The brethren are the True Israel.

Third, the criterion of judgment is not, as Dr. Oden alleges, help given to the needy, but rather response to the Kingdom of God. This is congruous with the entire synoptic tradition in which the Kingdom of God is central to all of Jesus' teaching and preaching. Manson writes:

> It is a striking feature of this . . . that those on the right are not commended for abstinence from the cruder sins, nor are those on the left condemned for indulgence in them. Again it is not that one group has been kind-hearted and the other callous. . . . Here the vital question is not whether a man has lived a moral life or whether he has been decently kind to other people. It is how he stood toward the Kingdom of God: was he on the side of the Kingdom or against it?[16]

The key to the criterion of judgment is the solidarity of Christ and his brethren. Though many they are one. Help given to one is help given to all—and above all to their Head. This is certainly the meaning of such passages as Matthew *10:*40-41, Mark *9:*41, and Luke *10:*16. Again, Manson comments:

> It is assumed that the help has been given to the disciples of Jesus, when they were engaged on their apostolic task, when they arrived in a strange town, hungry and thirsty or when they were worn out and ill through toil and travel, when they were imprisoned for preaching the Gospel—why else should the brethren of Christ be in prison? . . . The deeds of the righteous are not just casual acts of benevolence. They are acts by which the Mission of Jesus and His followers was helped and helped at some cost to the doers, even at some risk.[17]

In summary, rather than introducing a new and unique element in Jesus' teaching, Matthew *25:*31-46 is totally congruent with Jesus' message of the Kingdom as presented in the other Gospels. The destiny of persons is determined by their response to the Kingdom embodied in Jesus and mediated by the ministry of his discples. If anything, this passage teaches not the unrecognizable presence of Christ in the relationship between helper and helped, but rather the extension of the incarnation based on the union between Christ and His Church.

The doctrine of the Incognito Christ is extremely important to Dr. Oden. It explains how psychotherapy is able to mediate the accepting love of God through a nonbelieving therapist, because the Incognito Christ is present in the relationship even apart from faith. Oden's exegesis leads him to become almost euphoric in his eulogy of psychotherapy, despite the limited results which he claims one can reasonably expect from psychotherapy.

The conclusion of this lengthy tangent on exegesis is that while I am still prepared to seriously entertain Dr. Oden's central thesis, the Biblical support for an Incognito Christ to buttress that thesis is flimsy indeed. It is ironic that Oden comments in his third lecture about the alienation of American pastoral care from both exegesis and systematic theology when he so tragically illustrates the gulf.

Not only is Oden's exegesis questionable in the Matthew *25:*31-46 passage, but also in his use of the parable of the unmerciful servant in Matthew *18:*23-35. In that case he uses the parable to illustrate the ontological impossibility of guilt before God and the reception of pardon in such a way that it changes behavior. He writes off Jeremias' understanding of the parable without serious interaction with him. Again, he builds an important part of his effort at integration around a single, difficult to interpret, passage of scripture. His exegetical efforts decline even more noticeably when he cites I Corinthians *12:7* and holds that the gifts of the Spirit are given to the world rather than to the Church. Also Oden's understanding of the word "enlightened" in John *1:9* is probably better taken in the sense of casting light upon rather than inward illumination as he prefers.

The thing that appears incongruous to me is the strenuous effort Dr. Oden makes to utilize an innovative doctrine based on insubstantial exegesis when he could more easily account for the potential which lies within man on the basis of a high doctrine of the image of God in man and the unity of truth, whether its source lies in the givenness of revelation or the discoveries of the empirical scientific enterprise.

PRACTICAL OUTWORKING

Although Oden idealizes the therapist less in his lectures than in his books, he does see the therapist as God's agent mediating that accepting reality implicit in the cosmos. But what happens if the therapist mediates rejection, destruction, and disintegration rather than healing, either because of his technical ineptness or his personality. Of whom is he really an agent then? What about the therapist who restores social adjustment and intrapsychic comfort at the cost of a calloused conscience and a loss of faith? Or even in what sense is the therapist who mediates a mingling of the positive and negative mediating the accepting reality implicit in the cosmos? How can you say, "This is God at work" or "This is not God at work?" This is an area notable for its absence of development. I would invite Dr. Oden to wrestle with it.

In the final lecture, Dr. Oden sought to relate the eschatological hope expressed in theology to the human potential movement which he sees as "a prevailing mood of hope for a radical transformation of the human condition transcending what are viewed as the contemporary limits of man."[18] While Oden speaks of relating the evangelical hope to this recurring theme in Western Civilization, he is also aware of the sense in which they stand in tension.

The very first level of his four-level response begins with a brilliant summary in which he accuses the "pop theology" of the past decade of mismanaging the concept of hope by accommodating it to wave-of-the-future faddism. I concur with his prescription that the "antidote to pop theology is classical orthodox Christian thinking in dialogue with the emerging realities of history."[19] For more than a decade I have occupied much of the same terrain on which Dr. Oden now seeks to pitch his tent and have borne the barbs of ridicule of those bandwagon theologians whose every latest paperback was seen as groundbreaking and epochal. I share with Dr. Oden that loneliness of living on the boundary where radicals disparage you with the invective of "reactionary" and the conservatives can never quite bring themselves to trust you.

My own grounding in a progressive, but orthodox, understanding of Christianity in the late 1950's never allowed me to feel comfortable with either an irrationalist existentialism or a realized eschatology—both of which evacuate the future of any meaning. An affirmation of the value of reason through personal study of the writings of Edward John Carnell and an appreciation of the value of history derived from my major professor during college days have been a part of my heritage—as is the view that Christ's resurrection is proleptic of that Kingdom which will be consummated in the future.

This rootage prepared me to affirm the value of theological faddism as corrective of a pie-in-the-sky, futurist eschatology, a sterile classicism, a socially irresponsible devotionalism and/or an otherworldly pietism, while keeping me from any serious temptation to jump on any of the bandwagons. This approach, which gained in balance what it may have lacked in momentary enthusiasm, served well in my parish experience.

The second level of Oden's response draws heavily upon the work of Wolfhart Pannenberg. Believing that the dialogue between theology and psychotherapy has suffered by being de-eschatologized, Oden maintains that psychotherapy would benefit by being placed within the frame of reference of universal history.

He goes on, at the third level, to propose Cardinal Newman's concept of doctrinal development as a model for a theological method which relates tradition to renewal. Finally, at the fourth level, he pleads for a stabilizing and conserving direction in theology.

My response to levels two, three and four are intertwined because I see them as closely interrelated in Oden's thought.

The excursus of Newman is valuable, although it does not quite seem to fit with his announced theme as well as does the section on Pannenberg's universal historical thrust. The seminary which I attended was profoundly influenced by the Mercersburg Theology which drew heavily on the Oxford Tractarians and joined with them in insisting that the church remains the same only by changing, but that the changes must be a genuine exfoliation of the apostolic witness rather than a deviation from it. In this sense, Oden's analogy of the photographic plate, though static, is indeed well taken. While Newman, despite the excess baggage with which much of his thought is freighted, is an able exponent of the "mutability does not deny continuity" theme, Oden might have drawn upon Phil Schaff's *Principle of Protestantism,* published in 1645, for a development of the same idea from an evangelical, Protestant standpoint. Despite its age it is the current textbook for the core course on the Doctrine of the Church and Sacraments offered at Fuller Theological Seminary. I also believe that Oden's own thinking would be enriched by more serious interaction with Cullman who stresses that the tradition limited itself when it canonized the Scripture, thereby affirming that only that tradition is genuine which is a valid exfoliation of Scripture. This particular insight was recognized by the Roman Catholic Church in Vatican I when it renounced its historic, two-source theory of revelation. I feel that Oden needs to interact with this, not because he is unaware of it, but because much of what he says here does not seem to mesh with his own concept of a fourfold source of authority.

Despite some early indications to the contrary, I suspect that Oden's current romance with Pannenberg will bear little fruit in the dialogue of theology with psychotherapy. While my knowledge of his writings comes from secondary sources, of which Oden is one, Pannenberg appears too obtuse, too Germanic, and too psychologically unaware to excite American psychotherapists. I do not see him as a new model for dialogue—only as a corrective for a psychotherapy which has become too narrowly preoccupied with the present and the subjective. He calls psychotherapy to consider more seriously the interpersonal, the communal, the future so that man can be seen as a social being with a history, a present, and a future as well as a merely biological being. But even if it serves only as a corrective rather than as a basis for a new synthesis, Pannenberg's work will have rendered a service to psychology. What appears to me to make dialogue difficult is that I do not see much in Pannenberg that makes him open to the insights of psychotherapy.

CONCLUSION

Looking back over those hectic days in early January, 1972, I can cite several areas of growth in my own understanding as well as numerous instances where my thinking was clarified. First, I now believe that the key for integration in psychotherapy and theology lies along the lines of a high doctrine of the image of God in man as the source of human potential rather than the questionable doctrine of the Incognito Christ.

Secondly, I better understand both the inadequacies and strengths of my own Rogerian, via Hiltner orientation to pastoral care.

Thirdly, I clearly see Oden's correlation between Rogers and Barth as the finest attempt to integrate those schools of thought yet undertaken. For me, this means that Oden's books merit not only rereading, but also continued serious consideration.

Fourth, I was confirmed regarding some of my own misgivings about professionalism (be it in psychotherapy or the pastoral ministry) as well as strengthened in my inclination toward the support of the laicization of therapy.

Fifth, I received a sharpened diagnostic tool in the form of Oden's historical sketch delineating the conflation of moralistic pietism and optimistic liberalism which characterizes most mainline white, Anglo-Saxon Protestant (WASP) churches.

Sixth, I was helped to see the value of a living pietism for our time.

Seventh, I concurred heartily and with a deeper level of awareness with Oden's statement that orthodox Christian thinking in dialogue with the emerging realities of history is a relevant option in our day.

The Oden lectures are history. I am glad I was a part of that history. I have been enriched by the dialogue not only with Dr. Oden, but with the dialogue among the students and faculties of the three schools at Fuller. At times I stood in awe of the breadth of Dr. Oden's knowledge. At other times I was annoyed by an abstractness that sometimes bordered on the ethereal and the feeling that a vital contact with psychotherapy belonged more to Dr. Oden's past than to his present. But most of all what came through was the creativity of his thought and his genuineness and devotion to the task he has set before him.

NOTES

1. Oden, Thomas C.: The Diffusion of Therapeutic Agency, p. 2.
2. Oden, Thomas C.: *The Structure of Awareness.* Nashville, Abingdon, 1969, pp. 117-120.
3. Oden, Thomas C.: *Kerygma and Counseling.* Philadelphia, Westminster, 1966.
4. Oden, Thomas C.: *Contemporary Theology and Psychotherapy.* Philadelphia, Westminster, 1967.
5. Oden, Thomas C.: The Post-Therapeutic Situation, p. 31.
6. *Ibid.*
7. *Ibid.,* p. 37.
8. Oden, Thomas C.: *The Structure of Awareness, op. cit.* pp. 87-88.
9. Carnell, Edward John: *The Case for Orthodox Theology.* Philadelphia, Westminster, 1959, pp. 57-63.
10. Oden, Thomas C.: *Contemporary Theology and Psychotherapy, op. cit.* p. 98.
11. *Ibid.,* p. 100.
12. *Ibid.,* p. 99.
13. *Ibid.,* p. 101.
14. *Ibid.,* pp. 98, 100.
15. Major, H. D. A., Manson, T. W. and Wright, C. J.: *The Mission and Message of Jesus,* p. 542.
16. *Ibid.,* p. 543.
17. *Ibid.*
18. Oden, Thomas C.: The Human Potential and Evangelical Hope, p. 66.
19. *Ibid.,* p. 71.

CHAPTER 6

THE THEOLOGICAL SUBSTRUCTURE OF ODEN'S THEOLOGY AND PSYCHOLOGY SYNTHESIS

CALVIN R. SCHOONHOVEN

In this chapter I should like to address myself to certain constant and programmatic concepts that inform the books and lecture presentations of Dr. Oden. These axiomatic and all-controlling theological themes are crucial to any adequate critical interaction with the thought presented. Prior to a discussion of these, however, I wish to heartily endorse the intent, motivation and high level of scholarly involvement reflected in the writings, lectures and discussions of Dr. Oden. He writes with precision and force, and his public presentations are searching, forthright and free of determined dogmatism. It has been a delight both to read him and listen to him.

My interaction with Thomas Oden's crucial thematic constructions can best be accomplished by gathering my thoughts about five theological postulations that I think it essential to affirm if we are adequately to interrelate theology and psychology. They are:

1. The nexus between experienced justification, conscious faith and deliberate obedience must be kept invariably intact.
2. There is sin and a form of guilt that has ontological ultimacy.
3. "Common grace" is a necessary theological category.
4. Conceptualization (*notitia*) is an indispensable factor in the faith process.
5. The ultimate concern of theology is not to effect service to man, but to affirm the glory and worth of God.

THE NEXUS BETWEEN JUSTIFICATION, FAITH AND OBEDIENCE MUST BE KEPT INTACT

The first postulation that we think essential is that the nexus between justification, faith and obedience must be kept invariably intact. In his three major writings on theology and psychotherapy,[1] Oden brings this linkage into jeopardy. The nature of this disjoining may be instructive to us and assist in our effort to avoid a like problem in our relating of theology and psychology. The most striking phrase employed to denote this fragmentation is "covenant ontology." This means that by his very existence man is included in God's redemptive covenant. In no way can this covenant relationship be negated, and in no way can the grace and forgiveness of God be thwarted. The response of man neither makes real nor destroys the covenant. This is expressed in various phrases and contexts, some of which are the following: "No man can avoid being forgiven by the God who elects to love the world in Jesus Christ."[2] "Our essential affirmation is that man cannot be guilty before God, but only before the gods."[3] "The psychological experience of demonic guilt feelings is now perceived to be rooted in an ontological impossibility."[4] "God's pardoning verdict is not merely an empty possibility which we may or may not actualize, but *is* already in the most radical and universal sense an *actuality*."[5] The most extended statement in this regard is the following:

> But there is one thing that is not finally possible for man: to negate the forgiveness of God, to escape the gracious covenant which circumscribes his existence, to relieve himself of the constraint of the friendly divine verdict. He cannot run away from that, because he cannot run away from himself. God continues to renew the covenant even when it is from our side broken or neglected. That is the story of Israel and, symbolically, of all humanity. When the people of Israel are disobedient, the judging covenant Lord delivers them into captivity as an expression of his covenant love, in order to call them back into an awareness of their covenant freedom. From time to time the people indeed do actualize the covenant relation from their side in faith and obedience, but they remain a covenant people whether in idolatry, bondage, or faith. There is a sense in which, as the ancient fathers said, *grace is irresistible,* the sense that it cannot from our side be destroyed, even though it can from our side be unreceived.[6]

All this leads to the final and quite consistent discussion on universalism. Here is articulated a forthright statement on universal justification.

> I reject the notion of *universal salvation* (understanding salvation as the volitional participation of man in the saving deed of God), but at the same time thoroughly affirm the scriptural view of *universal justification*, which remembers that the divine pardon is addressed to all.[7]

According to this theological construction all men are justified, and this is to be conceived ontologically. Since this is an ontological category, there is no reason to speak of actualization or response. We cannot become something we already are. All are members of the covenant and all are in redemptive relationship with God.[8] Whether man lives wholistically or not, whether he acknowledges Christ or not, whether he feels alienated or not, whether he is evil or not, he is with God, in touch with God, accepted and loved by God. The force of this is stressed when man is depicted as not only valued despite his value negation, but valued unconditionally amid his value negations.[9] And it is to be assumed on the basis of these statements that if there is such a thing as an ultimate or realistic eschatology, if there is an eternal participation in the divine being, if there is to be the experience of the beatific vision, man will—even in his alienation, unfaith and unacknowledgment of God—enjoy and experience this. All are justified. This is the overarching category under which all reality must be subsumed.

The position is best articulated when it is affirmed by Oden:

> I make a sharp distinction between universal justification and limited salvation: God's pardoning verdict is declared and established for *all*, not some. It is unconditional, not contingent upon our response.[10]

Now, according to Oden, it is this universal, cosmic, accepting reality that makes possible effective therapy. To be healed, man must know that he is accepted. He must know that despite his hostilities, alienations, estrangement and incongruence he is loved, valued, esteemed, cared for. Indeed, the reason therapy is effective is because it is ground in an ultimate accepting reality that the Christian celebrates in the kerygma as the saving deed of God in Jesus Christ. Though this reality is unacknowledged, it yet provides the ground for the operational effectiveness of therapy. Empathy, congruence and non-possessive warmth [11] all find their counterpart and indeed their foundation in Christian theology. Thus therapy works because it is grounded in the accepting God.

Not only does therapy work because it is grounded in a universally accepting reality, but consciousness and conceptualization of this reality constitute in some way a benefit for the therapist and client in the therapy experience. Oden implies this when he states:

> With all due respect to the remarkable achievements of so-called "secular" psychotherapy, i.e. therapy without any consciousness of its ontological presupposition, we need to explore together whether it might be the case that the counselor who knows, understands, and celebrates the explicit witness of this special community to the self-disclosure of the accepting reality conceivably could attain clearer perception of human sickness and health than the therapist who only implicitly makes this assumption without examining it.[12]

After exploring this in his several books and studies it is questionable whether he makes the case that it is important for the therapist to know this. Indeed as we shall observe later, Oden comes close to denying that it is. And furthermore we shall see that according to the canons of logic he must not say this for it would be inconsistent with his metaphysical and epistemological presuppositions.

The problem with this position is the above noted assumption that irrespective of man's response he is indeed elected, redeemed, reconciled, justified, and pardoned. To put it in clear statement, the consistently maintained biblical connection between justification, faith and obedience is severed. There is no way this can be done without grave damage to the entire exegetical, theological enterprise. The nexus must be kept intact; justification, faith and obedience are inextricably together. Proof texts are not pleasing, but if the text picks up the overall inclination of the theology of the New Testament, they are in order. Such representative texts are found in Hebrews *5:9* where we read that eternal salvation is "to all who *obey* him" and John *3:36* where the writer affirms that "he who *believes* in the Son has eternal life; he who does not *obey* the Son shall not see life but the wrath of God rests upon him."[13] So, a man is not justified unless he responds to God, has faith in God, places his trust in God and subsequently obeys God. And without explicit trust, without this "volitional participation . . . in the saving deed of God," there simply is no experience of justification. This is only to give due weight to the biblical meaning of faith which is always used within a conceptual framework.

There are profound ramifications for theology and psychology when universal justification and limited salvation are maintained. It means that salvation is a term whose meaning is *totally* exhausted at the phenomenological, existential, experiential level. The saved are those who live in health and wholeness, and this wholeness is defined in terms of emotional maturity that reflects itself in interpersonal relationships.[14] Conscious participation in the supernatural, redemptive deed of God in Christ is not so important as affirming one's own self-worth within relationships on a purely immanential, nontheistic level. Indeed, there is here a forthright rejection of the two-sphere concept of the sacred/secular, the divine/human, and grace/nature. In a word, the vertical relationship is absorbed in and becomes real through the horizontal structures of life. Nothing really supernatural happens, because the category of the supernatural has been eliminated by right of absorption into the natural. And what we finally have here is not theology but anthropology, Feuerback "redivivus." For what is essentially important is not man's relationship with God (as one who is personally known because he has personally and supernaturally revealed himself), but rather man's salvation, his health, his personal integration, his maturity and his relationship with others that may be defined as good or bad, dependent upon the psychologist's explication or definition of these terms.

So much of contemporary theology is guided in its formulation by the concept that there is a spiritual impregnation of all reality by God. Celebration of life, affirmation of existence and universal grace are ideas that most appropriately relate God and created reality. God affirms ; he does not negate. Chardin, Tillich, Hillman, Bultmann, Barth, and the statements of Vatican II all in some way and at some point reflect this kind of thinking. Oden is the first so clearly to relate this theological formulation to psychology. Thus he holds that God informs all of life, that he impregnates the total cosmos, whether we realize this or not; but that when we do conceptualize this and volitionally participate in this reality of God, then there is greater potential for good or for psychic well-being.

But orthodoxy has consistently maintained that the proper relationship between God and the world is to be oriented about two foci, creation and salvation (justification). Insofar as these two relations are structured there is a total impress of God upon the world, but it is a totality that at the same time embodies distinctions, the distinction of judgment/condemnation as against justification/salvation. And that which is important and must be underscored is that both categories possess ontological ultimacy. This is why Christian theology has always been indissolubly connected with the Christian mission and witness. It is a correct biblical extrapolation that men are lost apart from Christ. And the concern has not been that witness be given to the end of psychic health, or better interpersonal relationships, but precisely because man's being was in eternal jeopardy.[15] Indeed, the response of faith was sought and obedience exhorted because this was the way both of actualized justification and meaningful salvation. The impingement of the eschatological realities of both salvation and judgement on present experience, finally and ultimately conceived, has had and we think must have decisive importance in pastoral counseling and psychology.

SIN AND GUILT POSSESS ONTOLOGICAL ULTIMACY

The second postulation is that there is sin and a form of guilt that has ontological ultimacy. In the light of the above it is obvious that much of Oden's integrative effort is centered on the Rogerian emphasis of providing an accepting environment where the individual can be free to express himself and experience growth. The idea of the growth potential of man through acceptance is, it seems, the place where the contact or interaction with theology becomes most acute. This is so because something definitive is said about man. Thus, we think, any resolution of the problem of the integration of theology and psychology must address itself to this problem. In Oden's system the problem is resolved by an easy adoption of the Rogerian emphasis on the healing virtues of love and acceptance, as these provide an environment for the intrinsic human drive or inclination toward health to actualize itself. It is not too difficult harmoniously to relate a theology that affirms the intrinsic goodness of man and a universally justifying God, with a psychology that postulates change for the good and psychic well-being on the basis of environmental categories of empathy, congruence and nonpossessive warmth. For such a psychology quite obviously also assumes man's capability to extricate *himself* (in the company of others) from his dilemma. The problem between psychology and theology, however, becomes exceedingly acute when theology maintains man's terrible inability, his evil inclination, his egoistic passion, his basic ontological condition of sinfulness; when it asserts with Augustine that original sin is not merely a psychological state but a metaphysical one; that man's fall was a fall in the order of Being. This correlated with the affirmation of a judging, wrathful God causes awesome problems for the harmonious relating of theology with most contemporary forms of human potential, and relational, client-centered psychotherapy.[16]

The point of this problem is sharpened by the observation that biblical teaching seems more to stress the evil nature of man. Great sections of Scripture are utilized to establish that man is so evil that there is no way out through his natural environment (Romans *3*). This is developed in such fashion that there is a frontal assault against any idea that man has ultimate worthfulness in himself, or that he should strive toward self-esteem. Indeed, the very opposite is true. The consistent teaching of apostles and prophets is that precisely the affirmation of self-worth, self-esteem and ego enrichment constitutes the real problem of man and not the solution (Philippians *3:*3,4). The kerygma further proclaims that the only way of escape is through a conscious, deliberate regeneration-conversion experience that is supernatural and that causes one to place all confidence in God. There is, then, no longer to be confidence in the flesh, but only trust in God. And this is the starting point and ending point of all theology. To stress this Paul asserts, "we are the true circumcision, who worship God in spirit, and glory in Christ Jesus, and put no confidence in the flesh" (Philippians *3:3.*)

We do not wish to say that man always reaches the full potential of his depraved nature, or that man has no remnants of good constitutive to him. For indeed the Bible consistently teaches the paradoxical character of human nature. In addition to his evil nature it must be said that he is also a "little lower than God; he is crowned with glory and honor and (under God) holds the dominion over the beasts and the whole world of nature" (Psalms 8:4-8). Man yet is in the image of God with all that this means.[17] However, it is impossible to erect a redemptive therapy on this foundation of the *imago dei.*

COMMON GRACE IS A NECESSARY THEOLOGICAL CATEGORY

According to Oden's evaluation of the therapy situation as that situation is informed by theology, the first order of the day is for the therapist to affirm the other person's worth. And the exhortation may be framed in the fashion in which Oden often frames it, "become who you are."[18] "The human predicament consists in man's failure *to be who he is.*"[19] But the biblical text seems more to exhort that we should "be who we are not." And this is so because in the flesh no one can please God. Man is "dead through trespasses and sin" (Ephesians *2:*1,5); he does not understand and seeks not after God (Romans *3:*11). His basic disposition, his habit, is other than to serve and glorify God. To affirm, then, that man is good by right of "beingness" is to maintain something that is difficult to square with the text and that may in the long haul work for man's detriment rather than his good. Let me explain this.

Working from the idea of a covenant ontology and putting into practice its demands we would place an individual or client in an atmosphere of total acceptance. And whether God were brought into the picture or not, ideally increasing organismic wholeness would be achieved; i.e. one would become emotionally more mature. Since the client in various ways is told of his goodness, he would want to be and to do good. This, I think, would be effective therapy in that any individual finds life less tense, less frustrating, less threatening when good is done rather than evil. This would be so even if the definition of good were incorrect. The important thing is that one thinks good is being done and that this be validated by observable results. Things work better and are handled more efficiently when the way of altruism, chasteness, fairness, beneficence and good will are followed. There is a peculiar kind of reinforcement that moves one along at this level reminiscent of ideas contained in Norman Vincent Peale's *The Power of Positive Thinking.*[20] However, we should have to say that here there is no relationship to redemption for there is no cognized, faith-in-God structure. There is a faith structure, but it is self-faith. The individual so treated may improve in emotional maturity, may be able to relate more meaningfully to his peers, and may be more balanced in his judgments. Indeed, he may be more psychically healthy. However, he is not at all closer to the Kingdom of God. Indeed, he may be considerably farther from it than before. Such a person may find adequate capability within himself to meet the problems of life as a result of this therapy with its insistence on intrinsic goodness. He then may feel no inadequacy or need for the supernatural empowerment of the Holy Spirit. Consequently, in the solving of man's problems of alienation, guilt, loneliness and despair there may be increased likelihood that the overtures of a radical Christology may not reach him in his self-sufficiency and total dependence upon self. He may finally come to the conclusion of Carl Rogers: namely, that there is within the human person all the potential for adequate and good health, and that when one has plumbed the depth of his own resources with the help of another, he finds there all that is necessary to meet the exigencies of his own contemporary experience. What we are saying here is that man can and will improve under certain therapies. Rogers' findings must be accepted as valid. When man has self-esteem, he probably will progress in wholeness. When one engages in the triad and does actualize empathy, congruence and nonpossessive warmth, he will probably experience more health. But this is not to be construed as redemption, salvation, reconciliation or justification. Rather, because not all men are in the church, because all men are not in Christ, because all men are not sympathetic to theism, the naturalistic, humanistic psychologist must start and end here. There are no other options for him. This is to assert that because there is a "kosmos" that is not redeemed, there must be such means to hold together the fabric of life so that the "elect" may have a matrix in which their redemption is actualized and that will thus prevent chaos from engulfing the elective, redemptive purpose of God.

Now, insofar as therapy does alleviate mental suffering, human frustrations and alienation, it is a good. Just as civil government is a good, and plumbing and ditch digging services are good, so too is this a good. It is a civil righteousness; it is the arena of creation, of common grace. And furthermore, whether done out of base motive or good motive, whether done for money or for the public betterment, it nonetheless is a good. But it is not theology, it has nothing to do with Christian teaching other than that of the general biblical teaching on vocation. This vocation like other vocations constitutes a grace for the church and makes up that matrix in which the redemptive purpose of God can be effected. We heartily agree with Calvin when he asserts that there "is a grace which is communal, does not pardon nor purify human nature, and does not effect the salvation of sinners. It curbs the destructive power of sin, maintains in a measure the moral order of the universe, thus making an orderly life possible, distributes in varying degrees gifts and talents among men, promotes the development of science and art, and showers untold blessings upon the children of men."[21] This is where helpful therapy fits.

But insofar as this good and effective therapy is not of faith in that it is carried on with motives that are not motives to glorify God, insofar as there is no theistic connection whatever, we have at the same time something that in the ultimate sense must be defined as sin.[22] So we are confronted with a paradox. At the same time that we affirm this therapeutic effort to be good as defined by man and even as defined by God according to the canons of good outlined for us in the biblical text, we must also affirm it to be sin, because there is no connection, i.e. no explicit, obvious, redemptive, motivational connection with the divine being, God revealed in Jesus Christ.

This, however, ultimately is not a stumbling block. It is possible to recognize this and yet continue with the effort of glorifying God and loving men. We simply recognize that in terms of soteriology, in terms of an ultimate saving relationship with God in Jesus Christ, psychotherapy does not necessarily move a person closer to the kingdom than does any other act of good perpetrated in the life of man. So that when we speak in terms of a radical Christian theology, when we make reference to the present and eschatological salvation of the individual and eternal matters, then we must say with the text of Scripture that in this regard all that is done apart from reference to God, apart from explicit love for God, apart from relationship with Jesus Christ, provides ultimately no salvation. There is then a strict and radical bifurcation between the discussion of salvation (of theology) and the therapeutic process. When we speak of soteriology in its present ontological and final eschatological sense, we are speaking of man transformed by explicit acknowledgment of the atoning work and saviorhood of Jesus Christ.

This, however, does not mean that there is an ultimate separation or bifurcation between this world of theology or explicit soteriology and the natural world, humanistically conceived, wherein at least in part healing is carried on through naturalistic psychology. For the concern of a radical divine soteriology overlaps with the concern of a humanistically oriented psychology insofar as both are concerned with man. Indeed Christian salvation has a present impact in terms of life integrated and whole which quite obviously at some point or points intersects with the concern of psychology.

Thus, whatsoever is not of faith is sin. Yet a radical, vertical theology and soteriology will utilize those developments in the natural sciences, including psychology, to sharpen the abilities of its own community to better reflect the power of the Holy Spirit. So that as incongruous as it may seem, what we end up saying again is, paradoxically, that the church, the body of Christ, utilizes the sin of the natural world (defined in ultimate terms) towards the improvement of its own community. And yet this should not be looked upon in a selfish or isolated fashion, for the church only utilizes that which is inherent in creation and dug out by the energy and effort of secular psychologists and scientists in order that it in turn may give back to the world, to all of creation, to man in his soaring wisdom, and to man in his lowly and stumbling ignorance that which is the the final capstone of all, connection with a truly redemptive person, Jesus Christ, the revelation of God. This knowledge and this relationship heightens, intensifies, and makes possible the mediation of final health and salvation.

CONCEPTUALIZATION (NOTITIA) IS AN INDISPENSABLE FACTOR IN THE FAITH PROCESS

What we are saying here relates to a philosophical theological problem in Oden's thought, the problem of epistemology. In the biblical framework and structure, "to know" plays a central role. This is not first an existential knowing, but rather an intellectual knowing: i.e. a conceptualization of truth.[23] Whatever else knowing means in Scripture, it surely means that there is a residual amount of the knowledge of the deed of God essential to respond to the call of God. We must "possess this knowledge" (I Corinthians *8:*7, Matthew *13:*10, Colossians *1:*24-2:6, Matthew *11:*27). This is quite contrary to Oden's diminution of conceptualization when he indicates that the "accepting reality need not be verbally expressed," and that one "cannot conceptualize until one has actualized."[24]

In all human experience and learning, progress is directly connected with knowledge. This is true whether in the sciences or the humanities, whether medicine or human experience (psychology). Indeed, conceptualizations do not guarantee any meaningful ethical conduct, for we can know and not do or act. But conceptualizations are nonetheless the prerequisite or the *sine qua non* of meaningful experience. If one is going to start, he must start here. The great problem that I have with Oden's attempted integration is that whereas in all other areas of his investigation epistemology is recognized and given due place, when it comes to the question of verbalizing the reality of God, there is hesitancy, and another sense of *knowing* can do the job as well. This other knowledge is the existential, phenomenological kind of knowledge that need not have any foundation in conceptualization. Thus, therapy can be and indeed is redemptive apart from any conceptual knowledge (noetic) of the redemptive deed. This constitutes, then, a truncated epistemology; indeed, an epistemology that is only half so. And it does not save the day to assert that in therapy we have communicated real experiential knowledge (noetic) of God the accepting reality. All that is communicated in therapy is that one individual is accepting, and this one individual can in no conceivable way be representative of God, not even in some kind of obscure experiential fashion. There is no way the therapist can become a surrogate for God. The ambiguities of human existence prevent this. Phenomenological and existential-experiential factors have quite thoroughly convinced the client that universal or cosmic reality is not accepting. Indeed, this is precisely why he needs therapy. There is no more validity for believing there to be an accepting reality in the universal scheme of things on the basis of one accepting therapist than there is for extrapolating a cosmic nonaccepting reality from the client's many prior, nonaccepting encounters. There is, however, a higher scale of validity for the latter extrapolation because negative encounters have been and probably will continue to be his normative experience.

To know that he is accepted by God, the client must in some subtle or direct manner be informed of this. In the therapy session something distinctly supernatural must transpire. Either by word or perhaps by attitude, life style and then word, the reality of the forgiving, redeeming Christ must be structured into the therapy experience. The way in which this should be done is difficult to ascertain and requires study that is not within a theologian's competence. But that it should be done is, it seems to me, indisputable if we affirm a more than superficial and casual connection between our theology and psychotherapy and take seriously the demands of biblical epistemology.

The older theologians spoke of the absolute, indispensable nature and logical priority of *notitia* (knowledge) and then the sequence of *assensus* (agreement) and *fiducia* (experiential trust). And this *notitia* was considered to be information conveyed to the mind and cognitively grasped. We think this a proper perspective and sequence.

Though ostensibly Oden has structured a thorough integration of theology and psychology, in reality the entire effort lacks coherance because as we have stated in another context, the absolutely indispensible connection between theology and psychology is broken; namely, the articulation of the saving deed of God is lacking which articulation or verbalization with consequent conceptualization *(notitia)* is absolutely indispensible not only to theology but also to Christian faith. Unless this ingredient is added in some way at some level, there is and will remain an unbridgeable chasm between the two disciplines on the conceptual-theoretical level, and no integration or interrelation can take place.

What we are trying to affirm here is that if the kerygma is not in some manner interleaved in the psychotherapeutic encounter, there is grave question as to how we can justify the exploration of any organic relationship between psychology and theology. Just as the community of believers can learn the dynamics of sensitivity from the psychological sciences, so too must these sciences feel the impingement of theology. And that impingement is precisely at the point of Christology. But it is an articulated Christology that brings into view the eschatological realities of judgment and grace and presses hard against humanistic psychology for an acknowledgement of universal history, the parameters of which are delineated in the cross and resurrection of Christ, and to which reality men are finally accountable.[25]

I am not suggesting that there be a return to a very simple, pastoral counseling procedure wherein there is a safe articulation of the gospel with little concern for the complexities of the human psyche. Nor am

I maintaining that this *notitia* be a part of the more scientifically structured and medically astute diagnostic and healing procedures carried on in many clinics.[26] Here the medical model is predominant and may largely preclude kerygmatic interleaving. There is an immediate need, and this need must be met by effecting behavioral changes through appropriate conditioning. It is just that there should be an underlying awareness that the remedies are palliative, that they are physical, mental, emotional, and that finally in some way and at some time the eschatological realities must break through and claim the spiritual attention and concern of all who would possess rest, peace and cosmic security.

What we are saying is that there must be stated, understood and reflected in experience the reality of the accepting God. Now if Oden would affirm this, he would be keeping intact the demands of epistemology. But when he maintains that man in the therapy session can experience the accepting God through the accepting therapist, the structure of logic and epistemology is broken and nothing more can be said. At least if articulation and conceptualization were given indispensable place in Oden's effort, we would have common ground upon which to proceed and would be keeping faith with appropriate canons of discourse. We could then discuss what must be articulated, what must be spoken, what really is the Gospel. But actually according to Oden's epistemology, not his intent, it finally makes no difference as to what the Gospel really is, for this Gospel is not necessarily articulated to the other in need.

Even according to Oden's scheme of theology, God's name need not be mentioned. Thus according to this epistemology the Christian therapist functions no differently from the non-Christian, and indeed the effectiveness of each is the same. Both exist under the reality of justification, and both have the same potential for psychic health. Indeed, according to Oden, the Christian may be able to actualize more fully an emotionally mature life because he knows of the saving deed of God; but finally this knowing is not indispensable to salvation, or to mental health. Indeed, as we have already pointed out, this would be inconsistent with his epistemology. The important matter is that on the human, secular level, apart from any conscious reference to God, the client be exposed to an environment that is accepting and that cares. When one has achieved this, he has achieved all that is necessary to become an emotionally mature person, and this seems to be the sum, substance and intent of the Gospel.

I am perplexed by an epistemological structure, such as Thomas Oden's, that can at the same time extol the need—or at least the resulting benefit—for the therapist to know of and, indeed, to conceptualize the universal accepting reality, but deny the need to know Christian truth for a redemptive experience in Christ.

THE ULTIMATE CONCERN OF THEOLOGY IS NOT TO EFFECT SERVICE TO MAN, BUT TO AFFIRM THE GLORY AND WORTH OF GOD.

In one place Oden extends great effort to show that the therapist functions not so differently from one engaged in other vocations. He affirms that wherever the neighbor is served, whether for gain or self-adulation, whether through one's ordinary vocational activity or in special deeds of care, God is being honored and his great command obliquely obeyed, even if unconsciously. "Thus," asserts Oden, "the unskilled laborer who works on the sewer line outside my office window, e.g., not only renders me and my colleagues a valuable (seemingly indispensable) service, but also thereby unknowingly honors God *through* his digging of that ditch."[27] He goes on to say, "There is a sense in which every lawful and legitimate vocation is a service to the neighbor and thus an honoring of God."[28] Then finally he asserts, in regard to the psychologist, "our point is scandalous: He may assess exorbitant fees, and be motivated by self-adulation, narcissistic pride and profit, but to the extent he is actually effective as a therapist so that through his empathy and congruence the power of neurotic compulsion is actually broken, he becomes unwittingly an oblique participant in the mediation of God's own healing love."[29]

Finally, then, Oden articulates the answer to the question as to what the difference is between psychotherapy and other vocational activities that admittedly give indispensable services to men. The answer: "Psychotherapy and it alone has made a deliberate science and a medical art out of the study of the process of mediating the very emphatic understanding and unconditional positive regard which from another perspective is the ultimate concern of theology."[30] So that which differentiates the therapist from the ditch-digger, the plumber, the secretary, the grocer, etc., is the fact that he helps and gives expert service in the area of personal and interpersonal relationships, which is for him a medical art. It is not a more exalted profession and its end, its ideal purpose or goal is no different from other vocations, i.e. to give service to the neighbor and thus, even though unknowingly and unwittingly, to give service to God. But that these statements possess ramifications of a shattering theological sort is indicated by the end statement of the above quotation where Oden asserts that empathetic understanding and unconditional positive regard from another perspective are the ultimate concerns of theology.[31] Embedded here is an interesting logical linkage. If it is true that psychotherapeutic effort is not to be qualitatively differentiated from other vocational services in that all are in the service of God as they serve man, then since the ultimate end of theology is the same as that of psychotherapy, it too merely struggles within the closed system of the unaided, natural human potentiality. Nothing really supernatural need be introduced *even by theology*. Service to man irrespective of motivational realities is that which sufficiently details the life that honors God. Since service to man without any divine reference is essential, it theoretically follows that (apart from the moral and ethical guidelines given) the conceptualized, articulated Christian underpinnings can be jettisoned. That this is so is inferred by Oden when he really questions whether as a matter of fact it is essentially important to witness, since in some sense it is better for the Christian to delight in God's redemptive impregnation of the kosmos. "The more difficult thing to express is that the Christian community celebrates the occurring love of God precisely in the midst of secular processes that are *unaware* of his unique self-disclosure."[32]

Somehow as we attempt to relate psychology and theology, we need to deal with the question as to where the purposes of both disciplines intersect or coincide. The worth and glory of God are the foremost concerns of theology. This obliquely results in peace and goodwill toward men. From a biblical perspective it is impossible to assert the reverse, that service and love to one's brother are the same as love for God, as though this were an *incognito* experience of a divine relationship.

The Christian is indeed one who rejoices in the good wherever it occurs, but his ultimate concern is not the good, the true and the beautiful that springs from a dilapidated attempt to buttress the ego by service expended for another, but good that really is good because it is consciously motivated by a divinely enabled desire to glorify God.

A FEW CLOSING REMARKS

The mood today is a mood of inclusiveness. It is the mood of existentialism. It wants a single, complete, unitary, synthetic view of man. There is strong feeling that the supernatural elements of life must really be felt, that there cannot be reliance upon an unfelt grace. In short, the superstructure must affect the substructure. The relationship between nature and grace cannot be thought of as two layers laid very carefully one on top of the other so that they interpenetrate as little as possible. This is important and helpful and quite in accord with theology at its best. Theology cannot be severed from life. This, however, does not necessarily mean that the Christian will be more altruistic, beneficent, compassionate or loving than the non-Christian. This is so because the energy of the non-Christian world to do good out of desire for self-adulation or to appease narcissistic pride is indeed immense. It does mean, however, that the motivational factor for good will be love for God, that there will be a reliance upon God, a rest in God, a trust in God, that will incline one to seek security only in Him. This in turn means that the interpenetration of grace and nature in life will be such that because grace has lifted the Christian from the hopelessness of life in this world and placed before him a very real eschatological hope, he will strive to treat others as ends in themselves and not manipulate for his own good. When one really relies upon God, then there is an intuitive bearing of fruit, indeed, the fruits of the Spirit. The superstructure of grace profoundly penetrates beneath so that there is a complete intertwining of pillar and fundament. The closed system of the natural being is broken into and, to change the metaphor, the entire loaf is leavened.

Thomas Oden has done all of us a great service by initiating a discussion on the relationship of psychotherapy and theology in such fashion as to underscore this need to observe life whollistically. We may disagree with his solutions in some very serious instances, but we must applaud the forthright manner in which he has introduced though-provoking concepts. He has stimulated reaction and helped all of us find at least temporary points of reference for the ongoing dialogue between the two disciplines. The crucial areas of interrelation have been hammered out, and though at times perhaps an excessive reaction to certain affirmations could induce a stagnation or ossification of the dialogue, yet we think this will not occur because of the essentially tentative nature of this dialogic engagement.

Much help has been rendered to me personally by the enlightening fashion in which the Rogerian triad and theology have been compared and to some degree harmonized. In the above pages I have pointed out the areas of my theological concern and the points at which I think some rather awesome problems exist, but at the same time I must also here record my great indebtedness to Thomas Oden for many insights I have received in regard to the parallel nature and intrinsic continuity of empathy, congruence and nonpossessive warmth with the Biblical doctrines of incarnation, God, revelation and love. This material has so impressed me that in many instances I have utilized it in various talks, discussions and lectures. Some of these insights are of basic programmatic significance for the continuing task of integration and enable us more clearly to understand the nature of the ministry of the church in this most crucial and strategic area of human concern.

NOTES

1. Oden, Thomas C.: *Kerygma and Counseling.* Philadelphia, Westminster, 1966. *Contemporary Theology and Psychotherapy,* Philadelphia, Westminster, 1967. *The Structure of Awareness,* Nashville, Abingdon, 1969.
2. Oden, *The Structure of Awareness, op. cit.,* p. 90.
3. *Ibid.,* p. 92.
4. *Ibid.,* p. 93.
5. *Ibid.,* p. 100.
6. *Ibid.,* p. 102.
7. *Ibid.,* p. 104.
8. That redemption is a universally applicable term is indicated when Oden discusses worldly theology and defines it as "a mode of reflection on the world that views the world exclusively as created, judged, and redeemed by God in Christ; that engages in a celebrative, Eucharistic relation with secularity on the assumption that Christ is the center of history; and that has learned to *view the whole created order strictly and consistently under the image of redemption* (this means me). Cf. Oden, *Contemporary Theology, op. cit.,* p. 44.
9. Oden, *Structure of Awareness, op. cit.,* p. 90.
10. *Ibid.,* p. 104.
11. These three qualities of experience are known as the therapeutic triad and make up the essential states of a successful therapeutic experience according to the Rogerian orientation. We use them here as the best example of what might broadly be termed the human potenial movement in psychology. They are representative of the contemporary emphasis in psychotherapy upon attitude and relationships rather than technique.

12. Oden, *Kerygma and Counseling, op. cit.,* p. 26.
13. Numerous other texts (Philippians *3:*9; Matthew *11:*27) could be cited to give basis for the above assertion because this concept of response is intertwined and implicitly or explicitly affirmed in the entirety of Scripture. In light of this it is, I think, quite improper for Oden to give an opposite meaning. He quotes such material as the parable of the unmerciful servant (Matthew *18*). It teaches that "it is impossible to be guilty before God" (Oden, *Structure of Awareness, op. cit.,* p. 117). Cosmic pardon is taught in I John *2:*1-2, "he is the expiation . . . for the sins of the whole world." II Corinthians *5:*18-19 teaches universal justification when it says, "God was in Christ reconciling the world (kosmos) to himself." The same is taught in Romans *3:*21-25, for *all* men, both Jews and Greeks, are under the power of sin. Just as all alike have sinned, so too *all are justified* by God's free grace alone. In discussing the above passage (Romans *3:*21-25), Oden quotes the section where the apostle affirms that "God designed him to be the means of expiating sin by his sacrificial death, *effective through faith* (italics Oden's). In commenting on this he states, "Far from saying that the veracity of God's deed depends upon our poor anr tenuous acceptance, this passage boldly resumes that God's justifying deed is given to all alike, without distinction, and that it becomes *effective* and *actualized* by those who appropriate it in faith" (italics mine). But this surely is not what Oden wishes to say (though forced to this by the text) because it has been his intent until now to assert that man does not actualize justication, but that he *is* justified. Indeed, it has been precisely the ontological impossibility of not being justified that he has underscored until now. In general we should have to say that there is in all three of Oden's volumes on theology and psychotherapy little serious interaction with the biblical text and where exegesis is attempted, it is hardly adequate. Compare *Contemporary Theology and Psychotherapy,* pp. 95-102 where Oden discusses Colossians *1* and Matthew *25* and sees here proof for universal justification. This paucity of exegesis is really quite surprising for in his third lecture entitled *The Human Potential and Evangelical Hope* (p. 71) he refers to Pannenberg as one who has given us "a much needed role model for building a systematic theology on the basis of solid biblical exegesis." He sharply criticizes American theology, "whose systematicians have never been conspicuously adept at biblical exegesis, and have recently assumed that they can pursue their theological inquiries quite apart from solid, historical, critical study of the Bible."
14. Of course Oden is quick to acknowledge that even in the church the "saved" experience the same psychic ambiguities as those outside the church.
15. Of course it has also been a biblical and a correct intuitive judgment of the church that if a man in faith responds to the saving act of God, his needs will be met. He then will possess a cosmic security, and out of this

personal fullness will be able to love others with the kind of organismic wholeness of which the psychologist speaks.

16. Not only does this theological proposition cause difficulties with an explicitly Rogerian form of psychotherapy but also for other psychological orientations. This is so because as Oden points out (*Kerygma and Counseling, op. cit.,* pp. 38, 39), whether Freudian, Rogerian, Adlerian, or ego psychology, "the ontology of acceptance is identifiable in any truly therapeutic process."
17. We wish to identify our thinking with Calvin when he stresses that of all the works of God man is the noblest and most admirable specimen of God's justice, wisdom and goodness. Even as a fallen being, man has an irrevocable dignity. Calvin speaks much and in a detailed manner about the horrible iniquity of man, but never to such an extent that he loses sight of the inherent dignity and glory of man which has foundation in the image of God.
18. Oden, *Structure of Awareness, op. cit.*, p. 105, 175.
19. *Ibid.,* p. 91.
20. Englewood Cliffs, P-H, 1952.
21. Calvin, John: *Institutes of the Christian Religion,* McNeil, John (Ed.), Battles, Ford, Lewis, Trans. Philadelphia, Westminster, 1960.
22. There are many verses that give support to this idea. God "sends" rain on the just and the unjust" (Matthew 5:45), and "whatsoever does not proceed from faith is sin" (Romans 14:23) are just two of the many. According to these statements any virtue engaged in apart from conscious connection with faith or trust in God must be categorized as sin.
23. However, we must agree that the Hebraic sense of "yada" [experience, obedience], is not lacking in the New Testament. Indeed, it is perhaps the predominant way in which *gnosis* is used. That knowledge in the conceptual sense is central, nonetheless, is clear even from the Christian appellation "believer."
24. Oden, Thomas C.: *Beyond Normalcy,* p. 30.
25. At this point we must note that Oden's last chapter, *The Human Potential and Evangelical Hope,* contains seminal thoughts that could go a long way toward reintroducing the supernatural dimension into the therapy situation. Here he discusses how Pannenberg's theology leads us beyond a realized eschatology that is connected with the introverted, subjectivized momentum of existentialism, to an affirmation of the need for a futuristic eschatology that impinges on the present proleptically in the Christ event. In this section he stresses the need for an apocalyptic understanding of history and rejection of the antihistorical bias in much contemporary psychotherapy. Then he asserts "Pathologies and therapies generally only have a limited view of the historical process. They look toward single events and transactions, and sometimes towards one single life history. But they need the completion of a larger historical perspective, which seeks to behold current suffering and alienation from the vantage point of its com-

pletion in the whole" (p. 79). Three critical concepts make up the uniqueness of this lecture as contrasted with the volumes above discussed: (1) Oden now sees rationalism in theology as needing more underlining, with a consequent de-emphasis on the subjectivist and personalist approach; (2) he wishes to rethink and affirm the need for a futuristic eschatology, without denying the impingement of the end time upon the now; and (3) the central place of the resurrection of Christ is affirmed because it is a prolepsis, a preactualization of the believer's resurrection experience. All three of these concepts are interrelated and could constitute a significant shift in Oden's thinking and require a rather substantial rewriting of his prior volumes. However, even in this lecture, there is no forthright statement that would indicate any substantial shift in his theology. Here we have guarded and tentative exploration.

26. We think here specifically of the Child Development Center at the Fuller Graduate School of Psychology where the most sophisticated technology is utilized to arrest various syndromes through behavior modification.
27. Oden, *Contemporary Theology and Psychotherapy, op. cit.*, p. 128.
28. *Ibid.*, p. 129
29. *Ibid.*, p. 130.
30. *Ibid.*, p. 136.
31. *Ibid.*
32. Oden, *Kerygma and Counseling, op. cit.*, p. 29.

CHAPTER 7

AN ANTHROPOLOGIST'S RESPONSE TO ODEN

CHARLES H. KRAFT

Introduction

I WOULD FIRST LIKE TO express my appreciation to the organizing committee for inviting me to participate in the Oden lectures and discussions. I found myself strongly identifying with Professor Oden in pursuit of answers to a complex of problems relating to the contemporary need within Western culture for understanding and expressions of Christianity that will be perceived as relevant. This is a major preoccupation of those of us primarily concerned with the communication of Christianity to peoples of non-Western cultures. It is, therefore, a challenge and privilege to relate some of the insights of my primary disciplinary specialty, anthropology, to certain of the concerns expressed by Professor Oden.

It is a growing impression of mine that the Western cultural philosophico-theological frame of reference in which theology as we know it has been studied, understood and proclaimed is no longer adequate as the sole framework for theological expression. There are at least two reasons for this: (1) Our understanding of life has forced us outside the limits which philosophy has traditionally defined for itself; and (2) since a primary purpose for gaining theological understanding is to develop greater effectiveness in the communication of Christian truth, such communication should be phrased and lived in terms demonstrably relevant to today's people, most of whom are not thinking in terms of a philosophico-theological frame of reference. Where the intellectual action really is for Western man today, as Oden points out, is in terms of the concepts provided by the behavioral sciences. It is, therefore, in this direction that Western theology must turn in order both to gain new understandings and to be perceived of as relevant to contemporary Western man.

It is not by accident that Oden's theological students no longer "dig the language of the church" but find it "easy and native . . . to get inside the language frame of psychology."[1] Our culture has changed, for both better and worse, and theology and the church have not kept up at the level where the eternal verities of theology either do or do not get across to people—at the level of perceived relevance, where people recognize that what is being recommended relates directly to the needs they feel and, therefore, is worth grabbing. Theology *is* relevant only in a potential sense. If it is not perceived as relevant by those who need it, all of its great potential is wasted. For relevance is as relevance is perceived.

That theology was meant to be perceived as relevant both in conceptualization and in communication I deduce from the fact that God has consistently chosen to employ man's tools, man's frame of reference, as the vehicle of his interaction with man. The theological concepts in terms of which God communicates himself are human concepts—out of the language and culture of those with whom he interacts. And this is still true even after a period of time when, in the process of God's employment of human concepts, these concepts are transformed into something greater than they were when the process first started. They are still human concepts even when employed by God. And the process of God's self-revelation in terms of these concepts is only effective when it involves both a sending God and a receiving man. In fact, communication science tells us, the way in which the receptor decodes the message is at least as important as, if not more important than, the way in which the sender encodes the message.[2] That is, what the hearer hears is at least as important as what the potential communicator says. Thus, when the would-be communicator discovers that the message is not getting through clearly to the hearers, it is incumbent upon the would-be communicator to adjust the manner in which he attempts to communicate so that the message is properly perceived at the other end.

I think, therefore, if my contention is true that modern Western man is more in tune with the conceptual framework of the behavioral sciences than with what has become a more and more isolated philosophico-historico-theological manner of conceptualization, God would have us give serious attention to the attempt to conceptualize and express theological truth in terms of the constructs of the behavioral sciences. This I regard as one of the more important of my own preoccupations, and it was with a great deal of excitement, therefore, that I read, listened to, and interacted with Professor Oden.

I must confess, though, that I often did not completely understand either what Oden was driving at or why he seemed to be having such difficulty making his point. I attribute this difficulty in understanding to one of two reasons: (1) my own unpreparedness with regard to specialized concepts and jargon of psychology; or (2) Oden's inability at points to completely free himself from either the philosophical frame of reference of Western theologizing or the intellectual faddism in which Western theology has too often indulged itself. The first of these problems (i.e. my own lack of acquaintance with psychological concepts and jargon) does not concern me greatly though I would beg the indulgence of the reader at those points when my deficits in this regard become disturbing to him. At the second point, however, I think Oden, with a more relevant behavioral scientific frame of reference already in his grasp, did not help his argument by continually reverting to formulations of theological truth expressed in terms of the philosophical frame of reference—truths that can and should be reexamined without apology in terms of the new perspective and phrased unashamedly in the language and concepts of the behavioral sciences which Oden has discovered to be the linguistic home territory of his students.[3]

But in spite of these concerns, I found myself experiencing at many points an exciting identification with either Oden's aims, his approach, or both, and it is in that spirit, perhaps to be equated with what he termed an "ecumenism of spirit," that I offer the following remarks.

The Perspective: Beyond Dialogue

I would like to comment on Professor Oden's presentations from a perspective that I would recommend for a contemporary approach to theology. It is not fully anthropological, though I consider myself an anthropologist, nor does it conform to the philosphical perspective of Western theology, though in my own nontraditional sense I would also claim to be a theologian. The perspective is, rather, one which I have tentatively labeled "Christian Ethnotheology"—a perspective which attempts to examine and interpret theological data from a point of view that not only takes God and man seriously, as traditional theology does, but also takes culture, society and psychology seriously, as the behavioral sciences do. The aim of this approach is not to compromise traditional theology, but to rechannel it in such a way as to increase its relevance to real life.

This is simply an attempt at translation, predicated on the assumption that the linguistic and conceptual frame of reference of modern, Western man has, as Oden points out, changed. This approach seeks to resist the tendency of any of the source disciplines to absolutize its own particular concerns or to press its own particular predilections if the findings of another discipline prove contradictory. It rather attempts to reconcile the contradictory findings in the conviction that in reality truth is one and that, therefore, a given insight is not to be rejected simply because it has been arrived at by psychologists rather than by theologians. I will suggest below certain modifications in our theology that I believe psychology and/or anthropology are calling us to consider.

Where, however, the insights of two or more disciplines are largely complementary rather than contradictory, Christian ethnotheology seeks to integrate both into a single frame of reference. Thus it finds it possible to accept most of what Western theology has to say about the character and power of God and the fact of man's sinfulness and to unite these insights with those of the behavioral sciences. Diagrammatically such concerns and the primary sources of information about them may be expressed as follows:

Christian ethnotheology, then, looks to traditional theology for information concerning the absolute God and transcultural truths concerning man. It specifically accepts the existence of the personal, omnipotent, omniscient, omnipresent, triune, transcendent, self-reveal-

Figure 7-1.

ing God of Judeo-Christianity. It further accepts the remainder of the cardinal doctrines of Christian theology (generalized) concerning, for example, the fallenness of man, his need for redemption, God's concern for and activity (revelational and otherwise) toward the meeting of that need, the primacy of human faith-response to God in the salvation process, the activity of the Holy Spirit in revealing (and/or illuminating) and in the leading of Christians (individuals and groups) toward spiritual maturity, the need for groupness in response to God and man (e.g. the Hebrew nation, the Church), the primacy of the Incarnation and of the Bible in God's revelation to man, etc. Christian ethnotheology does not, however, feel bound in every case to interpretations or formulations of these doctrines in purely Western cultural terms or in philosophico-theological terms.

This integrating approach looks unashamedly to disciplines such as anthropology, sociology and psychology for greater insight into matters of human limitation, perception, and potential. The fact that human perception is not only pervasively influenced by sin and finiteness (as traditional theology states) but is also pervasively influenced by culture, subculture, and individual psychology is an important contribution of the behavioral sciences to ethnotheology.

This approach, then, is an attempt to go beyond the mere dialoguing between two (or more) disciplines that Oden discusses to the more important matter of actually facilitating the communication of theological truth. That is, the real aim cannot simply be that the representatives of the disciplines talk *to each other* (and, probably, fight with each other) but, rather, that someone begins to employ whatever insights God permits us to the carrying out of the real task before us: the mediation of God to man. And for this purpose the question that Professor Oden found so difficult to deal with as to the difference (if any) between believing and unbelieving therapists fades into insignificance behind the greater concern: How do we as believers maximize the employment of whatever cultural, disciplinary, theoretical or other channels that are available to us to mediate God to men.

And at this point a distinction focused on by anthropology is helpful. The key terms ordinarily used are *form* and *function,* though the latter may be looked at from several perspectives including objective social and/or psychological function, internal perceived meaning (of individuals or groups) and the use to which a thing is put. Theology, the church, psychotherapy or an axe, for example, each have distinct forms in terms of which they may be described. These forms have various properties which may be very much in focus both in the minds of the users of the forms and of the recipients of whatever use is made of them. But this use of the forms, their meanings and the overall part they play in the cultural and psychological frame of reference of the user are labeled *function* and regarded as quite distinct from the form.

The same is true of disciplines such as theology or psychotherapy or of institutions such as the Church or the family. In setting up such disciplines or institutions there must first be a determination of the intended function and then the choice of the appropriate form(s) to employ in order to serve the desired function. In the case of an already existing form numerous questions must continually be asked. Is this form serving the desired function, or has it somehow ceased to serve that function? Has its perceived meaning changed over the years as word forms in language continually change their meanings, adding new connotations and losing some of their previous connotations? For one of the disturbing things about cultural forms (whether they be institutions or words) is that as culture changes, the meanings attached to these cultural forms change. And this gives rise to a very interesting principle of translation and "transculturation" of cultural institutions from culture to culture or from a previous to a present stage of the same culture: it is frequently necessary to alter or update the form of a concept or an institution in order to preserve its intended meaning.[4]

This, I believe, is the same truth that Oden quotes Cardinal Newman as pointing to by suggesting that "it is only through change . . . only through the living process of constant rebirth and reappropriation of the Christian tradition that the church can be the same."[5] Perhaps, though, had Newman's frame of reference had available to it such a dynamic concept of behavioral science as the form-function dichotomy, he would not have had to resort to the static photographic analogy to make his point. For such a model has to account for any new interpretation of truth in terms of a discovery of something already stored in the forms themselves rather than in terms of a dynamic, living, Spirit-led, continual updating of the cultural wineskins in recognition of the fact that the wine of Christianity refuses to stop fermenting, for God's Spirit is still guiding his Church into new truth (John 16:13). It is the behavioral science that provide us with this more dynamic model and, I suppose, it is a function of my own involvement in a behavioral science frame of reference that I had difficulty understanding why Professor Oden or anyone else could get so excited about Newman's exposition, except for the fact that he arrived at this insight so long ago.

Contemporary Western theology, set in the philosophico-theological mold, is a form of theologizing which may once have communicated the intended meaning. But it has long since ceased to do so (at least to the unbrainwashed). If we take as the true purpose of theology what Oden claims it to be: "to put to work human language in an attempt to point to how God has made himself known in history"[6] or "to make self-consistent and intelligible the life of faith in the Christian community" in ultimate accountability "to the reality to which Scripture points, viz. the judgment, grace and care of God in and for the whole of history,"[7] then certain changes are warranted. And I believe that it is somewhere in this area that the purpose of theologizing lies. For theology, like the Word and acts of God which form its data base, is intended to effectively communicate God to man. It must, therefore, give just as much attention to the adequacy with which its message is getting across to men as it gives to the accuracy of its understandings of the revelation.

With regard to psychotherapy the situation does not appear to be quite the same, at least on the surface. The language, concepts and overall frame of reference of psychotherapy are, as Oden points out, eminently "in," even with theological students, leading to the temptation on the part of both client and therapist to expect more of this form than it can deliver. Yet psychotherapy, like theology, is but a form that can be put to any of a number of functions, only some of which can be regarded as appropriate from a Christian perspective. That is, psychotherapy functioning as a substitute for God, rather than employed to mediate him, must be regarded as an inappropriate use of the psychotherapeutic form by those whose dedication to Christianity predisposes them to see all of life and labor as instruments to be employed to God's glory.

Thus, whether or not God is active in psychotherapy as administered by an unbeliever is an interesting but irrelevant question (although I would claim that the answer must be positive since—"incognito Christ" or not—my understanding of the Bible is that God is constantly at work in all that goes on in the world). The only relevant question for me in this regard is how well the primary and specific Christian objective is being realized. For the end of psychotherapy for the Christian psychotherapist cannot simply be the bringing about of psychological health—health which, in the absence of a relationship with the eternal God, can only serve to make more bearable man's ultimate estrangement from God. For health and life themselves are but means, not ends. The Christian psychotherapist must be concerned with at least specifying the Christian tenets to which he has committed himself if he is not to be acting unconcernedly vis-a-vis his client who, like he, needs ultimate as well as penultimate answers. But for this purpose Christian psychotherapists need to hear from a theology that not only talks about what Christian objectives are, and how to live and speak convincingly concerning them, but talks in such a way that what it says is perceived as relevant by Christian psychotherapists.

We can make genuine progress in relating theology and the behavioral sciences, if only we will stop regarding Christian theologizing as the sacred preserve of those still enmeshed in the philosophico-theological matrix and begin to experiment with more contemporary formulations of the same truths—not, of course, neglecting to consult with and elicit the assistance of the traditional theologian. For many of them are likewise unhappy about the current state of affairs and as anxious to experiment as any of us. But we must get beyond mere dialogue.

Mediating God's Love and Concern Across Interpersonal Boundaries

Of the numerous subjects discussed by Professor Oden I would like to comment on several which connect in some way with my own thinking relative to the development of the ethnotheological understanding of the God-man interaction briefly outlined above. These areas gravitate around the central theme of how to mediate God's love and concern across boundaries—boundaries that are ultimately interpersonal but that may intermediately be interdisciplinary, intercultural, or between God and man.

Oden speaks much about empathy [8] and likens it to the Incarnation. I accept this analogy and, in fact, seek, below, to make even greater use of the analogy that he does. Oden's struggle to understand man's ability to empathize in terms of such philosophico-theological constructs as Barth's "incognito Christ," however, bothers me since I feel that it is based on the perpetuation of an overly negative, Western theological evaluation of human potential to which such approaches as psychotherapy give the lie. Why not, rather, take seriously the observation of psychology (and of a great deal of other social scientific investigation and popular experience) that human beings possess the ability to partially enter into the experience of other human beings, and then modify our theological understanding of what constitutes basic humanness?

I would contend that since truth is one, insights into the nature and potential of human beings derived from Western psychology are at least as likely to be valid as are those of Western theology, provided only that they do not conflict with the clear teachings of Scripture (an evaluation just as difficult for theologians to make as for Spirit-led psychologists). I cannot, therefore, regard as acceptable the hypothesis (apparently held by Oren) that one of the effects of man's fall was the total eradication of the ability to empathize with other human beings and also, to an extent, with God, especially in the light of (1) the findings of the behavioral sciences, and (2) the Incarnation.

The effective crossing of barriers, however, involves a number of principles pointed to by various of the behavioral sciences as present in effective man to man relationships, and all employed by God in the Incarnation. Those to be dealt with here are (1) the necessity for the initiator of the communication/therapy process to adopt the frame of reference of the receptor/client in order to make effective contact with him; (2) the responsibility of the receptor/client to determine what his response will be; (3) the greater effectiveness of the communication/therapy process if the primary type of learning that takes place is by means of discovery on the part of the receptor/client; and (4) the need for some sort of permanent therapeutic community for the receptor/client to fit into if the process initiated between the potential helper and the helped is to continue. It is probably no longer possible to categorically identify each of these insights with but a single social science. Personally, however, I identify the first with communication theory, the second with anthropology, the third with learning or education theory and the fourth with a combination of anthropological insight and reflection on what the Church is intended to be (i.e. an ethnotheological insight).

1. The Receptor-Client's Frame of Reference

With respect to the first principle, it seems axiomatic that a potential helper must know where the one he seeks to help is before he can assist him. But he must go much further than simply knowing: he must make the effort to genuinely enter into the experience of that person—into that person's cultural and psychological frame of reference—in a way that is acceptable and intelligible to that person if the barrier(s) between them are to be effectively crossed. For as mentioned above, how the hearer understands the communication is every bit as important as and probably more important than the way the potential communicator attempts to get the message across.[9] And this principle applies whether the potential helper is a psychologist working within the same culture as his client, a missionary working across cultures, or the Son of God crossing the supraculture-culture boundary.

But this crossing has several complicating factors to it in addition to the major psychological and cultural ones on which we have already focused. Chief among these factors are those surrounding the role adopted by the potential helper and the stereotype of the helper held by the potential recipient of the help. It is at this point that I feel that Oden's remarks concerning the inadequacy of the doctor-patient model for the therapist (though not original with him) are particularly well-taken. For, it seems, it is not until the potential helper, be he therapist, missionary, preacher, professor or God, is perceived by the one in need of assistance as fully human (i.e. fully within his frame of reference) that the setting for true assistance has been attained. "You don't seem like a university professor," one young student once remarked as we chatted in my university office. This was an indication that we had reached that stage in our relationship when she could relate to me as a *human being,* not simply in terms of her stereotype of a professor.

Likewise, I read between the lines of the prologue to the First Epistle of John a large degree of amazement at a God who broke through the stereotype, turned his back on an assigned prestige and a prescribed role in favor of a position *within* the frame of reference of those, like John, whom he sought to reach—a position in which he was forced to *earn* as an intelligible equal, rather than to *demand* as a largely unintelligible superior, any respect accorded him. And this type of stereotype-smashing both communicates and genuinely helps.

No, I would say, the therapist does not help his therapy by imitating the doctor-patient model, and the best indication that we have of this is the fact that when God—who presumably could have employed any model he chose but who, we can assume, chose the most effective—sought to really help, he employed a model that put him fully within the frame of reference of those he came to assist. And the really redemptive part of the process was what those close to him got to know about him because he allowed them to get close. As John says, they saw and heard and *touched* him. Apparently the helping relationship, both as employed by God and as defined by at least some psychotherapists, is most effective when the helper gets beyond whatever the stereotype of his role is and allows himself to be perceived by the receptor as human, vulnerable, touchable and, therefore, credible.

Our understanding of why breaking out of the "doctor" or "God" stereotype facilitates genuine communication, and thus therapy, is furthered by the insights of a relatively recent field of study known as Information Theory. Two of the major tenets of this perspective are: (1) The less expected the action or word is, the greater the amount of information it conveys; and (2) the more specifically a matter is presented, the greater the amount of information conveyed. A doctor (or God) is expected to fit into a particular stereotype. If he breaks out of it, he communicates more than if he stays within it. If, in addition, he *specifically* reveals himself to the one he is trying to assist, the potential for more effective communication and assistance is greatly increased. The Gospels are full of illustrations both of Jesus' unpredictability and of the specificity of the real life stories and experiences in terms of which he communicated eternal truths.

It should, however, be noted that in order to cross a stereotype barrier the helper (be he therapist, missionary, preacher, professor, doctor or Son of God) must first have legitimate title to the prestige and other accoutrements that the stereotype assigns him. If the therapist has not come, through or apart from (and sometimes in spite of) training, both to personal wholeness and to an understanding of how to assist others toward wholeness, he will not be well-advised to risk the position to which the stereotype assigns him for the sake of a position that requires him to *earn* (rather than to demand) whatever prestige he attains in each new interpersonal encounter. For such an unqualified person or one who lacks the personal security to take such a risk, the professionalism allowed by the stereotype provides a ready refuge from reality. That is, one cannot turn his back on a position that he has not already attained. Nor could Jesus turn his back on his "Godness" for the purpose of truly identifying with man (as Philippians *2* states that he did) if he did not have a rightful claim to retain his position as God.

This insight, I believe, has something to say both to the problem that Professor Oden raised with regard to the acceptance or rejection of the doctor-patient model and also to the concommitant problem which Professor Oden labeled professionalism. It would seem advisable for prospective therapists to get professional credentials and, in the process, come to psychological wholeness themselves as a necessary precondition to their adopting an incarnational approach to therapy. Only then will they have the credentials that will enable them to effectively communicate by (at least at times) turning their backs as it were on that which they have the right to have ascribed to them by their clients—professionalism and psychological health—and genuinely sharing themselves, even their specific hangups, with their clients. Jesus gave him*self,* shared him*self* and *won* the admiration and allegiance that he had a right to demand. And this set the stage for a truly therapeutic relationship.

2. *The Responsibility of the Receptor/Client to Determine His Response*

But whether or not the openness, acceptance, and self-sharing of the communicator/therapist will result in therapy for the receptor/client is up to the receptor. If the process of barrier crossing is to be anything but an academic exercise, it must involve not only the *offer* of self on the part of the communicator but the *acceptance* of that offer on the part of the receptor. And this positive response is no less important to the therapeutic process than the offer itself, as innumerable psychologists, missionaries and others who seek to cross interpersonal barriers of one kind or another for the purpose of bringing about change can testify.

Thus it is that the best situation for effective interpersonal barrier crossing involves both the demonstration of relevance (i.e. empathy) on the part of the communicator and the feeling of need for whatever the communicator has to offer on the part of the receptor. If, then, the felt need, there already or stimulated by the communicator, is for behavioral change, the change itself can be accomplished only by the recipient of the message, never by the communicator/therapist. The anthropologist H. G. Barnett,[10] strongly influenced by a psychological perspective on social change, distinguishes at this point between the function of an advocate and that of an innovator. The advocate of change may come either from outside a culture or from within. An innovator, however—the one who actually effects the change—must always be an insider since the locus of culture change is within the culture whether the new idea is produced by the insider or suggested by an outsider. It is only the insider who can take the new idea in whole or in part and integrate it into the culture of which he is a part in such a way that from henceforth that innovation forms a part of the structure of that culture.

On the psychological level the therapist can analyze, understand, empathize, and identify with the client, but ultimately his part of the operation is limited by the fact that he cannot actually effect the required changes in the client's psychological structure. He can only advocate, recommend, persuade, and plead for such change. Any actual changes—any innovations—must be made by the client himself.

Professor Oden, of course, recognizes this fact when he states that "the real healing competencies reside in the ordinary people in the group, not in the professional therapeutic agent."[11] They could reside nowhere else. But what he seems to ignore in stressing his concept of laicization is the fact that though people possess great potential for a variety of things, much of this potential is not actualized due to lack of stimulus. Thus, though the therapeutic behavioral change takes place within the client and *may* result from response to the fairly weak stimulus of an untrained friend, in many cases the stimulus must be greater than can be provided by one whose imagination and experience have not been expanded through training. That such behavioral change is often most effectively carried out in the context of groupness is undeniable (see below), but not always—as Oden implies—because of the therapy applied by the group but, rather, because the group provides the *support* needed by the individual while he himself effects the therapy recommended to him.

Note, however, the attribution of responsibility to the receptor of the therapy—a point on which certain theological positions are weak. The behavioral sciences demonstrate, I believe, that human beings are capable (even without the Holy Spirit's aid) of positive change for the better. It is, therefore, incumbent upon us to accept this evidence and to modify our theological understandings accordingly. A theological understanding of man that regards him as essentially passive in either the determination of his response to God or in the effecting of positive behavioral change after conversion must, I believe, in response to the findings of the behavioral sciences, be abandoned. Certainly a theological view of man more consonant with that of the behavioral sciences would enable Oden to escape some of what seems to me to be unwarranted dependence upon the concept of the incognito Christ to explain man's ability to help himself.

3. The Greater Effectiveness of the Communication/Therapy Process If the Receptor/Client Learns by Discovery

The third of the principles which contributes to effective interpersonal barrier crossing relates to the means by which the potential acceptor comes to new insight and understanding. In our day of mass dissemination of predigested information it is easy to overlook the significance of the fact that the value to the learner of the material learned is usually directly proportionate to the degree of discovery-learning involved in the learning process. Predigested information has its value, but its influence on the learner is seldom either as deep or as pervasive as that of insights that the learner feels he has discovered himself.

"Keith, I didn't know you were a Christian," said a secretary to Keith Miller on one occasion as continued association with and observation of him bore fruit in discovery learning. "You are the Christ, the Son of the living God," said Peter long ago in expression of a similar discovery encouraged by a teacher who respected the intelligence of the people he sought to help to such an extent that he taught in terms of real and true-to-life analogies which challenged his hearers to *discover* truth and thus to learn how to learn. This he did in preference to a method that would have predigested truth into theological propositions and taught his hearers to depend upon their teacher rather than to think for themselves. This latter approach might well have resulted in a greater uniformity of theological understanding but only at the expense of (1) a lower degree of real learning, since the discovery element would have been lessened considerably; and (2) the demonstration of a much lower estimate of man on the part of the Son of God.

Again this insight of behavioral science and the observation that Jesus sought to stimulate discovery-learning speak eloquently for the revision upward of our theological estimate of man. Both Jesus and the behavioral sciences seem to evaluate man and his potential more highly than Western theologies do.

If a man of faith, functioning as a therapist, genuinely shares himself in his integrated Christian wholeness with his client, he invites discovery on the part of that client of the source of his integrated wholeness. Witness is, of course, far more than a verbal and cognitive thing, but person to person, boundary-crossing sharing does not stop short of verbalizing even cognitive understandings when interaction with the client and the possibility of thereby leading him to discovery invite it. But even then it is my opinion that the type of verbalization employed should, after Jesus' example, be in the kind of meaningful linguistic and conceptual forms that stimulate further inquiry and further discovery, rather than in those that hinder or squelch discovery. Often a retreat to what appears to the hearer to be authoritative theological propositions results in the latter.

The discovery approach to arriving at truth is but the focus of a longer process of experiential growth which I see as involving at least the following stages which apply whether the experience under consideration is ordinary learning or something more special like Christian conversion or one of the multitude of similar learnings in the process of psycho-spiritual maturation. Since Professor Oden specifically asked me to comment on my understanding of this process, and since I feel this tangent is germane both to the matter of interpersonal boundary crossing and to the task of Christian psychotherapy, I will briefly

Cycle I: **Stimulus → Expanding Awareness → Encounter → Discovery → Decision → Peak Experience → (Tentative) Incorporation (→)**

Cycle II: **Feedback → Adjustment → Final Incorporation**

Figure 7-2.

outline it here. The stages,[12] arranged in two cycles, are (as I see them):

The stimulus may be anything, informational or otherwise, that breaks through to the learner from any source (such as therapy, preaching, personal experience) whether external or internal to the learner. This stimulus is reflected upon by the learner, producing an expansion of his awareness and an encounter with it in which he is forced to deal either positively, negatively or evasively. Though even a rejection of the demands of the new awareness may result in discovery of some kind, acceptance of it is more likely to issue in the kind of discovery that produces a decision and a revitalized peak in the learner's experience. This is often followed by at least a tentative attempt to incorporate the discovery into his frame of reference. In this attempt at incorporation, however, he may be only partially successful, or he may only partially incorporate due to a variety of factors. These factors often center around what turns out to be an incorrect or incomplete understanding of either the nature or the implications of the stimulus. And, of course, at this point, as at any of the others in the cycle, most potentially useful learning simply gets lost because we are not ready for it, we forget it, or it is displaced by some other learning.

At this point, however, the learner, if he is fortunate, may be able to begin a second cycle by obtaining assistance in the form of sympathetic feedback, preferably from the source of the original stimulus, if that source be human (e.g. therapist, teacher, preacher, etc.), on the basis of which the learner makes whatever adjustments may be required prior to the final incorporation and integration of the learning into his life.

As I have said, such discovery-learning typically takes place at innumerable times in a person's life and he may label one of these times as his "Christian conversion." But this discovery and peak experience may not be perceived as differing in kind from other such learnings

—only in content. Nor need it always differ in intensity from other such learnings. Indeed, it is the experience of most of us who were converted early in life that there have been many peak experiences subsequent to that which we label the conversion experience that have in most ways been both more meaningful to us and more life-changing.

The aim of such a learning process, whether stimulated by therapy, teaching, interpersonal relationships, or whatever, is of course, maturity. Two points of particular importance for theology are, it seems to me: (1) that maturation whether psychological, intellectual or spiritual is always a process. One never arrives at the point where he can be regarded as mature in any but a relative sense; and (2) though many of the stages in the Christian conversion and maturation process can also be assigned distinct theological labels (such as the "wooing of the Spirit," "illumination," "redemption" and "sanctification") designed to enable us to focus on the activity of God in the process, this in no way obviates our right to describe and come to understand the whole process from the standpoint of human perception of what is transpiring. For there is no greater sanctity to one set of terms or concepts over the other even though the one approach claims to view the process from God's perspective while the other views it from man's. Indeed, it is only as we analyze the process in terms of our human perception that many of us gain the insights we need to effectively advocate such Christian innovation on the part of others across interpersonal barriers between us and them.

4. *The Need for Therapeutic Community*

But it is in the fourth point that all this really comes together—in the need for a therapeutic community. Professor Oden spoke wistfully of his desire to see the therapeutic process become self-perpetuating through the activity of those thus brought to wholeness in the society of which they are a part, and of its own search for "stability, continuity . . . a sustained viable self-understanding . . . and an understanding of the wholeness of the historical process,"[13] in short, for a permanent therapeutic community. For man is a cultural being and seems to function best when he is part of a group affording security.

When, as the Leightons,[14] and Hughes [15] and others have shown, groupness breaks down and cultural disintegration takes place, psychological problems increase. An important part, therefore, of the therapeutic process, if the wholeness is to endure, must be the building once again of integrated cultural matrices within which the processes leading toward psychological wholeness can be continued and encouraged. A proper aim of psychotherapy, then, is "ethnotherapy"—the reintegration and revitalization of groups.

This is a powerful concept, often considered to be beyond the reach of psychology proper, but again with a theological analog—the Church. Oden reaches out for such a concept when he describes clients as "searching for intimacy, for community: looking for some way of putting it all together, of integrating their lives into some cohesive unity,"[16] and suggests that what such expressions indicate is that today's clients expect through psychotherapy to be led into genuine community in which are achieved "relationships of intimacy and trust which they find missing in ordinary social transactions."[17] He suggests, however, that the expectations aroused by modern psychotherapy are often more than it can deliver in this regard and asks, "How are we to care for persons who have experienced seriously all that effective psychotherapy can do, and still find themselves struggling?"[18]

Oden's solution envisions the secularization of the therapeutic process as therapeutically-helped persons reach out to help others in need. But, I feel, following an unfortunate terminological choice on the part of certain theologians, he misses (as they missed) the real significance of what is being called for. For, if I read correctly between the lines, what the secularizing theologians are really expressing is not a need for the secularization of Christianity but for the revitalization of a disintegrating society through the deprofessionalization and de-institutionalization of Christianity. But they, like Freud, are not free to see society as a positive thing since Western culture teaches us all to regard social structuring as "repressive"[19] and so feel compelled to look elsewhere for answers. But what they (and, I believe, Oden, when he speaks of the deprofessionalization of psychotherapy) are really hinting at is the need for social revitalization. But they all, in their culturally inculcated, anti-social-structure mentality seem to miss the important point that dissatisfaction with the present type of leaders does not warrant the contention that leaders are unnecessary, much less demonsrate that things would be better without leaders. What Oden (and the secularizationists) haltingly call for, I believe, is a movement with leaders who are psychologically and spiritually healthy, aware and contemporary in their approach. Such leadership seems to be all too rarely found within traditional Christianity.

In recommending a real solution, however, we would do well to look at a very significant study by the anthropologist Anthony F. C. Wallace entitled "Revitalization Movements."[20] In this study the author points to a process of cultural change through which many societies go, and from which many societies never recover. This process involves a progression from what Wallace terms a "steady state" through a state of "individual stress" into a "period of cultural distortion." The steady state is a state of maximum integration and well-roundedness for the society and for a high percentage of the individuals within the society. It is a time in which the society (which may be a whole culture or any definable social group within it) fulfills for the vast majority of its people a maximum of the security, regulation, personal and group definition, and overall organization of life that a well functioning society is intended to provide for its participants, with a minimum of the stress that characterizes a malfunctioning society.

As time goes by and the culture changes, however, he describes a tendency toward the decreasing efficiency of "certain stress-reduction techniques" resulting in a "continuous diminution in its efficiency in satisfying needs." This Wallace terms a "period of increased individual stress" which in turn issues in a "period of cultural distortion" in which some members of the society are able to function reasonably well while others are driven to "psychodynamically regressive innovations."

> The regressive response empirically exhibits itself in increasing incidence of such things as alcoholism, extreme passivity and indolence, the development of highly ambivalent dependency relationships, intragroup violence, disregard of kinship and sexual mores, irresponsibility in public officials, states of depression and self-reproach, and probably a variety of psychosomatic and neurotic disorders. . . .
>
> In this phase, the culture is internally distorted: the elements are not harmoniously related but are mutually inconsistent and interfering. For this reason alone, stress continues to rise. "Regressive" behavior, as defined by the society, will arouse considerable guilt and hence increase stress level or at least maintain it at a high point: and the general process of piecemeal cultural substitution will multiply situations of

mutual conflict and misunderstanding, which in turn increase stress-level again.[21]

As such a process continues, driving people to the breaking point, not infrequently disillusionment, demoralization and social deterioration of the kind we see all around us set in, leading to the disintegration of the society. As happened with many of the American Indian cultures, people may lose the will to go on, and then they begin to stop reproducing themselves. Or a society may be defeated in war and its customs suppressed either through assimilation or dispersion, resulting in the extinction of the society.

Or, says Wallace, the whole process may move in another direction entirely—into a state which he calls a "period of revitalization," a state which he sees as a rather frequently occurring but "special kind of culture change phenomenon." A revitalization movement, says Wallace, "is defined as a deliberate, organized, conscious effort by members of a society to construct a more satisfying culture."[22] This reconstruction which, he claims, has happened literally thousands of times in documented human history, typically involves the "innovation of whole cultural systems, or at least substantial portions of such systems,"[23] systems which tend to be religious in character.

It is, I believe, this type of ethnotherapeutic revitalization that our rapidly distinegrating society is calling for and which the Christian Church once again brought into touch with reality and recalled to true groupness must speak. The institutional church, to the extent that it has contributed to the production of social and individual stress—including guilt, self-reproach and general lack of harmony and consistency in behavioral expectations—has, of course, joined in the intensification of the problem. But essential Christianity, whether expressed in the institutional church, in Christ-motivated psychotherapy or in any other cultural institution could again provide the kind of revitalizational spark that it produced in a similar social situation in the first century. It would need to be presented in such a way that it would answer questions that people are concerned about now, e.g. questions about interpersonal relationships and groupness and the expression and experiencing of love and acceptance.

The Christian groupness here envisioned as an agency of social revitalization is, however, more than a temporary thing—more than an occasional or weekly get-together, be it for church or for group therapy. Such groups would be characterized by the kind of permanence, mutual commitment and functional interrelationships that are the hallmark of well-integrated social groups operating in Wallace's steady state. In such a society the kind of psychotherapeutic laicization of which Oden speaks is a reality, for people meet each other's needs before they grow into psychological illnesses, within the confines of a naturally accepting, committed, caring and permanent group, and without either the expectation of ostracization if one deviates a bit from the group norms or the necessity of being forced to go to a stranger for counsel.

The kind of group here recommended is specifically not oriented toward two-sphere thinking of either the Church versus outside society or the encounter group versus outside society types. It does not indulge itself in escape from existing cultural patterns but specifically dedicates itself to the transformation of them—the kind of transformation that Christianity originally brought about when it turned certain sectors of the Greco-Roman world upside down *from within*. This kind of community takes existing cultural forms and invests them with Christian meanings, starting consciously with the transformation/revitalization process. This kind of community is already developing in many parts of the world, including Africa, where God has taught many of us what the Church was meant to be. For, returning from such exposure to true, well-integrated groupness we begin to see afresh in the pages of the New Testament a dynamic *community* experimenting with what Christian groupness meant in Greco-Roman society. We see there not a community that has already become perfect but one which, in the words of John V. Taylor, "is manifestly alive and moving and becoming. If there is a falling back into sin, it is known to friends, challenged and forgiven: the partially blind have others' eyes to guide them forward: the timid draw on others' courage, the stale are renewed by others' inspiration."[24] Such a church community we need today.

Groups such as these should probably start (as both Israel and the Church started) as small face to face groups of persons well on their way toward psychological, sociological and spiritual wholeness, people dedicated totally to carrying out both within and outside of the group the functions and meanings of Christianity in whatever cultural forms they carry over from their previous society or develop anew. Effective face to face groups in our culture seldom number more than thirty, although, over a period of years, they could get much larger through biological growth. I find the kind of self-perpetuating therapy that Oden calls for, difficult to conceive of apart from such groups. The fact that members of numerous, small, well-integrated cultures in all parts of the world betray little if any serious psychological dysfunction testifies to the fact that the result Oden seeks lies in this direction.

Such groups could orient themselves either existentially or eschatologically or with some measure of each—depending on the felt needs of the members of the given group. The point at issue should not be the fulfilling of the American cultural imperative to arrive at a single "best" outlook on life, but to select from these and perhaps other options offered by Christianity those which turn on the members of the group. Diversity in the forms of conceptualization and expression of Christianity should, however, always be allowed, accepted and encouraged in recognition of the fact that the cohesion of the group is predicated not on sameness of cultural, especially cognitive, forms but on commitment to Christian functions. For it is only the functions and meanings of life that may be labeled "Christian" since any of the forms—even religious, theological and psychotherapeutic forms—are equally operable by Christian and non-Christian. It is the use to which these forms are put, the motives of those who employ them, and the content or meaning with which they are invested that may properly be labeled Christian or non-Christian—a point that Jesus came back to again and again in his dealings with the Pharisees.

Now, since my concern here is the effective crossing of interpersonal boundaries, let me make more explicit the part that I see real, integrated, psychologically healthy groupness playing in this process:

1. The major function of such a group/church would be to provide a permanent, well-functioning sociocultural matrix in which psychological and spiritual health can be experienced and can grow.

2. The groups would need to be small enough to be face to face, integrated enough to provide a maximum of security (defined as low stress) for each member, especially at times of high vulnerability (such as at natural and unnatural times of crisis), but open enough to allow for and encourage individual and subgroup diversity in an atmosphere of trust.

3. The groups should be permanent, living together, spending considerable time together, and dynamically moving together along the path of growth to maturity.

4. But just as each member of the group should come to see himself as living not only for himself but also for the group, so each group, in fulfillment of its life with and for Christ, needs to see itself as living for the sake of other persons and groups as well as for itself. And one result of this self-image will be the dedication of such groups to the establishment of other such groups which function in such a way that they are genuinely Christian and genuinely therapeutic whether or not they bear much formal resemblance either to what we now know as churches or to any present-day therapy group.

Conclusion

In conclusion, I would like to say again that I have been greatly stimulated by the Oden lectures. I wish to express my gratitude both to Professor Oden and to the organizing committee for the opportunity to participate. I sincerely hope that the reactions that these lectures and discussions have stimulated in me are as valuable to my readers as they have been to me.

NOTES

1. Oden, Thomas C.: The Post-Therapeutic Situation, *op. cit.,* p. 45.
2. Nida, E. A.: *Message and Mission.* New York, Harper and Row, 1960, p. 34.
3. Oden, Thomas C.: The Post-Therapeutic Situation, *op. cit.,* p. 45.
4. Nida, E. A. and Taber, C. R.: *The Theory and Practice of Translation.* Leiden, Brill, 1969, p. 5.
5. Oden, Thomas C.: The Human Potential and Evangelical Hope, *op. cit.,* p. 92.
6. Oden, Thomas C.: The Post-Therapeutic Situation, *op. cit.,* p. 67.
7. Oden, Thomas C.: The Human Potential and Evangelical Hope, *op. cit.,* p. 81.
8. Oden, Thomas C.: *Kerygma and Counseling, op. cit.,* pp. 1-17.
9. Nida, *Message and Mission, op. cit.,* p. 34.
10. Barnett, H. G.: *Innovation: The Basis of Cultural Change.* New York, McGraw, 1953.
11. Oden, Thomas C.: The Diffusion of Therapeutic Agency, *op. cit.,* p. 10.
12. My scheme is an elaboration of that suggested in Tippett, A. R.: *Verdict Theology in Missionary Theory.* Lincoln, Lincoln Christian Coll P, 1969.

13. Oden, Thomas C.: The Human Potential and Evangelical Hope, *op. cit.*, p. 24.
14. Leighton, A. A.: *My Name is Legion.* New York, Basic, 1959.
15. Hughes, C. C.: *People of Cove and Woodlot.* New York, Basic, 1960.
16. Oden, Thomas C.: The Diffusion of Therapeutic Agency, *op. cit.*, p. 39.
17. *Ibid.*, pp. 38-39.
18. *Ibid.*, p. 35.
19. Oden, Thomas C.: The Post-Therapeutic Situation, *op. cit.*, p. 27.
20. Wallace, A. F. C.: Revitalization movements. *Am Anthropologist*, *58*:264-281, 1956.
21. *Ibid.*, pp. 269-270.
22. *Ibid.*, p. 265.
23. *Ibid.*, p. 264.
24. Taylor, John V.: *The Primal Vision: Christian Presence Amid African Religion.* London, SCM Pr, 1963, p. 131.

CHAPTER 8

AN OPEN LETTER TO TOM ODEN

WARREN WALKER

Dear Tom:

YOUR VISIT WITH US at the Fuller Graduate School of Psychology was stimulating and rewarding to me, and I hope it was for you, too. The most significant effect your visit had upon me was to plow up the theological part of me that has lain fallow for many years and to nurture the growth of theological strategies that interrelate with, and are in constant tension with, psychological realities. Personally I need the kind of stimulation you provided.

I am at a crossroad in my life. I have reached the place where my theological training no longer has the immediacy it had when I was in the formal ministry. Clinical psychology now occupies my time and interest. I feel as if there is some danger that I will go so far in the direction of psychology that I will become more and more out of touch with theology. I do not want to do that. I want to be continually enriched and stimulated by both disciplines, and I want to capitalize on the training and rich experiences I have had in both areas.

For the first time I am beginning to realize the extent to which psychology and theology need each other. As I see it, I was rather enamored of theology while I was in the seminary and for about half of the seven years I was in the pastorate; but theology lost its rosy hue for me when I came up against the ethereal, moralistic quality of most theology as it is dispensed by the church. Having lost my enthusiasm for theology, I have become progressively more enthusiastic about psychology during the last four years; but now I am beginning to see that there are some serious shortcomings to psychology, especially when I confront that part of it that is amoral or is caught up in a barren scientism that is lacking in the richness and depth which theology can provide.

I am excited now by the high degree to which the two disciplines can enrich each other. Each can check the excesses and strengthen the weaknesses of the other. I am becoming very much enamored with psychology and theology held in dynamic tension, each informing and explicating the other.[1] You have been instrumental in this new enthusiasm in me, and I thank you for your contribution.

Looking Back From The End Of History

One thing in this current series that turned me on was your presentation of Wolfhart Pannenberg's thoughts on history and hope. It is only in the fulfillment of universal history that we can see the meaning of human histories. The full power of creation and the Christ Event are indiscernible without a regard for the eschaton and the fulfillment of which Christ is the promise. I sense that this concept has the potential of being a fertile stimulant to psychotherapy and psychology, although I have only a faint notion of how it might be applied. I do, however, experience one area where eschatological hope is necessary in order to do effective psychotherapy. I work with some clients whose lives are terribly twisted and broken. In some, even the hope of slight improvement is unrealistic. They are so horribly deformed or so emotionally impaired that hope for ease of the anguish in their lifetime is beyond my capacity.

I find myself withdrawing from such persons. It is hard for me to look at deformity and twistedness. I do not think it would be possible for me to be present to them and allow myself to feel their feelings and think their thoughts without some perspective that is broader than the span of a lifetime. When I am working with such clients, the faith that their case histories are part of a universal history that will have a rectifying, restorative end is absolutely essential. I am sure I could not help but continue to withdraw if it were not for such a faith, and I am finding that I am able to be close to such persons when I do have such a faith.

The other application to psychology of the concept of universal history is the use of the concept to gain a more accurate perspective on history. It is helpful to project myself forward through the years to the time of the fitting end of history, and then to look back to the present and forward to the immediate future. Such a perspective certainly does have a humbling effect! You and I are only an instant in universal history, and our wildest schemes seem relatively unimportant. My experience has also been that such a perspective engenders something which I believe is powerfully therapeutic: a kind of "let-be" attitude. I tend to take myself a little too seriously and to get wrapped up in my plans and schemes and to become extremely demanding of my friends and enemies. It is, then, when this is going on, beneficial to me to take a look at myself in the broad perspective. It counters my compulsive, "clutch-at-life" tendencies. It also makes me more accepting of my neighbors. After all, they too are passing through, most of the time doing the best they can with what they have.

Recent History

Getting even a glimpse, no matter how dim, of universal history in mind provides a perspective on recent history. H. Richard Niebuhr said one time that "Every theology is an overcorrection of a previous theology."[2] I certainly view theology in that light, and I would guess that you do, too; but in the light of this, I am troubled by the way you criticize what you call "pop theologies." Take for example your statements about Harvey Cox. Don't you see his work as a needed corrective to the theologies which sacredized so much that they led to the generating of a dualism? I suppose Cox's work was an overcorrection too, but it was primarily a corrective; and it played an important role in the development of American theology.

Certainly pop theology—all theology for that matter—needs criticizing, but I have problems with the way you criticize it. When you say that a most serious drawback of the work is that it miscalculated history, you are implying that the motivation of the theologians was to calculate history. The further implication is that they developed their theories primarily to get themselves on the historical bandwagon. This is getting you pretty close to doing something I know you do not want to do—impugning their motives.

Your Own Work As Correction And Over-Correction

Taking the cue from Niebuhr, I view your work in the integration of psychology and theology as correction and overcorrection. On the one hand, I see it as corrective to the "two separate spheres" frame of reference of the confessional theologians such as Barth. On the other hand, I see your work as corrective to the pale Christology of the pastoral theologians. You have successfully shown that a theology can be post-liberal and still be highly correlated with psychotherapy. I view this as an important contribution. I believe you have successfully made the case for the need of dialogue between theologians and psychologists, and that you have shown that one does not have to sacrifice sound theology on the one hand, nor probing theories of personality and potent therapeutic learnings on the other, in order to sustain that dialogue.

But I wonder if you have not now overcorrected on the side of a rigorous theology, and I suspect that you are steering yourself away from deep, sustained dialogue with psychology. Haven't you tended to back away from that no man's land between the boundaries of psychology and religion and inadvertantly become firmly entrenched in the theological camp? You seem to view psychology from afar as if it were an alien land into which you make occasional forays but in which you do not feel as comfortable as you do in the realm of theology. I feel the need for you to emotionally pry yourself away from the seminary and to locate in a situation where you can immerse yourself more fully in psychology and psychotherapy.

Here are some of the things which occasion my remarks.

1. In this current series I do not see you as prizing any psychologist or psychotherapist in the way that you prized Carl Rogers in previous works. You demonstrate a very high regard for Wolfhart Pannenberg and Cardinal Newman, but there is no parallel appreciation of the current giants of psychology.

2. You have a jaundiced view of Freud. It is certainly true that there are destructive aspects of Freud's work, but why point them out without also alluding to his role as pioneer in the exploration of mind and feeling? Freud did not encourage people to defer from society. Rather, his operational definition of a well-functioning man was one who could love and work. His efforts were aimed at helping the individual to adapt to society. Freud left us an unweeded garden to be sure, but what a garden![3]

3. Your use of Szasz to critique psychiatry is heavy-handed. His major point, that the concept, "mental illness," is misused, is well taken; but you apparently are not aware of the fact that the medical model has long been waning in psychiatry and psychology. There are many within the field who have known the truth of what Szasz overstates for years.

If Szasz had not been so zealous and polemical with his statements, a vast majority of people in the psychiatric professions would have joined with him in correcting the misuses of the concept of mental illness; but the difference between the word "misuse" and the word "myth" is the difference between progressive cooperation on the one hand and foot-dragging on the other. If you are seeking allies in the psychiatric professions—and I think you should be if you are not—you will have to use a critical scalpel rather than the kind of two-edged sword Szasz supplies.

4. You do not demonstrate that you are in touch with the progressive aspects of psychology in the same way that you are in touch with progressive• movements in theology. For example, many of the statements you make about the inefficiency of one-to-one therapy and the laicization of therapy have long been made by people in the area of community mental health. They are, for example, thinking in terms of social learning centers rather than mental hospitals. They are also working to develop the model of "consultant" for professionals. I think it would be well for you to be in better touch with your allies in the fielr of psychology.

5. You ignore the whole behavioral school of thought. B. F. Skinner, along with other experimental psychologists, has either made an important contribution to humanity or posed a serious threat to it. Is Skinner boon or bane? More particularly, what aspects of Skinner's thought are consonant with sound theology and what parts are totally incompatible? There is a great debate raging between humanistic psychology and behaviorism and some of us are willing to look to both schools for benefits to the human enterprise. What do you think?

6. Your method is theoretical rather than research oriented. This may well be the best method for you to use to accomplish your objectives, but it would be easier for psychologists to enter into dialogue with you if you were more tuned to existing research findings on some matters you discuss, and if you would more consistently go in the direction of the concrete rather than the general.

On The Boundary

I would like to see you return to the no man's land between therapy and theology. It is precisely in those uncharted regions between the two disciplines that a faith may be developed which will inform both disciplines. As I see it, your espousal of the human potential movement and your proposal that psychotherapy be laicized is an example of your return to the boundary line. You seem, at this point, to be forsaking the safer retreats of the church and theological systems and once again declaring your convictions from the boundary line between psychotherapy and theology. That is the kind of thing

I acclaim in you. You take considerable risk in speaking up for the encounter groups in the church even at this late stage in their development. Certainly it would be safer for you to suggest that pastors develop well-proscribed bible study groups in the churches; but rather, you encourage pastors to wade out into the unknown waters of human interaction and, no matter how imperfectly, to develop groups which will encourage deep human communication. I agree with that.[4]

The Role of the Professional

I think you are on the right track when you take to task the role of the professional. Your denunciation of the idolatrous nature of professionals is indeed relevant to the practice of psychotherapy, and I hope many professionals will hear it. Part of me, the part involved with training to be a professional psychologist, wants to say, "So what! There is a lot of idolatry in the professionalism of the clergy, too"; but I would prefer to refrain from this kind of defensive reaction and to think positively with you about the role of the professional in the face of the laicization of therapy. Won't that role be unashamedly that of trainer and facilitator? The style of the most beneficial small groups I have experienced has been shaped by *professionals.* Carl Rogers, for example, has developed a whole style of facilitating small groups from which many have benefitted. Real professionals are complementary to, not antagonistic toward, the human potential movement. They are equipped to develop methods for groups and then do careful research on the causes and effects of group interactions. Such research can encourage constructive methods and help weed out destructive methods.

I have another response to your statements about the drawbacks of psychotherapy by professionals and the limitations of the one-to-one model of psychotherapy. I believe there is a sense in which the one-to-one model for therapy includes within it the basis for training and facilitation. I think it would be a mistake to be securely tied to this model, because it needs improving and expanding. I certainly want to be in on developing new and more efficient delivery systems for psychological services, including groups, training centers, preventive measures, etc.; but at the same time, I have a high regard for one-to-one therapy, primarily because it is so communicable.

I experienced a year of client-centered therapy several years ago.

It was a righly rewarding experience, and I believe it facilitated considerable change in my life-style; but what I am struck with is the degree to which I was able to informally do the work of the therapist with others while I was receiving therapy as a client. This happened long before I began formal training in therapy. I experienced nonpossessive warmth, accurate empathy, and congruence in the therapist, and I was increasingly enabled to be warmly and genuinely empathic to others. There is a trainer-trainee factor in client-centered therapy that has a wider scope than is apparent.

What I am saying is that for me, the therapist at once brought therapy and facilitated the therapeutic resources within me. Again, we are confronted with the facilitator model for the therapist. It seems to me that this is exactly the model from which churches could most benefit. Laymen in the church are often better at facilitating the human resources of fellow parishioners than their pastors. I demythologize the feeding of the five thousand,[5] and see it as a paradigm for the therapeutic process in the church. The people gather together and share what they have. The most significant thing they can give to each other is not fishes and loaves, but their caring, understanding, genuine selves. Christ is the one who symbolically facilitates the deep sharing in the church. He induces people to give of each other to each other.

I wish more congregations could be caring communities. A step in that direction is to aid pastors to develop their skills more fully. Seminaries should extensively train students in group facilitation; but behavioral science skills can also be brought to bear even more directly upon the churches. One of the things which we are developing at the Fuller Graduate School of Psychology is a church consultation service. Consultation takes many forms, but one of the more interesting to me is that of training church people in the therapeutic triad, i.e. empathy, warmth, and congruence. It may be possible for trained therapists to bring forth therapeutic qualities in members of the congregation so that they not only are able to bear one another's burdens, but actually do the real work of the church—increase among men the love of God and neighbor.

Beyond Therapy

Some of the work to be done after therapy (actually before, during, and after therapy) will be to explore the great questions of life, "death, meaning, honesty, spiritual awareness, and moral courage."[6] You seem to be puzzled that people are still asking these questions after therapy. I am not. Nor am I puzzled that people ask these questions after theological training or after church membership. These questions of course, are part of the never-ending human quest for the meaning of existence. I do not know any therapist who believes differently. There is no need to view therapy with alarm because these questions are not answered in therapy. It would be better to simply join therapists and psychologists in doing a better job of asking the right questions in depth.

I see the major contribution of your work at this point as the aiding of the psychologist and the psychotherapist in formulating an ontology for dealing with the great life questions; not necessarily to answer them but to explicate them and to challenge easy answers. Perhaps you could do so in such a way that even those not in the Christian community could benefit. There is a great need in this area. I hope you will have a hearing with the psychologist or psychotherapist who is not self-consciously Christian, but who is certainly responsive to the great human questions. I think you may have jeopardized that hearing by your failure to be firmly in touch with psychology in the ways that I have suggested earlier; but I hope you will return to the task of hammering out a psychological ontology.

The Indisputable Need For Theology In Psychotherapy

My own experience is that a sound theology is an absolute need of persons in trouble. When I was the pastor of a church, the daughter of one of my parishioners died at the very time when the parents were experiencing radical theological doubt and distrust. What struck me was that, although their disbelief took away from them the possibility of doing sound theology, they nevertheless did theology, twisted and destructive as it was. They began entertaining the notion that fate was punishing them for previous mistakes, and that they should never get involved with anyone because fate would likely take another loved one from them. My experience here was that sound theology was absolutely essential in this crisis, and that twisted theology rushed into the vacuum left by doubt. What this means to me is that theology is not a luxury reserved only for those persons who happen to be within the Christian tradition but a necessity for every person who lives and attempts to understand what is happening to him. It is this kind of rock-bottom, nitty-gritty situation to which theology must be addressed.

Behavior Modification

There is another task to which I would like to challenge you, and that has to do with taking behaviorism seriously. My first impulse, perhaps like yours, is to reject it out of hand; but I find as I read Skinner's book, *Beyond Freedom and Dignity,*[7] that much of what he has to say about the misuse of the concepts of freedom and dignity make considerable sense. Also, it is hard to find fault with his case against punishment. Skinner is either a worthy foe or a significant ally. Are there not some theological correlates to Skinner's theories?

Some of these correlates may be found in your book, *Kerygma and Counseling.*[8] There you demonstrate the importance of Charles Hartshorne's analogical concepts of God. You seem to buy into the social analogy and the mind-body analogy, but you betray grave reservations about the impersonal-artist analogy. I think this analogy of the work of God in the world might be dusted off to provide some useful parallels to the work and thought of Skinner. Is it too audacious to suggest that God's creating process is partially a shaping process? If so, perhaps operant conditioning is a way of understanding and defining that process. I certainly can see a lot of theological problems with this kind of brainstorming, and I only suggest it as a tentative possibility hoping that you will be inclined to apply your theological skills to behavioral theory.

Therapy As Preparatio Evangelica

Finally Tom, I think you have neglected to explicate something which may be implicit in your statements, but which needs to be clearly delineated: that in an important sense, the Gospel is exactly what lies beyond therapy. I resonate to your statements regarding the incognito Christ in the therapeutic process and ontic acceptance as the underlying assumption of therapy, but I think there is a sense in which therapy is preparation for the Gospel rather than any sort of corollary to it. In the therapeutic process, the individual is helped to peel back the layers of inauthentic questions and cheap answers, and to ask the deep questions of meaning, morality, and love. He is sometimes enabled in therapy to give up the little gods that fail and to experience a confrontation with the slayer of the gods. It is precisely at this point when he is jarred loose from his idolatries and has gotten in touch with the abundant loving, trusting process within himself that he may be able to perceive revelation. Precisely then is he able to respond within his total person to the Christ Event.[9]

Gratefully yours,
Warren Walker

NOTES

1. John Davis, a fellow student at the Fuller Graduate School of Psychology, has helped me to formulate my thinking on this point. An example of the way in which psychotherapy can correct the ethereal nature of some theology comes out of my seminary clinical training. In my report of an interview with a patient. I wrote, "This man needs to know that God loves him." The training phychiatrist wrote in the margin, "Right Walker, let God love the S.O.B., because you sure can't!"
2. Remarks made at Harvard Divinity School, April, 1963.
3. Transactional analysis, for which Dr. Oden has affinity, is squarely based upon Freudian theory. Parent, Adult, and Child seem parallel to Superego, Ego, and Id, despite Berne's statements to the contrary.
4. One evening at a sensory awareness group meeting, we were tenderly and gently touching a partner's face with our hands. I happened to be assigned to a person for whom I had some bitter feelings. Experiencing her uniqueness and responding to her total personality as symbolized by her face, I remember suddenly thinking, "No one who has held a person's head in his hands could ever do violence to a person." I hope everyone, especially those in the church, can at least have the opportunity to experience something like this.
5. Mark *6:*35-44.
6. Oden, Thomas C.: The Diffusion of Therapeutic Agency, *op. cit.*, p. 35.
7. Skinner, B. F.: *Beyond Freedom and Dignity*. New York, Knopf, 1971.
8. Oden, Thomas C.: *Kerygma and Counseling, op. cit.*, pp. 139-141.
9. I owe thanks here to Richard Sharpe, a student at the Fuller Graduate School of Psychology, for helping me to develop this point.

CHAPTER 9

RESPONSE

THOMAS C. ODEN

WITHOUT SLACKING THE critical thrust of our conversation, it seems fitting for me to note personally that I cannot remember a lecture and discussion series in which I have participated in which I have been more affirmed and felt that I was being taken more seriously as a scholar, churchman and human being than during these discussions at The Graduate School of Psychology at Fuller. I am grateful that there has been a serious and sustained attempt to understand the intention of my project, a patience with its ambiguities, and a sense of shared responsibility and joy in beholding the profundity of the task of dialogue between therapy and theology.

So abundant and varied are the issues which have been afforded me by my partners in dialogue that I am compelled to be extremely selective as I develop those which seem most urgent or promising for further pursuit. Rather than attempting to address the whole spectrum of questions and criticisms addressed to me by my respondents, it seems wiser to concentrate on a few and try to do justice to them. So I shall work deliberately on three themes: (1) a response to those who feel that the empirical evidence for my argument on laicization is weak; (2) a more discriminating clarification of types of therapeutic agency and their relation to degrees of professionalization; and (3) some unsolved theological dilemmas which have been noted.

I. THE WEIGHT OF EVIDENCE

In responding to critics of my laicization thesis, it is evident that I need to spell out more clearly the body of supportive empirical evidence which lies behind my conclusion that nonprofessionals can be effective therapeutic agents, and that there is no direct and reliable correlation between professionalization and effective therapeutic agency. A principal criticism, noted by both clinicians Warren and Walker, is that I have neglected research findings which would bear on my discussion. Thus I shall begin by reviewing in this section some of the hard evidence which does stand in support of hypotheses and conclusions, which otherwise might appear to be flimsy, improbable, or isolated personal opinion.

ON AVERAGE PSYCHOTHERAPY

Is average psychotherapeutic care effective so as to warrant further professionalization? A growing body of evidence makes a positive answer increasingly dubious.

a) As early as 1952 H. J. Eysenck [1] set forth evidence that professional psychotherapy on the average is no better than spontaneous remission rates or rates of improvement achieved simply through the passing of time. Although Eysenck's assumptions and definitions were questioned, a great deal of evidence has been amassed since that time, the overall result of which is reported by Truax and Carkhuff as follows: "After a careful review of the relevant research literature, it now appears that Eysenck was essentially correct in saying that *average* counseling and psychotherapy as currently practiced do not result in average client improvement greater than that observed in clients who receive no special counseling or psychotherapeutic treatment."[2]

b) In *Persuasion and Healing,* Jerome Frank noted that the improvement rate for patients receiving psychotherapy was about two thirds, but that same rate was found for comparable clients who had no psychotherapy.[3]

c) The study by Barron and Leary in 1955 of 150 neurotic patients who were tested before and after treatment on Minnesota Multiphasic Personality Inventory neuroticism scales showed no significant differences in improvement among patients who had received group psychotherapy, individual psychotherapy and no psychotherapy.[4] "Significant" in this instance means that "there is statistical reason to believe that differences are large enough so that there are less than five chances out of 100 that chance alone could explain the differences."[5]

d) Of the 650 delinquency-prone boys studied by Teuber and Powers, the evidence indicated a slightly more positive improvement rate among boys who had not received counseling than among boys who had received psychoanalytically-oriented and client-centered counseling.[6]

e) Similarly in separate studies comparing spontaneous remission rates to rates of improvement under psychotherapy by Cartwright and Vogel,[7] Mink and Isaacson,[8] Gliedman, Nash, Imber, Stone and Frank,[9] Walker and Kelley,[10] Barendregt,[11] no overall differences in outcome were reported in favor of those receiving psychotherapy.

f) Anker and Walsh studied improvement rates among hospitalized patients receiving group therapy as opposed to drama group activity and found consistently greater improvement for the drama activity group than for those in therapy.[12]

g) In 1959, Frank, Gliedman, Imber, Stone and Nash compared minimal-contact patients with longer term patients who received psychotherapy for more than six months. The improvement rate favored the dropouts. On the discomfort scale measure, no significant differences were found between minimal-contact patients and those in long-term therapy.[13]

h) Five-year follow-up on spontaneous recovery rates shows that 80 per cent experience remission of symptoms without treatment.[14] Eysenck's study in 1960 of 7,000 cases treated by psychotherapy showed a 64 per cent improvement rate,[15] and similar studies by Levitt showed a 67 per cent improvement rate at termination of therapy in 3,399 cases.[16]

These data have been very difficult for psychotherapeutic professionals to accept. Wishing it were not so, many therapists have chosen either to ignore the findings or to assume that there must be something wrong with them. But a growing mass of evidence seems to support the conclusion that "the average effects of therapeutic intervention (with the average therapist or counselor) are approximately equivalent to the random effects of normal living without treatment (control groups, or what has been traditionally labeled 'spontaneous' improvement)."[17]

These negative findings are made even more ironic by additional evidence that counselors in the above studies often considered their therapeutic relationships highly effective. Despite all of this evidence, there seems to be a continuing momentum of confidence among therapists that any therapy will be better than none, and that their therapeutic skills are worthy of the patient's trust, fees and the whole apparatus of professional confidence.

Furthermore, on the basis of extensive studies by Levitt (1957), Eysenck (1960), Mink and Isaacson (1959) and others, "*the evidence now available suggests that, on the average, psychotherapy may be harmful as often as helpful, with an average effect comparable to receiving no help.*"[18]

Thus my questions about the efficacy of psychotherapeutic professionalization are not without a substantial body of evidence which displays statistical reliability and care in data-gathering. The implausibility of psychotherapeutic professionalization under the medical model is no longer merely a matter of speculation, dissatisfied clients or casual ivory-towered debate. Rather it is a serious moral question which each professional psychotherapist must ask for himself and with his colleagues: Do my own vested interests and professional identity prevent me from seeing the limitations of psychotherapy, and buoy continued confidence in a dubious way in my own capacity to be of help to others, and in the legitimizing of my professional identity?

SOME NEEDED ACKNOWLEDGEMENTS

It has never been my intent to undermine well-functioning psychotherapeutic professional care. For the record, let me repeat, and reaffirm, several positions which seem consistent with the direction of this inquiry:

1) I have supported, and continue to support, fee-basis counseling. (I have even gone so far as to attempt to provide some theological legitimation for fee-basis counseling!)

(2) I have supported, and continue to support, structures for professional psychotherapeutic training which would in fact improve therapeutic agency.

(3) Far from believing that the psychotherapeutic tradition is senile or dying (as Fromm and others have worried), I have stated my opinion that the therapeutic tradition is currently in a very vital moment of transmutation and will continue to have significant impacts upon the formation of the human spirit.

(4) I have no doubt that severely disturbed persons may need professional psychotherapeutic care, including hospitalization, under

the medical model, even though that model is counterproductive to therapeutic agency in most relationships.

What then do I want from psychotherapeutic professionals (if not the end of professionalization)?

(1) Less *hubris*, (pride) more self-critical awareness of their limitations.

(2) A clear acknowledgement that therapeutic agency (hinging on empathic capacity, nonpossessive warmth and genuineness) is found among many nonprofessionals, and may not always be found in professionals.

(3) A clear rejection of the medical model as an ideological frame of reference and sociological prototype for behavior-change. This means the end of talk about patients, mental illness, treatment, and much of the objectivized and pejorative language of pathology, as well as an elitist and gnostic conception of knowledge, and much, much more.

(4) Less defensiveness about lay or paraprofessional therapeutic agents or counselor's aids whose track record is not all that bad, and intensive group processes where the laicization of therapy is currently occurring and pursuing significant experimentation which deserves the support, encouragement and guidance of psychotherapeutic professionals.

THE DEFINITION OF THERAPEUTIC AGENCY

Further, it remains desirable that we present solid evidence which will support not only these negative, but positive, aspects of the laicization proposal, viz. that (1) the ingredients of therapeutic agency are definable, and (2) that therapeutic agency can be achieved and embodied by nonprofessionals and paraprofessionals competent in the therapeutic triad but with little formal knowledge of developmental psychology or psychotherapy.

A substantial body of evidence exists to indicate that.

> Therapists or counselors who are accurately empathetic, nonpossessively warm in attitude, and genuine are indeed effective; the greater the degree to which these elements were present in the therapeutic encounter, the greater was the resulting constructive personality change. These findings seem to hold for a wide variety of therapists and counselors *regardless of their training or theoretic orientation;* and for a wide variety of

> clients or patients, including college underachievers, juvenile delinquents, hospitalized schizophrenics, college counselors, mild to severe outpatient neurotics, and the mixed variety of hospitalized patients. Further, the evidence suggests that these findings hold in a variety of therapeutic contexts and in both individual and group psychotherapy or counseling.[19]

Among the supportive evidence for this conclusion are the following studies:

1. The Johns Hopkins study done by Whitehorn and Betz (1963) compared seven psychiatrists whose schizophrenic patient improvement rate was 75 per cent with seven other psychiatrists of similar training whose improvement rate was only 27 per cent. The differences in therapists were not in theoretical orientation or training but in attitudinal and relational styles. Betz's descriptions of successful therapists were consistent with the therapeutic triad which appears so frequently in most major theoretical formulations of effective therapy: empathic understanding, nonpossessive warmth and authenticity.[20]

2. The Wisconsin Schizophrenic Project, reported in 1963 by Truax, and others,[21] compared patients receiving individual psychotherapy with carefully matched control patients in an effort to analyze the effects of high conditions of the therapeutic triad versus low conditions upon outcome:

> Patients had been randomly assigned to either therapy or control conditions within the matched pairs, and a complete battery of psychological tests was given initially and later. The therapy patients received a minimum of thirty individual therapy sessions and a maximum of 280 sessions throughout a three-and-one-half-year period. . . . Samples of tape-recorded interviews were independently rated for therapist accurate empathy, nonpossessive warmth, and genuineness. . . . The findings indicated that patients receiving high levels of the three conditions showed an overall gain in psychological functioning; but those patients who received rather low levels showed a *loss* in psychological functioning.[22]

3. Using the Barron Ego Strength Scale, developed to predict psychotherapeutic outcome, Wargo (1962), in a related study, concluded that the therapists who present to the client low degrees of unconditional positive regard, accurate empathy, and congruence may be facilitating loss of ego strength.[23] "It may well be, then, that no therapy is better than low conditions therapy, but that high conditions

constitute the essential ingredient in constructive personality change during therapy."[24]

4. Gendlin and Geist (1962), also related to the Wisconsin project, concluded that extremely low levels of genuineness or congruence in the therapist tend to invalidate the effects of higher levels of empathy and positive regard offered by him.[25]

5. Using the Accurate Empathy scale, Bergin and Solomon (1963) showed that accurate empathy, as rated from tape-recorded therapy, was significantly related to ability to produce constructive therapeutic outcomes.[26]

6. Studying a group of psychoanalytically oriented therapists, Strupp, Wallach, Wogan and Jenkins (1963) showed substantial correlations between therapists' ratings of successful therapy and client ratings of therapist's feelings of nonpossessive warmth and liking for the client.[27]

Among other studies which further substantiate the evidence for the central therapeutic ingredients as empathy, nonpossessive warmth and genuineness are Halkides (1958), Truax and Carkhuff (1963), Barrett-Lennard (1962), Combs and Soper (1963), Cartwright and Lerner (1963), and the outpatient study at Johns Hopkins by Truax and others (1966).[28] Thus when I speak of therapeutic agency in my laicization thesis, clearly what I mean is a high level of conditions of the above therapeutic triad. I am not speaking merely of my opinion of therapeutic agency, but of a substantial body of evidence concerning the nature of therapeutic agency.

TRAINING IN THERAPEUTIC AGENCY

The next question is: Can such therapeutic agency be taught to nonprofessionals? Is it possible to develop a training program wherein persons with little or no background of training in developmental psychology and psychotherapy can be rather quickly taught the elements of the therapeutic triad in such a way as to enable them to function as therapeutic agents in a ratable and scientifically demonstrable manner? Does such a body of evidence that such training is possible exist?

Many psychotherapeutic training supervisors are convinced that therapists are born, not made. I myself am tempted to agree with this view, as one who has taught pastoral care and counseling, and deliberately attempted to train ministers in the therapeutic triad. At least it is clear to me on the basis of that experience that some individuals are much less trainable than others, and some only need more accurate conceptualities for doing what they seem to do naturally. A body of evidence is beginning to build, however, to indicate that "even potential therapists who, in their usual human encounters, are nonempathic, hostile and defensive could learn to communicate these [therapeutic triad] ingredients in the specific encounter of the therapy hour."[29]

Again it is Carkhuff and Truax who have done the pioneering work in this area. They have designed research instruments to measure the therapeutic triad and have fashioned a training program seeking to improve the conditions of therapeutic agency in both professionals and lay persons integrating both didactic and experiential learning in the training process. The results were remarkable. They compared levels of therapeutic agency of lay persons to postgraduate clinical psychology trainees, and then to a group of experienced therapists such as Drs. Carl Whitaker, Rollo May, Albert Ellis, and Carl Rogers:

> After slightly less than 100 hours of training . . . the levels of accurate empathy communicated to patients were not significantly different between the three groups in terms of the level of nonpossessive warmth. . . . With respect to therapist's genuineness, however, the experienced therapists showed a significantly higher level of genuineness or self-congruence than did the lay trainees. . . . The level of patient depth of self-exploration achieved in therapy did not differ significantly between the three groups. . . . These data suggest that the ingredients can be learned, by both professional and nonprofessional persons.[30]

In a related study, lay persons who had received the same 100-hour training were assigned as group therapists to eight groups of hospitalized patients. Their tape-recorded levels of effectiveness in the therapeutic triad were rated and subsequently compared with levels of effectiveness of nineteen experienced therapists in group therapy with hospitalized patients. "Thirty-one per cent of the sample of experienced professional therapists provided levels of accurate empathy and nonpossessive warmth *at or below the level* provided by the beginning lay therapists."[31]

Summarizing, Truax and Carkhuff say: "The currently available evidence, then, suggests that these ingredients of accurate empathy, nonpossessive warmth and therapist genuineness are 'teachable'; and that even nonprofessional persons lacking expert knowledge of psychopathology and personality dynamics can, under supervision, produce positive changes in chronic hospitalized patient populations."[32]

THERAPEUTIC AGENCY WITHOUT TRAINING AND SUPERVISION

The remaining question is: Can lay agents function with high levels of the therapeutic triad *without* such training and *without* such supervision? I would hypothesize that *there exists among certain persons in the lay population a high level of capacity for therapeutic agency prior to training.* This is not to deny the importance of training, but rather to point to propensities prior to training which in fact do not depend upon training for their therapeutic efficacy.

Such a hypothesis has not been investigated by the psychotherapeutic researchers. Despite the immense contributions of Truax and Carkhuff and other studies mentioned above, there seems to be no willingness yet to take the inquiry an obvious next step further and ask to what extent therapeutic agency is present in lay populations completely apart from professional training and supervision. When I ask myself the reason why there is resistance to such an inquiry, the most apparent answer is that it would challenge the very foundations of the professionalization process in psychotherapy, which has deeply rooted vested interests against the entire line of inquiry which we are here pursuing.

Frankly, it seems to me to be an incredibly proud and self-serving assumption to presuppose, as I think current psychotherapeutic research has been presupposing, that therapeutic agency can only occur when channeled through their own approved tracks of professional training and supervision. The most obvious historical similarity to this assumption is the medieval conception of sacramental grace as being available only through the approved channels of clerical-ecclesial professionalization. It was against just that sort of *hubris* that the laicization process developed, first in the Reformation and finally more powerfully in Protestant and Jewish pietism.

My critics have correctly pointed out that if psychotherapy is counterproductive, then that is contradictory to what I say about it being desirable for lay people to be involved in therapeutic agency.

The distinction we need to clarify is that between *average* psychotherapy, which is apparently not very helpful, and excellent therapeutic agency. I do not recommend average psychotherapy to laymen for laicization. But what I see happening are lay therapeutic agents functioning quite effectively (i.e. above average) without credentials, and quite apart from the apparatus of professionalization.

In the 1970 *Annual Review of Psychology* on "Psychotherapeutic Processes," Gendlin and Rychlack wisely summarize:

> Many other studies also report descriptively, or by today's crude criteria, that undergraduates, mothers, and others practice and evaluate as well as highly trained psychologists. Are subprofessionals and new 'mental health professionals' a threat to the statue of fully educated psychotherapists? We think not. Supervisory and intellectual leadership will be needed more than ever, and in more places and programs. . . . The findings which imply that people do as well or better at practice after a short training period as university trained psychologists could be read to mean that it is eminently possible and economical to give future researchers and thinkers on human behavior the necessary practice training which would enable them to be part of the new field of mental health practice.[33]

CAN PSYCHIATRIC PATIENTS BE THERAPEUTIC AGENTS?

Among the most intense research projects which have been carried out over a period of several years on leaderless or autonomous groups is that which has been done at the Human Relations Training Laboratory of the Veterans Administration Hospital at Houston, a program in which I was privileged to work as an associate of Philip G. Hanson and Walter M. O'Connell in 1970. Since I spent several months as a research associate in the laboratory process, coming away strongly persuaded of the effectiveness of lay therapeutic agency, I will describe it in some detail.

The Human Relations Training Laboratory is an open psychiatric ward with a social-psychological orientation. Patients are divided into small groups called "Development Groups" (D Groups) of seven to nine men each. These groups are self-directed and meet daily. No staff member is present while the D Group is in session. In lieu of a professional therapist, self-administered behavioral rating scales, check lists and attitude measures are provided to guide the group's attention to processes and behaviors that are important to individual learning and group development. Each group is free to develop its own agenda and procedures for handling the agenda. Through instruction, many functions of a trainer or therapist are taken over by the group members themselves. These activities may include interpreting ongoing behavior, gate-keeping, reflection of feeling, process analysis, problem solving, experimentation, feedback, and generalization.[34]

> Participants are told to take all problems to the group, and individual consultations are discouraged without the group's consent. While each group has a staff consultant, his assistance is sought only after the group has thoroughly exhausted its own resources and has reached an impasse. The consultant may attend the group session only by invitation, and that invitation has to be by group consensus. Many groups complete their program without having once called the consultant in for help.[35]

All this occurs in a psychiatric ward, where one third of the patients have been diagnosed as "psychotic," and where some are still experiencing "psychotic symptoms."

In group psychotherapy the therapist

> uses his professional experience and educational background in making group members aware of what is happening in the group. He interprets behavior, brings to the attention of the group discrepancies between what is being said and what is being felt, and describes group dynamics in terms of the kinds of interactions taking place. The autonomous groups in the laboratory learn to do these things for themselves through lectures, exercises, and problems . . . role playing and group training exercises. The autonomous groups are given the tools to analyze their own groups and function as self-facilitators. In this way the group members themselves become sensitive to group process rather than relying on the trainer for this important function.

The problem of professionalism in the therapeutic relationship is centered on the tendency and temptation to create and nurture dependency:

> The autonomous group, on the other hand, is thrown upon its own resources. The group must organize and set up its own procedures. . . . Instruments cannot give approval or response as can a trainer or a therapist. In addition, the group gains benefit from inventing and discovering ways to direct its own operations.

It is less accurate to term the Houston project an experiment in leaderless groups than in *autonomous* group processes, where the actual process of therapeutic agency is turned over to the patients themselves. That does not deny the function of professional therapists, but here the therapists are seen essentially as designers of an instrumented process and as special consultants when the process (only rarely) breaks down. It can be described in part as an automated, lay therapeutic self-help program:

> Through the instruments, participants can rate any number of variables concerning their group in terms of the type of group structure . . . group atmosphere, degree of leveling and participation of each participant, group accomplishment, and cohesion. Instruments are constructed which also provide quantitative means of providing individual feedback in a direct, face-to-face relationship with one another.

The results of this laboratory process have been reported in several studies. After twelve years of operation, we have a right to ask how effective is such a program in which therapeutic agency is dispersed in the patient population, as opposed to a program which professionals directly control and for which they provide surveillance. Studies in 1963, 1965 and subsequently have indicated that participants were less tense, less depressed, less dependent and more willing to assume responsibility.[36] A nine-month, follow-up study comparing participants from the laboratory with conventional group therapy programs showed that more were employed at the time of follow-up with a longer mean time of employment. Using the MMPI, the Symptom Rating Scale, a Discomfort-Relief Quotient and several other instruments, participants showed "significantly less symptomatic behavior and fewer feelings of discomfort or tension after the laboratory."[37]

The Houston laboratory provides some of our most stunning and significant evidence for the viability of lay therapeutic agency. Even psychiatric patients, when properly trained and supervised, can be effective therapeutic agents.

II. TYPES OF THERAPEUTIC AGENCY VIS-A-VIS DEGREES OF PROFESSIONALIZATION

My views on the laicization of therapy have been criticized on several grounds: (a) as unappreciative of professional training;

(b) as dangerous; (c) as too optimistic about lay therapeutic agency; and (d) as unaware of current work being done in counselor aid training. In order to answer some of these objections, it has been necessary for me to search for more appropriate distinctions between higher and lower levels of therapeutic agency in relation to degrees of professionalization.

NINE CATEGORIES

The following list may help us to be more discriminating in distinguishing between levels of therapeutic agency and levels of professionalization. In a circle which includes the whole population, there are at least nine distinguishable categories:

1. lay therapeutic agents with high levels of warmth, empathy and genuineness
2. paraprofessional, lay counselor aids with high levels of the triad
3. professional psychotherapists with similar high levels
4. laymen with moderate levels of warmth, empathy and genuineness
5. paraprofessionals who function with average levels of the triad
6. professional therapists with only moderate levels
7. lay persons with low levels of warmth, empathy and genuineness
8. paraprofessionals with low conditions of therapeutic agency
9. professionals with similarly low conditions.

The vertically slanted columns in Figure 9-1 indicate degrees of training in psychotherapy:

(1) lay—untrained nonprofessionals without special educational or clinical credentials in psychology or psychotherapy

(2) paraprofessionals—partially trained lay counselor aids who have received some intentional, though limited, training to help them improve their skills as therapeutic agents

(3) professional psychotherapists—including clinical psychologists, psychiatrists, psychoanalysts and counselors who are fully licensed with recognized educational credentials and clinical training to receive fees for their services.

Our categorization follows studies by Rogers, Carkhuff, Truax and others, showing that when an incongruent person comes into significant contact with another person who is empathetic, congruent and positively regarding (the therapeutic triad), and when the other is

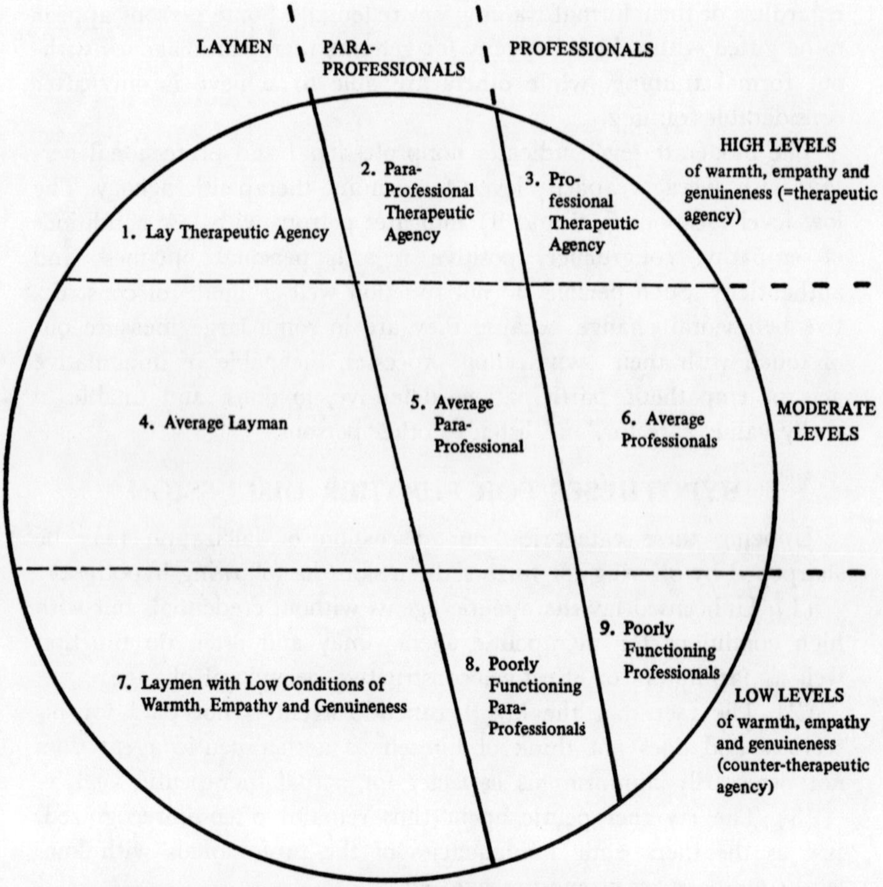

TYPES OF THERAPEUTIC AGENCY VIS-A-VIS PROFESSIONALIZATION

Figure 9-1.

perceived as such, then constructive change will invariably occur. Without "necessary *and sufficient* conditions" of therapeutic agency, constructive behavioral change cannot be expected to occur.[38]

Accordingly, the horizontal strata of Figure 9-1 distinguish three levels of therapeutic agency. The highest level, comprising sections 1, 2 and 3, indicates persons, whether lay, paraprofessional or professional, who function with high levels of personal openness, congruence, positive regard, empathy, and personal authenticity. Such persons have a high capacity for constructive behavioral change regardless of their formal training or credentials. Some persons appear to be gifted with a high capacity for genuine therapeutic agency without formal training, while others are able to achieve it only after considerable training.

The moderate level indicates nonprofessional and professional persons with average capacity levels for genuine therapeutic agency. The low level (sections 7, 8 and 9) indicates persons with low conditions of empathy, congruence, positive regard, personal openness and authenticity. Such persons do not function well as agents of constructive behavioral change because they are in some large measure out of touch with their own feeling processes, incapable of imaginative acts of empathetic participation, defensive, anxious, and unable to really value, care for, and listen to other persons.

HYPOTHESES FOR FURTHER DISCUSSION

Utilizing these categories, our discussion of laicization may be sharpened by offering for further discussion the following hypotheses:

(1) Unlicensed lay therapeutic agents without credentials but with high conditions for therapeutic agency may and often do function well as facilitators of effective, constructive behavioral change.

(2) The fact that the lay therapeutic agent is not paid for his services and does not think of himself as a therapeutic agent does not necessarily diminish his capacity for actual therapeutic agency.

(3) The lay therapeutic agent thus remains often unrecognized, just as the therapeutic inadequacies of the professionals with low-level conditions are often unrecognized.

(4) Professionals may help laymen move from lower to higher levels of therapeutic agency through well-conceived didactic and experiential training sequences designed to nurture the therapeutic triad. Well-functioning professionals have an important and evident responsibility for such lay training.

(5) The populist category of "ordinary persons," which I have previously employed, is not especially helpful in this connection, since it fails to acknowledge crucial differences between the above categories 1, 2, 4, 5, 7 and 8.

(6) Those in need of therapy, and the governmental and funding agencies who seek to care for their interests, cannot reasonably expect all professional psychotherapists to be genuine therapeutic agents with

high levels of the therapeutic triad, although we have a right to expect that their training will tend generally to move them up on the scale of therapeutic agency.

(7) Many well-functioning professionals, we hypothesize, function well because of resources and talents which are not necessarily the product of their training process. There are natively gifted persons with therapeutic potential, and we hope that they will elect to go the professional route, rather than those with less native gifts.

(8) The placebo effect accounts for a substantial amount of the effectiveness of professional therapists operating with low levels of therapeutic conditions, since clients come with the expectation that they will receive help. Nonetheless, the laicization process is making the placebo effect increasingly less plausible and less operable.

(9) The *intent* of therapeutic professional training is to take persons in the lay category and move them toward professionalization, *and* to move them up the scale of therapeutic agency. Evidence indicates that the actual result of professionalization is often only to move them toward professionalization, not always upward on the scale of therapeutic agency.

(10) Sometimes the professionalization process becomes regrettably counterproductive by moving individuals *down* on the scale of actual therapeutic agency.

(11) The ordinary public expectation is that there is a direct correlation between actual therapeutic agency and professionalization. Evidence indicates that some lay therapeutic agents are more effective than some professionals. If so, the process of assigning credentials to psychotherapeutic professionals is put severely into question.

(12) Clients have a right to expect from professional therapists a high degree of actual therapeutic agency.

THE DANGERS OF OVERCORRECTION

Neil Warren has wisely argued that psychotherapists need not become overly professionalized, but must not fail to professionalize either. Such a balanced statement is consistent with my assessment of the situation, and perhaps the weak side of my statement still is its seeming neglect of the importance of well-conceived professionalization, even though I have vigorously supported that need at several points.

Warren does not seem to understand my purpose for the proposal on laicization. My purpose is to raise some of the questions which he has begun, very responsibly I think, to work toward answering. I am not as convinced as Warren is, however, that all laymen need training by professionals. My argument is based on the assumption that laymen have exceptional abilities for therapeutic agency even without training, and thus I resist the position of saying only those laymen are capable of therapeutic agency who have gone through a particular process of training that professionals have designed and feel good about. That, to me, is also a professional exaggeration.

Warren is disturbed about my instruction to ministerial candidates to do their own experimentation in their parish. He is concerned about irresponsible experimentation which is in fact dangerous, and he cites Bergin's studies of the dangers and possibilities of deterioration in psychotherapy. I, too, am alert to those potential dangers in psychotherapy, but I am not asking ministerial candidates or pastors to experiment with technical psychotherapy. I am asking them only to search for authentic human community and improved human communications. I agree that psychotherapy is particularly vulnerable to the demonic. The context in which I made my statement about doing experimentation in the parish was not focused upon psychotherapeutic experimentation but upon doing experimentation in group processes and the nurturing of authentic community. The Protestant community-building concern has been a solid tradition for over 200 years. It is not something new which might depend upon psychotherapeutic interpretation for its viability.

I want to take seriously Walker's and Mulholland's charge that I have now overcorrected and am steering myself away from a deep, sustained dialogue with psychology. I hope that is not the case and, in fact, I hope a more serious confrontational style will deepen that engagement with psychology. A serious criticism of psychology, however, has been overdue from various perspectives, including that of religion. If such criticism blocks, rather than elicits dialogue, I can only regret it, since what I am hoping for is a deeper and more realistic dialogue. I do not feel myself all that firmly entrenched in a single language, certainly not theological language, nor do I see psychology as an alien land. Rather, I see myself fundamentally as a modern man for whom it sometimes takes great effort to make functional or meaningful the historic language of the church.

III. THEOLOGY, ESCHATOLOGY AND THERAPEUTIC AGENCY

Finally, a cluster of theological questions have been posed by the respondents which deserve a careful, studied response. Each of the respondents were in their own way grappling with questions concerned with the integration of theology and therapy. Again, so diversified and broadly gauged were these issues that I must be selective in dealing with those which seem most manageable, but without any pretense that all of them shall be dealt with as deeply as they deserve.

A) THE LINGUISTIC HERITAGE. Charles Kraft correctly notes that the traditionally defined, Western philosophical-theological frame of reference can no longer be assumed uncritically as a starting point for dialogue between theology and the behavioral sciences. Furthermore, I respect his bold appeal to adopt outright the language and thought-forms of contemporary behavioral sciences as models for Christian reflection and proclamation. But I myself do not wish to try to *completely* free myself from the philosophical frame of reference of Western theologizing, since I think that (1) western tradition still has much wisdom to bequeath to modern man; (2) however much modern man may attempt to surpass or negate his own western history, he cannot finally escape the fact that it is and remains his history; and (3) both the disciplines of theology and psychology are dependent upon language models which have roots clearly predating the modern behavioral sciences. Although I respect the appeal for clarity and relevance of language, I do not think the dialogical task is well served simply by dismissing or attempting to dismiss *in toto* a deeply rooted, linguistic tradition involving paradigmatic words like *therapeia, ministerium, salvus, anthropos, humanum* and *psyche*.

B) THE TRANSITION TO PRACTICE. I am especially grateful for the concern expressed by both Warren and Walker that some more specific pragmatic content be given to the eschatological perspective on psychotherapy which I have proposed. The relativizing of contemporary history, the "let-be" attitude, the possibility of seeing the brokenness of clients in a larger frame of reference than the mere here-and-now personal transaction—all of these are positive achievements. Likewise I am grateful for Charles Kraft's ethno-theological critique of the medical model based on information theory and communications research. It was a refreshing perspective.

I share Warren's perplexities about how a universal understanding of history may become more functional to therapeutic agency. He is correct to complain that I have not given much practical assistance to making this perception operational. I, too, have only recently moved out of the existentialist-subjectivist stance. The wholeness of universal history, however promising and significant, remains a strange new world and difficult to translate into operational terms.

C) THE HUMAN PREDICAMENT. I want to take sin as seriously as Calvin Schoonhoven and orthodox Christian theology have taken it, but not in such a way as to miss the evangelical proclamation that God himself has taken man's sin so seriously that he has taken it upon himself, rendered a radical divine verdict against it in favor of Jesus Christ and for all humanity. Disobedience consists in the human attempt vainly to make real and historical that which is an ontological impossibility, and herein is the essence of sin. But the fact that man is a sinner does not give sin an ontological status in creation. In the light of the resurrection of Jesus, man's continued attempt to negate the creative and redemptive word of God is understood as absurd (Romans 6). We must not give sin a legitimate place in creation. Creation is God's gift, and sin cannot be deduced from man's nature, which, as the creation of God, is and remains good. To the question of why man is disobedient to God's grace, the most appropriate answer is silence.[39]

In making this evangelical affirmation, I do not think that I have capitulated to an optimistic, Quixotic, romanticist view of man's nature as essentially and existentially good. In fact, I have attempted elsewhere to show that humanistic psychology sometimes exhibits a profound bondage, similar to the Augustinian-Calvinist conception of the bondage of the will (see *The Structure of Awareness,* Sections 8-11, 23, 33, 39, 41-43, 45; *Kerygma and Counseling,* Chapter 3; and *The Intensive Group Experience,* Chapter 3). It is not I who talk about the intrinsic goodness of man or assume man's unlimited capability to extricate himself from his alienation. In fact it is precisely "man's terrible inability, his evil inclination, his egoistic passion," of which Schoonhoven speaks, which is most powerfully exposed and beheld in effective therapeutic agency. Consequently, there is probably nothing in Schoonhoven's statement that surprised and disturbed me more than to see him link my views with the truncated, bourgeois, manipulative, positively thinking romanticism of Norman Vincent Peale. The Christian gospel of God's judging love in Jesus Christ is not positive thinking nor defiant human optimism in the face of death, nor success-oriented strategies of replacing bad thoughts with good ones, but rather it is good news of God's will to overcome the desperate depths of human sin.

D) THE NEGLECT OF CHRISTOLOGY. Accordingly, my intent is not to nurture a dialogue with therapy on the basis of anthropology, but rather upon the basis of Christology and eschatology. Thus Schoonhoven apparently mistakes my intent when he suggests that I wish to build a theological interpretation of therapy on the foundation of the *imago dei.* In this same connection, I see Schoonhoven surprisingly aligning himself with an older tradition of liberal theology when he pleads for the notion of common grace as the necessary category for dialogue with therapy, and drops or neglects any eschatological or Christological basis for the dialogue. It is far from my intention to deny the presence of common grace, although I would identify its historic lineage in a different way than Schoonhoven has. (It seems only Calvinist in a derivative sense, and much more heavily dependent upon the scholastic tradition prior to Calvin, which later was transmuted by the tradition of Schleiermacher). Schoonhoven's advice seems to be to give up the attempt to relate therapeutic agency to Christology, and simply reserve space in the doctrines of creation, man, civil righteousness and common grace to deal with therapy. I am inclined to see this as a constricted frame of reference which has been badly overused during the period of the hegemony of liberal theology. Schoonhoven reneges on the essential task to which I understand myself to be committed when he concedes that "there is a strict and radical bifurcation between the discussion of salvation (of theology) and the therapeutic process."

It is clearly not my intention to be read as saying that salvation is a term whose meaning is totally exhausted at the phenomenological level. In fact, I have specifically attempted to distinguish between salvation and psychotherapy (in the first chapter of *Kerygma and Counseling*). It is not I who says conscious participation in the deed of God in Christ is not as important as affirming one's own worth. I myself would prefer not to put those as an either/or option. Schoonhoven is correct to see that a rejection of two-sphere thinking implies that the vertical relationship becomes real through the horizontal structures of life. But it is not a consequence of that to say that no act of God occurs or that the vertical is thereby absent. Rather the self-revelation of God is fully present amid and made known through historical structures.

E) THE HYPOTHESIS OF A NONACCEPTING REALITY. I found myself puzzled, prodded and most seriously challenged by Schoonhoven's statement that "there is no more validity for believing there to be an accepting reality in the universal scheme of things on the basis of one or more accepting therapists than there is for extrapolating a cosmic nonaccepting reality from the client's many prior nonaccepting encounters. There is, however, a higher scale of validity for the latter extrapolation because negative encounters have been and probably will continue to be his normative experience." To this, I wish to make three separate, but related, responses: (1) It is not my intention to argue that we can infer a universal accepting reality from the limited, fragmented, occasional glimpses of human acceptance that may or may not occur. Quite the contrary, evangelical proclamation helps us to grasp and celebrate the fact that we are free to believe in the reality of acceptance precisely on the basis of its eventful self-disclosure in history. Thus evangelical reasoning moves analogously from the acceptance of God in Jesus Christ to the accepting act of the therapist, and not characteristically the other way around. Thus, my own analogical direction is quite the opposite of what Schoonhoven seems to be assuming.

(2) Even if it were the case that one persistently had nonaccepting encounters from men (as we in fact do frequently and persistently) still the gospel announces the acceptance of God precisely in spite of the nonacceptance of the neighbor. Unconditional love, which is only vaguely knowable amid our human loving, is (according to Christian witness) an *event* in Jesus Christ. It is now possible to respond not only to the abstract concept of unconditional love, but to a history in which humanity is in fact unconditionally loved. The Christian community celebrates the presence of God's love precisely amid the absence of human loves! The faithful man is being called to embody the love of God through his own interpersonal encounters. Nonetheless, however inadequately God's love is received, it continues to be offered as a gift (Hosea *11:*8, Luke *15:*11 ff., Ephesians *2:*4 ff).

(3) Evangelical thought is not locked into the position of looking for the providential or redemptive action of God only in the midst of positive encounters. The cross teaches us to look for the action of God precisely amid the severest negative encounters. To argue for the nonacceptance of cosmic reality, however, is already to make a decision about the nature of reality, against which evangelical thought must place the credibility of its own witness.

F) THE BELIEVING THERAPIST. In discussing the distinction between the believing and nonbelieving therapist, which as Warren knows is crucial yet somewhat hidden in my analysis, I am not satisfied with the tendency of Warren and Kraft to give up on attempting the distinction. Warren is correct to say that the distinction is heavy with speculation, and is of a low order of reliability, provided one defines reliability in strictly quantifiable, scientific terms. Admitting the difficulty of defining the term Christian, I would prefer not to come to the conclusion that it is at all levels simply undefinable.

If so, the burden is on me to clarify the definition of a believing or Christian therapist, as distinguished from one who is not a believer. An effective therapist who is a believer is one who functions with high levels of therapeutic agency, mediates the reality of acceptance, and knows and celebrates the love of God in Jesus Christ as the final self-disclosure of that accepting reality. Quite simply, a Christian is one whose life is shaped by the proclamation of the church that Jesus is the Christ. Admittedly it might be possible for a therapist to function with minimal empathy and with neurotic defenses, and still at a superficial level believe that Jesus is the Christ. In that case we would have an ineffective therapist functioning low on the scale of therapeutic agency, who nonetheless was a Christian, at least nominally. Both Christian and non-Christian individuals can be found in any of the nine categories of our previous diagram. But to the degree that his Christian experience and confession is real, he would tend to move up on the scale of therapeutic agency.

Warren usefully distinguishes between a life-changing encounter with Christ and a superficial kerygmatic encounter. I agree that the only encounter with Christ that deserves to be credited as such is one that changes life, transmutes behavior and attitudes, and revises one's understanding of the historical process. So when we believe that Jesus is the Christ, we refer to something more than a superficial brush with Christian consciousness. I would hypothesize that persons who in fact have had such a life-changing encounter tend to move up on the scale of genuine therapeutic agency. The life of the person who has affirmed that Jesus is the Christ in the most profound way is changed so as to recognize that he is loved in history with an unconditional love which frees him to love others. He is forgiven in such a way that he is capable of sharing that forgiveness with others. He is the recipient of God's care so that he can care for others in a liberated way.

G) CHRIST AS EVENT. With Neil Warren I would also want to contest an understanding of the Christ event which would say that Christ only points to the truth which is (without Christ) always there, (even though he perceives me as saying just that). Rather, the fact that Christ is an event already makes the Christ something more than merely a sign. Although Jesus Christ does point to the way things are, to the inner character of reality, things would not be the way they are without the self-revelation of God. Instead of saying that the love of God is fully present in history without the Christ, rather Christian worship remembers the triune God who eternally loves the world in the particular way which is clarified in the history of Jesus, so that the world in its own time comes to know of that love which has always been there.

Similarly, when Mulholland suggests that the not guilty verdict is ontological, I do not want him also to infer from it that it is not at the same time historical through an event. The verdict of forgiveness has an implication for universal being precisely because it is historical in the Christ event as an eschatological event. A covenant ontology thus points to rather than away from history.

H) BIRFURCATING INDICATIVE AND IMPERATIVE. One of the most persistent criticisms with which I have had to deal is that I have drifted off into a vague and uncritical soteriological universalism. Aware that I have dealt already with the theme of universalism at some length (in *The Structure of Awareness,* Sections 16-18; *Kerygma and Counseling,* Chapter 5; and *Contemporary Theology and Psychotherapy,* Chapters 7 and 9), I would like to add only these few comments: Schoonhoven correctly distinguishes my rejection of universal justification, but it puzzles me that, after quoting these views extensively, he still does not recognize that I, too, like him, want to hold together justification and obedience, as well as

God's grace and human responsiveness. He rightly urges that conscious faith and deliberate obedience be kept invariably intact. My only confusion is that he could fault me on this point, which I have taken such care to state in a way that is quite consistent with his formulation. If the comparison may be pardoned, I feel a bit as Luther must have felt when critics read only the first half of his treatise on Christian Liberty, only to assume that he was antinomian, then laid aside the treatise without reading its last half which deals with good works as the fruit of faith. Referring again to my previous discussion of the parable of the unmerciful servant (*The Structure of Awareness,* Section 17) and chapter IV on indicative and imperative in Radical Obedience, I can only reaffirm with Schoonhoven the texts of Hebrews *5:*9 and John *3:*36 on the intimate relationship of salvation and obedience. This does not weaken, however, the evangelical witness to the universality of God's justifying grace toward the whole of human history, as seen in such texts as Romans *5,* Colossians *1* and the epistles of John.

I) Reclaiming the conceptual task. Schoonhoven has rightly stated the proper sequence of *notitia,* (knowledge), and *assensus,* (assent), and *fiducia,* (trust) both in Christian thought and the therapeutic process. But I would interpret that sequence rather differently, in this way: Trust is never simply based upon a lack of knowledge of the trustworthiness of the trusted one. Trust is only possible on the basis of the disclosed trustworthiness of the other, whether in psychotherapy or evangelical witness. In therapeutic agency, it is only when one perceives and cognizes (*notitia*) one's neighbor as trustworthy (in the form of internal congruence, unconditional positive regard and accurate empathy) that assent (*assensus*) can be given, from the depths of one's being, to the relationship, and an act of trust *(fiducia)* can take form. This implies, as Warren rightly notes, "a delicate balance between cognition and affect."

When I place the priority on event before language, Schoonhoven infers some objectionable diminution or neglect of the importance of verbalization and conceptualization. It is not a neglect of the verbal or conceptual when evangelical theology argues that God himself speaks through history and addresses man through events in such a way that our own human speech can in fragmented ways reflect and become accountable to God's own speech.

CONCLUSION

Having been given the privilege of the last word in this conversation, I am aware that there are many points in this response to which I know my colleagues and critics would wish to have a rejoinder. Although this colloquy is now coming to an end, surely we can have no illusions that these concluding comments represent any closure of the issues or any sort of definitive statement. Rather we have succeeded in what I take to be the larger intent of our discussion, viz. to open up further the agenda of unfinished business between religious understanding and the science of behavioral change.

Where do we go from here? (1) Toward careful empirical studies of therapeutic agency and religious community. I know of nothing which the dialogue needs more urgently than well-conceived, controlled empirical data gathering, which could achieve credibility with the community of clinical psychology, concerning the relation of the religious commitment and community to therapeutic change and actual therapeutic outcomes. The fact that few psychology departments are attempting such studies makes it all the more imperative for those who are making such attempts to do it well and convincingly.

(2) Therapists who are trying to understand themselves theologically are being called to try to perceive their therapeutic operations as existing within the context of a larger universal history, inclusive finally of some hope toward the end of history.

(3) We need urgently to develop specific training programs, structures and processes, both didactically and experientially, to improve and channel the immense resources of lay therapeutic agency.

NOTES

1. Eysenck, H. J.: The effects of psychotherapy: An evaluation. *J Consult Psychol, 16:*319-324, 1952; for critical responses see DeSharmes, R., Levy, J. and Wettheimer, M.: A note on attempted evaluations of Psychotherapy. *J Clin Psychol, 10:*233-235, 1954; Rosenzweig, S.: A transvaluation of psychotherapy: A reply to Hans Eysenck. *J Abnormal Soc Psychol, 49:*298-304, 1954.
2. Traux, Charles B. and Carkhuff, Robert R.: *Toward Effective Counseling and Psychotherapy: Training and Practice* (hereafter TECP). Chicago, Aldine, 1967, p. 5.
3. Frank, J. D.: *Persuasion and Healing.* Baltimore, Johns Hopkins, 1961; cf.

Frank, J. D.: The role of hope in psychotherapy. *Kentucky Centennial Psychotherapy Symp,* Lexington, University of Kentucky, 1965.

4. Barron, F. and Leary, T.: Changes in psychoneurotic patients with and without psychotherapy. *J Consult Psychol, 19:*239-245, 1955. The therapy in this case was dynamically oriented and neopsychoanalytic.
5. Truax and Carkhuff, *TECP, op. cit.,* p. 7n.
6. Teuber, H. L. and Powers, E.: Evaluating Therapy in a Delinquency Prevention Program. *Association for Research of Nervous and Mental Disorders.* Baltimore, Williams & Wilkins, 1953, vol. 31, pp. 138-147.
7. Cartwright, Rosalind D. and Vogel, J. L.: A comparison of changes in psychoneurotic patients during matched periods of therapy and no therapy. *J Consult Psychol, 24:*121-127, 1960.
8. Mink, O. G. and Isaacson, H. L.: A comparison of effectiveness of nondirective therapy and clinical counseling in the junior high school. *School Counselor, 6:*12-14, 1959.
9. Gliedman, L. H., Nash, E. H., Imber, S. D., Stone, A. R., and Frank, J. D.: Reduction of symptoms by pharmacologically inert substances and by short-term psychotherapy. *AMA Arch Neurol Psychiatr, 79:*345-351, 1957.
10. Walker, R. G. and Kelley, F. E.: Short-term psychotherapy with hospitalized schizophrenic patients. *Acta Psychiatr Neurol Scand, 35:*34-36, 1960.
11. Barendregt, J. T. (Ed.): *Research in Psychodiagnostics.* Paris, Mouton, 1961.
12. Anker, J. M. and Walsh, R. P.: Group psychotherapy, a special activity program, and group structure in the treatment of chronic schizophrenia. *J Consult Psychol, 25:*476-481, 1961.
13. Frank, J. D., Gliedman, L. H., Imber, S. D., Stone, A. R. and Nash, E. H.: Patient's expectancies and relearning as factors determining improvement in psychotherapy. *Am J Psychiatr, 115:*961-968, 1959.
14. Truax and Carkhuff, *TECP,* p. 11.
15. Eysenck, H. J. (Ed.): *Behavior Therapy and the Neuroses.* Elmsford, Pergamon, 1960. Eysenck argues that the average spontaneous recovery rates in nontreated neurotics are 66 to 72 per cent, better than the 64 per cent rate treated by eclectic psychoanalytic psychotherapy.
16. Levitt, B. E.: The results of psychotherapy with children: An evaluation. *J Consult Psychol, 21:*189-196, 1957.
17. Truax and Carkhuff, *TECP,* p. 12.
18. Levitt, *op. cit.;* Truax and Carkhuff, *TECP,* p. 21. Many of the pretenses of therapeutic professionalism are exposed in Jefferson M. Fish's *Placebo Therapy: A Practical Guide to Social Influence in Psychotherapy.* San Francisco, Jossey-Bass, 1973, an excellent recent follow-up to our theme in Chapter 1.
19. Truax and Carkhuff, *TECP,* p. 100.
20. Whitehorn, J. C. and Betz, Barbara J.: A study of psychotherapeutic relationships between physicians and schizophrenic patients. *Am J Psychiatr, 3:*321-331, 1954; Betz, Barbara J.: Bases of therapeutic leadership in psy-

chotherapy with the schizophrenic patient. *Am J Psychother, 17:*196-212, 1963; Betz, Barbara J.: Differential success rates of psychotherapists with "process" and "non-process" schizophrenic patients. *Am J Psychiatr, 11:* 1090-1091, 1963.

21. Truax, C. B.: Effective ingredients in psychotherapy. *J Counsel Psychol, 10:*256-263, no. 3, 1963; Truax, C. B., Liccione, J. and Rosenberg, M.: Psychological Test Evaluations of Personality Change in High Conditions Therapy, Low Conditions Therapy and Control Patients. *Brief Research Reports.* Madison, Wisconsin Psychiatric Institute, U of Wis, 1962, p. 10; see further documentation in Truax and Carkhuff, *TECP,* pp. 83 ff.
22. Truax and Carkhuff, *TECP,* p. 96.
23. Wargo, D. G.: The Barron Ego Strength and LH[4] Scales as Predictors and Indicators of Change in Psychotherapy. *Brief Research Reports.* Madison, Wisconsin Psychiatric Institute, U of Wis, 1962, p. 21.
24. Truax and Carkhuff, *TECP,* p. 97.
25. Gendlin, E. T. and Geist, Marilyn: The Relationship of Therapist Congruence to Psychological Test Evaluations of Personality Change. *Brief Research Reports.* Madison, Wisconsin Psychiatric Institute, U of Wis, 1962, p. 24.
26. Bergin, A. E. and Solomon, Sandra: Personality and Performance Correlates of Empathic Understanding in Psychotherapy. Paper read at American Psychological Association, Philadelphia, September, 1963.
27. Strupp, H. H., Wallach, M. S., Wogan, M. and Jenkins, Joan: Psychotherapists' assessments of former patients. *J Nerv Ment Dis, 137:*222-230, 1963.
28. For a report of these studies, see Truax and Carkhuff, *TECP,* Ch. 3.
29. Truax and Carkhuff, *TECP,* p. 108.
30. Carkhuff, R. R. and Truax, C. B.: Training in counseling and psychotherapy: An evaluation of an integrated didactic and experiential approach. *J Consult Psychol, 29:*333-336, 1965; Lay mental health counseling: The effects of lay group counseling. *J Consult Psychol, 29:*426-431, 1965; Truax and Carkhuff, *TECP,* pp. 108-109; Truax, C. B., Carkhuff, R. R. and Douds, J.: Toward an integration of the didactic and experiential approaches to training in counseling and psychotherapy. *J Counsel Psychol, 11:*240-247, 1964.
31. Truax and Carkhuff, *TECP,* p. 111.
32. *Ibid.,* p. 111.
33. Gendlin, E. T. and Rychlak, J. F.: Psychotherapeutic Processes. Palo Alto, Annual Review of Psychology, Annual Reviews, 1970, vol. 21, pp. 170-171.
34. Rothaus, Paul, Hanson, Philip G., Johnson, Dale L., Lyle, Francis A., and Moyer, Raymond: The anticipation and management of crisis on an open psychiatric ward. *J Appl Behav Sci, 2:*431 ff., no. 4, 1966.
35. Hanson, Philip G., Rothaus, Paul, Johnson, Dale L. and Lyle, Francis A. Autonomous groups in human relations training for psychiatric patients.

Appl Behav Sci, *2*:305 ff., no. 4, 1966 (quotations in subsequent three paragraphs also from this essay).

36. Rothaus, Paul, Morton, R., Johnson, Dale, Cleveland, S., and Lyle, Francis A.: Human relations training for psychiatric patients. *Arch Gen Psychiatry, 8*:572-581, 1963; Johnson, Dale, Hanson, Philip, Rothaus, Paul, Morton, R., Lyle, Francis A. and Moyer, Raymond: Follow-up Evaluation of Human Relations Training for Psychiatric Patients. In Schein, E. H. and Bennis, W. G. (Eds.): *Personal and Organizational Change Through Group Methods.* Somerset, Wiley, 1965, pp. 152-168.
37. Murillo, N.: Conceptual Approaches to Dependency Assessment, unpublished doctoral dissertation. Houston, University of Houston, 1965.
38. For the original statement of this thesis, see Rogers, C. R.: The necessary and sufficient conditions of therapeutic personality change. *J Consult Psychol, 21*:95-103, 1957. For further development, see Truax and Carkhuff, *TECP.*
39. For a detailed development of this position, see Barth, Karl: *Church Dogmatics, The Doctrine of Creation.* Edinburgh, T & T Clark, 1960, vol. III, sec. 43, p. 2.

[illegible]

[illegible] Psychol. [illegible]

29. For a detailed development of this position see [illegible] *The Doctrine of Creation*, Edinburgh, T. & T. Clark, 1960, vol. III [illegible]

INDEX

L

M

N

O

P

R

S

T

U

V

W